M. V. Volkenstein

Physical Approaches to Biological Evolution

M. V. Volkenstein

Physical Approaches to Biological Evolution

With a Foreword by Manfred Eigen

With 142 Figures

Springer-Verlag
Berlin Heidelberg New York
London Paris Tokyo
Hong Kong Barcelona
Budapest

Professor Mikhail V. Volkenstein †

Translator:

Artavaz Beknazarov

1 Kulakov Street, Block 1, Apt. 24,
123181 Moscow, Russia

Cover picture: Computer DNA Model (Photo: Dan McCoy)
© ZEFA/STOCKMARKET, Düsseldorf

ISBN-13:978-3-642-78790-4

Library of Congress Cataloging-in-Publication Data. Vol'kensteĭn, M. V. (Mikhail Vladimirovich), 1912- . Physical approaches to biological evolution / M. V. Volkenstein; translated by A. Beknazarov. p. cm. Includes bibliographical references and index. ISBN-13:978-3-642-78790-4   e-ISBN-13:978-3-642-78788-1
DOI: 10.1007/978-3-642-78788-1   1. Evolution (Biology)
2. Molecular evolution. 3. Biophysics. I. Title. QH366.2.V625   1994   575'.001'53–dc20   94-6894

© Springer-Verlag Berlin Heidelberg 1994
Softcover reprint of the hardcover 1st edition 1994

Cover Design: Struve & Partner, Atelier für Grafik-Design, Heidelberg
Production Editor: A. Kübler, Heidelberg
Typesetting: ASCO Trade Typesetting Limited, Hong Kong
SPIN: 10059996   55/3140 - 5 4 3 2 1 0 - Printed on acid-free paper

Foreword

"Mr. Wolkenstein's *Physical Approaches to Biological Evolution*, whether or not it proves to give the ultimate truth on the matters with which it deals, certainly deserves, by its breadth and scope and profundity, to be considered an important event in the philosophical world."

This is a quotation from an introduction written by Bertrand Russell for Ludwig Wittgenstein's *Tractatus Logico-Philosophicus*. I exchanged only name and subject. As for the rest, I could continue quoting Russell, but I would rather say something myself.

As Wittgenstein did with formal logic, Wolkenstein rectifies our views on how to approach the logic of life from a formal theoretical basis. Many biologists do not believe that their subject lends itself to the scrutiny of physical theory. They certainly admit that one can simulate biological phenomena by models that can be expressed in a mathematical form. However, they do not believe that biology can be given a theoretical foundation that is defined within the general framework of physics. Rather, they insist on a holistic approach, banning any reduction to fundamental principles subject to physical theory.

This is a misconception, not of biology, but of physical theory. The aim of theory is not to *describe* reality in every detail, but rather to *understand* the principles that shape reality. This can be accomplished only by abstracting from reality. In other words, theory is not simply a substitute or an alternative to the (previously) descriptive and (nowadays) largely experimental approach to biology. This is similar to chemistry where quantum mechanics provided the fundamental theoretical understanding; the solutions to the problems in chemistry still require a specific "chemical" approach. This will be true for biology as well. The theoretical problems of biology are of a typical nature. The mechanisms and the boundary conditions are intertwined in a unique way; otherwise, the situation does not differ greatly from physics, which only recently started to deal with "complex adaptive systems" that are typically found in biology.

Physicists, on the other hand, have not always appreciated that a theory of biology has to start from biological facts. They often thought that biology is just another field to which they could immediately apply their equations. While this may be true for certain models, for example those involving diffusion and reaction processes, physical theories did not offer easy access to an understanding of "life itself". How to handle the complexity of life in a reproducible way, how the information is generated that leads to the organization of this complexity at the molecular (genetic), the cellular (developmental) or the cell-network (e.g. neurobiological) level, are questions that had no precedent in conventional physics. When facing these problems physicists often came up with proposals devoid of any relation to biological facts. This aroused the suspicion of biologists and led them to reject *any* theoretical approach. Max Delbrück (himself a theoretical physicist) was perhaps the first to realize this dilemma and he decided first to become a true biologist and only then to try to solve their problems. He was the founder of (a physically intelligible) phage genetics. Another physicist who became a biologist, Walter Gilbert, recently declared (in an editorial in the Journal *Nature*) that the time is now ripe for a change of paradigm in biology. I do not think that paradigms are really a basis for gaining any new knowledge, but in this connection I find the term justified. The time is ripe to order the wealth of biological data within the framework of a general theoretical scaffolding.

Exactly this has been accomplished — for the first time in such depth — in Mikhail Wolkenstein's book. I knew Mikhail for many years, first only by exchanging letters. He referred to the Kafkaesque situation in his country, and although he was not permitted to visit us we managed to establish a true exchange of ideas. Despite all the obstacles that were put in his way, he loved his country (not the system) and he was deeply concerned. And when we were finally able to meet personally and speak freely to each other we realized what we had missed in all those years. Unfortunately, the remaining time was short. He died a few days before he was supposed to receive heart surgery (which Otto Creutzfeldt and I had arranged for him at a clinic at Bad Nauheim). During his visits I became acquainted with the book he was preparing. However, I was truly surprised when, after his death, I received the manuscript. I had not expected such a comprehensive treatise on theoretical biology.

The book is written from quite a superior point of view. It takes issue with opinions that neglect either the physical or the biological prerequisites. The author is quite outspoken and does not shrink from naming those scientists with whom he does not agree. Throughout the book he addresses the fundamental problems of biology. The geneticist Theodore Dobshansky once said "Nothing in biology makes sense except in the light of natural selection." This is the conclusion of a biologist who confirmed Darwin's theory by doing ingenious experiments. In his book Mikhail Wolkenstein tries to find a *theoretical basis* for Dobshansky's conclusion that characterizes biology so aptly. Let me conclude this foreword by once more quoting Bertrand Russell:

"As one with a long experience of the difficulties of *a theory of life* (*logic*) and of the deceptiveness of theories which seem irrefutable, I find myself unable to be sure of the rightness of a theory, merely on the ground that I cannot see any point on which it is wrong. But to have constructed a theory of *life* (*logic*) which is not at any point obviously wrong is to have achieved a work of extraordinary difficulty and importance. This merit, in any opinion, belongs to Mr. Wolkenstein's (Wittgenstein's) book, and makes it one which no serious *biologist* (*philosopher*) can afford to neglect."

Göttingen, June 1994   *Manfred Eigen*

Preface

Contemporary natural sciences have become historical. History exists only in open far-from-equilibrium systems. In such systems, dissipative or synergetic, new information is created.

The "historicalization" of natural sciences was started by Darwin, who was the first to show how ordered, organized development of the biosphere emerges from chaotic variability. It is not for no reason that the great physicist Ludwig Boltzmann called the nineteenth century "The century of Darwin".

The understanding of certain areas of modern physics destroys the "Berlin Wall" that separated biology from physics. Narrow specialization is being replaced by united natural sciences.

Physical approaches to the basic problems of biology are especially topical today. An attempt has been made in this book to approach biological evolution from two directions: from the viewpoint of atomic and molecular physics and from the viewpoint of the physics of dissipative systems and information theory. This "physicalization", let us call it, is only in its infancy at present.

In the 1930s the union of Darwin's theory with genetics led to the creation of the synthetic theory of evolution. Today science is faced with new problems – the theory of evolution is being united with molecular biology, synergetics, and information theory. The theory developed by Darwin continues to be of fundamental significance.

Some sections of this book have been written on the basis of the works published by me and my coworkers. Special credit goes to my colleagues Mikhail Lifshits and the late Boris Belintsev. Before his untimely death, Boris's bright talent shed light on an entire domain of science. The book also presents work that was performed and published in collaboration with Igor Golovanov, Boris Goldstein, Natalie Yesipova, Michael Conrad, Vladimir Nauchitel, Oleg Ptitsyn, Theodore Rass, Vladimir Sobolev, Dmitry Chernavsky, Yuri Sharonov and Werner Ebeling. I express my deep gratitude to all of them, and also to numerous scientists for advice and material provided by them. Special mention should be made of Nikolai Vorontsov, Albert Goldbeter, Steven Gould, William De Grado, Ken Dill, Gabriel Dover, Motoo Kimura, Bernd-Olaf Küppers, Max Perutz, Ilya Prigogine, Hans Frauenfelder, Hermann Haken, Syrus Chothia, Eugeny Shakhnovich, Harold Scheraga, Peter Schuster, and Manfred Eigen. Of course, none of these scientists is responsible for what has been written by me.

I wish to take this opportunity to pay tribute to the translator, Artavaz Beknazarov, who has put much effort into translation of this book into English and who has patiently and painstakingly gone through the difficult job of dealing with the diversity of subjects discussed in the book.

Moscow   *M.V. Volkenstein*

Mikhail V. Volkenstein
(1912–1992)*

Mikhail Volkenstein died in Moscow on 16 February, 1992. His life story might inspire any scientist to try and explore a remote field of knowledge. He was one of those who managed to instill physical views on novel objects such as multi-atom molecules, polymers, proteins, and complex biological systems. This is not an easy goal, since it implies an ability to notice the major physical patterns of complicated systems and to envisage simple ways of studying them. Mikhail Volkenstein drew pleasure from this rare ability, which always allowed him to work in the "hot spots" where physics intruded into chemistry and biology.

Coming from the well-known physical school of L.I. Mandelshtam and G.S. Landsberg, Volkenstein started his research career in 1933 with the then just evolving molecular spectroscopy. His main achievement in this field is the theory of intensities of the vibration spectra of complex molecules. The physical basis of this theory is simple and fruitful: the valence-optical scheme, whereby the changes in the dipole moment and polarizability of a vibrating molecule are given as the sum of the appropriate characteristics of its valence bonds.

In 1948, Volkenstein ventured into a new field – polymers – and moved to the recently established Institute of High-Molecular-Weight Compounds in Leningrad. He immediately formulated the basic goal of this field: to predict the flexibility and other physical properties of polymers from the chemical structure of their chains. Volkenstein put forth a simple physical model whereby a polymeric chain is envisaged as a linear cooperative system with several discrete states of each monomer. This permitted him and his students to apply the ideas of the mathematical theory of such systems – developed long before and awaiting its hour – and thus obtain a relatively simple solution of this seemingly intractable problem. This was the first instance in molecular physics of the physical properties of a complex system being calculated from its chemical structure.

Volkenstein's treatment of the polymeric chain as a linear cooperative system with a set of discrete states proved very important for biophysics and molecular biology. The widely known theories of the helix-globule transition in polypeptides and DNA are based on the same idea. Moreover, an extension of this method to the behaviour of a polymeric chain in an external field, under-

---

* This obituary was originally published in the Russian journal "Molekulyarnaya Biologia" – Molecular Biology.

taken by Volkenstein in 1953, has now become the point of departure in modern attempts to predict the protein tertiary structure from its amino acid sequence. It is noteworthy that this prediction problem can be regarded as a continuation of the related problem of a flexible polymeric chain, formulated by Volkenstein in the early 1950s.

In the early 1960s Volkenstein came to the conclusion that polymer science would develop along two main lines: into technology and into biology. He chose biology, and in 1967 took positions in the Institute of Molecular Biology in Moscow and in the Institute of Biophysics in Pushchino. With his clear physical perception, he delved into the most intricate problems of biophysics and made great contributions to this field, including the meaning of the genetic code, the interrelationship between the electronic and the conformational (i.e. nuclear) degrees of freedom in enzyme catalysis, the neutral theory of evolution (in particular, at the molecular level), a view of the origin of species as a phase transition, and many others.

Mikhail Volkenstein was quite an extraordinary man. With his really encyclopaedic knowledge, he was an intellectual of the type more often encountered in the Renaissance than nowadays. He fluently spoke the main European languages, he was freely conversant in physics, chemistry, zoology, botany, history, literature, and art. He painted, wrote poetry and science fiction, played chess, collected stamps and butterflies. With his wife Stella, who outlived him by only a week, he had in Leningrad and then in Moscow a hospitable home where Russian and foreign intelligentsia could meet. He was very active in social life, and played an important part in the Russian democratic movement.

His life was not an easy one. He tragically lost his first daughter and in the early 1950s, was forced to leave the Leningrad University; only in the last years of his life was he allowed to go abroad. Yet he was harmonious and happy, he never envied anyone, and it was impossible to hurt him. He was a charming man, and we shall never forget him.

O.B. Ptitsyn

Contents

Part I

Biological Evolution

1. Introduction

1.1 Living Nature

There is a pleasure in the pathless woods,
There is a rapture on the lonely shore,
There is society where none intrudes,
By the deep Sea, and music in its roar.
I love not man the less, but Nature more,
From these our interviews, in which I steal
From all I may be, or have been before,
To mingle with the Universe, and feel
What I can ne'er express, yet cannot all conceal

Byron, "Childe Harold's Pilgrimage", CLXXVII

In these beautiful lines Byron writes about Nature, about the sea, the wood. The forest is a community of living organisms, both plants and animals. We can perceive the wood poetically; Nature is a source of joy and inspiration. We can also look at the wood with an artist's eyes and see the dazzling whiteness of birch trunks illuminated by the sunlight, the briskness of a red squirrel .... .

One can, however, contemplate the essence of what one has seen, without limiting oneself to emotional perception.

Let us compare two famous books about great voyages: *The Frigate 'Pallada'* by I.A. Goncharov, a Russian writer of the last century, and *Diary of the Voyage of H.M.S. "Beagle"* by Charles Darwin.

Goncharov, a clever and subtle observer and writer, saw and understood the life of other peoples, the historical and social conditions of their existence, and the characters of people. *Pallada* contains poetical, picturesque descriptions of the sea and the sky, and of their fantastic beauty in tropical regions. But at the same time Goncharov writes:

"The sea hedge is a half-plant, half-animal: it grows and, perhaps, breathes."

"I saw many giant butterflies, and something resembling wasps, blue, hairy and with a spot on the head."

"Apart from the pine-tree, various trees grew close to the countryside, trees that I had seen nowhere before .... . Unfortunately, there were no amateur-naturalists among us, nobody whom we could have asked about these trees."

Goncharov described the plant and animal life, but like the majority of people he knew nothing about it.

In contrast to *Pallada*, the *Beagle* played a very important role in the history of culture. It was this voyage that served as a source of Darwin's theory, one of the highest achievements of human thought. As is known, the key idea of the theory came to the young naturalist when he visited the Galapagos Islands:

"The most curious fact is the perfect gradation in the size of the beaks in the different species of *Geospiza*, from one as large as that of a hawfinch to that of a chaffinch and ... even to that of a warbler .... Seeing this gradation and diversity of structure in one small, intimately related group of birds, one might really fancy that, from an original paucity of birds in this archipelago, one species had been taken and modified for different ends."

The Galapagos finches of the species *Geospiza* are known today as Darwin's finches.

Darwin writes emotionally and interestingly about people, history and social relations and especially about nature. This is how he described a day spent in a forest:

"Bahia, or San Salvador. Brazil, Feb. 29th. The day has passed delightfully. Delight itself, however, is a weak term to express the feelings of a naturalist who, for the first time, has wandered by himself in a Brazilian forest. The elegance of the grasses, the novelty of the parasitical plants, the beauty of the flowers, the glossy green of the foliage, but above all the general luxuriance of the vegetation, filled me with admiration. A most paradoxical mixture of sound and silence pervades the shady parts of the wood. The noise from the insects is so loud, that it may be heard even in a vessel anchored several hundred yards from the shore; yet within the recesses of the forest a universal silence appears to reign. To a person fond of natural history, such a day as this brings with it a deeper pleasure than he can ever hope to experience again."

The conscious attitude toward living nature depends on the ability to see, to ask questions, to seek and, if you are lucky, to find an answer. The encounter with a new flora and fauna is a joy which is the greater the more you know about them.

In the winter of 1967 I found myself in the West of Cuba Island. I wandered in a forest for an hour and a half. Having heard birds' calls and songs, I walked into the thicket, ignoring the thorns, and saw a large bird of the family *Trogonidae* (called a trogon), a grey bird with a bright red tail. Then I saw two quarrelling Cuban parrots and what I thought was a blue butterfly, which proved to be a hummingbird. Suddenly a large lizard crossed my path; it was an anole with a blue-red colouration. When I went farther, I saw leaf-cutting ants marching in a line, each carrying a piece of a leaf. They would then grow fungi on this substrate.

At a later time I saw the tundra vegetation along the coast-line of the Okhotsk sea on Kamchatka (at the latitude of Kiev). Two species of the dwarfish wild roses, with crimson and pink flowers, grew in close proximity to taller brown-cap boleti.

The questions that come to mind in the forest, in the field, on the sea shore, or when you dive into the sea are:

What is it?
What is its structure?
How does it live?
How did it emerge?
How does it relate to and interact with other plants and animals?

Scientific biology begins with systematics. The diversity of the animal and plant kingdoms is enormous. In this lies one of the main distinctions between biology and the physics of inanimate nature, whose most important objects are not so numerous. For example, the number of chemical elements, as well as the number of elementary particles, is of the order of 10^2. The number of various animal species is four orders of magnitude greater. And this is only 0.1 percent of the number of species that have existed on Earth. According to present-day estimates, 99.9 percent of the species have become extinct.

Accordingly, biology is based on zoology, botany and, of course, palaeontology, which are all descriptive sciences. Physicists often neglect these sciences and consider them to be boring and fruitless. One eminent scientist expressed his opinion about the classification of spectral lines: "But this is pure zoology!" The great Rutherford, to give another example, referred to biology as "postage-stamp collecting". But without the description of the diverse objects observed, no further progress can be made.

Theoretical biology began with Carolis Linnaeus, the founder of modern taxonomy, who established the classification of living things in a particularly methodical way. The Linnaeus system, as was found out later, reflects the real hierarchy determined by the evolutionary development of the biosphere. Linnaeus himself was a creationist, i.e. he believed that living things were created in the beginning by God.

The modern classification of living organisms is given here (in a somewhat simplified way): there is the superkingdom (prokaryotes and eukaryotes), the kingdom (prokaryotes, plants fungi, and animals), the subkingdom (unicellular and multicellular for animals); then, bypassing the intermediate taxa, we arrive at a phylum (23 phyla for animals) or a class (22 classes for plants). The phylum is divided into classes, classes are divided into orders, followed by a family, genus and species. Thus, man belongs to the animal kingdom (*Animalia*) the subkingdom of multicellular animals (*Metazoa*), the chordate phylum (*Chordata*), the vertebrate subphylum (*Vertebrata*), the mammalian class (*Mammalia*), the order of primates (*Primates*), the suborder of the higher primates (*Anthropoidea*), the superfamily of apes (*Hominoidea*), the family of hominoids (*Hominidae*), the genus *Homo*, the species sapiens (*Homo sapiens*). Finally, the subspecies *Homo sapiens sapiens* is different from the Neanderthal *Homo sapiens neanderthalensis*.

The great discovery made by Linnaeus is the possibility of characterizing an animal organism by two words designating the genus and the species. A species is biological reality.

Systematics is a branch of science which is far from being completed and is still developing. At present more than 1.5×10^6 species have been described, the actual number being perhaps three times more. More than a million species are animals (among these there are about 850,000 species of insects), about 350,000 species are plants, and about 40,000 species are fungi. The number of species of prokaryotes is a few thousands. New invertebrate species are being discovered constantly. As well as species, even new phyla are sometimes discovered: in 1955 and 1957 A.V. Ivanov offered a well-grounded characterization of the pogonophora phylum. At the same time, systematics undergoes change and is made more and more accurate. Thus, in the last century hares were believed to belong to the order of rodents (*Rodentia*). However, in 1912 the zoologists established that they differ significantly from rodents, and today hares, rabbits, and pikas are classified as belonging to the special order *Lagomorpha*.

Systematics rests on the discreteness of species and, more importantly, on that of higher taxa. A species is a group of individuals that form geographically and ecologically representative populations possessing common morphophysiological characters, capable of freely interbreeding under natural conditions ... occupying on the whole a common continuous or partly broken habitat; each species is separated from other species ... and there is practically complete biological isolation (i.e. there is no interbreeding) [1.1]. Reproductive isolation is the most important feature of a species for sexually reproducing organisms. After interbreeding with members of another species, hybrid offspring may be produced (a mule is the hybrid of a donkey and a horse), but the offspring is sterile in most cases. This is not the case with prokaryotes and Protozoa, which reproduce asexually. Nonetheless, here too the notion of species remains valid; species differ in morphophysiological characters [1.2].

Taxonomy, which begins with a kingdom and ends up with a species, proceeds from the general to the particular. At each level all members of a particular taxon are characterized by similar traits and are in this sense equivalent. The cells of all eukaroytes have a nucleus, in contrast to prokaryotes. At the kingdom level the flatworm *Planaria* is equivalent to an elephant, at the phylum level an elephant is equivalent to a lancelet. Human beings and elephants belong to the same class; man and lemurs are classified as belonging to the same order. As we go down the hierarchial ladder the description aimed at identifying a living creature becomes more and more detailed. The determinants of animals and plants are constructed according to the bifurcation principle: whether a particular life-form possesses a particular character or not. This bifurcation eventually reflects the real divergence of species.

Biological systematics is confined to species, or at most to subspecies. Further division practically goes beyond the scope of biology. There exist races (for example, the squirrel *Sciurus vulgaris* in Bulgaria have, in contrast to our squirrels, a chocolate colour). All human beings, irrespective of the skin colour, belong to the same species, though the bushman strongly differs in external appearance from the Russian. The differences between human races extend to the molecular level. Good men in Australia, wishing to take care of the aborigines, organized a boarding school for their children. The children were well cared for and provided with good food, with milk being included in their diet. But the children began to get seriously sick and it seemed they would die. It turned out that they could not tolerate milk; their bodies were devoid of the enzyme lactase, which is required for the digestion of lactose (milk sugar). Nonetheless, both the white Australians and the black aborigenes belong to the same species *Homo sapiens*, with its 46 chromosomes in each somatic cell.

The intrusion of man sharply and rapidly increased the number of races and breeds of living organisms, as human beings proved to be more ingenious than Nature. All breeds of dogs, beginning with the mastiff and ending with the lap-dog, belong to the same species; a species whose chromosome number matches that of the wolf *Canis lupus*.

The emergence of new races and breeds is determined not only by man's breeding new animals and plants, which he requires for agriculture or for pleasure (for example, breeds of doves or decorative aquarium fishes). The chemical means produced by man for killing destructive insects have led in a short period of time to the appearance of new races resistant to DDT. This insecticide proved to be a factor in natural selection. The same refers to the fight against pathogenic micro-organisms; strains of pneumococci which are resistant to penicillin can arise, due to selection of the corresponding mutants with the aid of this antibiotic.

The ultimate unit of biological hierarchy is an individual. This does not require explanations when we are speaking of a human being or a higher animal; we know how diversified the characters of dogs or cats are. It is possible, however, that earthworms, which seem to us to be identical, are also individuals, which is "understandable" to them but not to us .... .

Science classified human beings still further; an asthenic, pyknic, and athletic according to Kretschmer [1.3], or as phlegmatic, sanguine, melancholic [1.4].

Systematics (classification) forms the basis of any branch of science. The greatest discovery in chemistry (the Periodic System of the Elements) was made purely on this basis. D.I. Mendeleev was a genius at classification.

But the unravelling of the natural scientific basis of systematics, of its actual causes, requires a more detailed study. By arranging the chemical elements in the sequence of their atomic weights, Mendeleev discovered the periodic law and came close to understanding the ultimate causes of this periodicity. He established the most important attribute of an element, its atomic number. As was later found out, this number represents the number of protons in the atomic nucleus and the equivalent number of electrons in the atom. The periodicity discovered by Mendeleev was fully accounted for in quantum mechanics at a later time.

The finding of the true basis of the classification of species is a much more formidable task. Its solution is associated with the development of the theory of evolution, which is still far from being completed.

1.2 The Theory of Evolution

The great physicist Ludwig Edward Boltzmann said that if he were asked to describe the 19th century, he would answer that it was the century of Darwin.

In other words, the century of evolutionary theory is the century of the study not so much of established structures as of the process of becoming. "From being to becoming", as Prigogine put it [1.5].

Today we know that the universe, the galaxies, the Earth and the biosphere change with time. Some structures disappear and other new structures emerge. Geology and palaeontology provide more and more detailed information about the evolution of the biosphere.

Biological evolution is a part of the evolution of the universe and of the solar system. Biological evolution is localized in an infinitesimally small part of the world space, in a narrow area of water, land, atmosphere at the surface of the earth. We have no data at our disposal on the existence of life outside the earth [1.6]. On the contrary, the time span of biological evolution is commensurate with the time of existence of the universe, which is estimated at about 1.5×10^{10} years. Life on earth arose about 3.9×10^9 years ago, as evidenced by the ancient fossilized imprints of bacteria and primitive algae. The age of the Earth is estimated at 4.5×10^9 years.

Biological evolution was thus preceded by the emergence and evolution of the galaxies and the solar system, by the geological and chemical evolution of the Earth. The monograph of Ebeling and Feistel [1.7] contains tables of evolutionary events, in which the time is shortened by 1.5×10^{10} times, i.e. the time period that elapsed from the moment of the "big bang" up to the present time is represented by one year (see Table 1.1).

Table 1.1 presents the periods of appearance of large new groups of organisms. Only once is the disappearance of one such group noted: the extinction of dinosaurs in the Mesozoic era. Incidentally, if the statement that 99.9 percent of the species that ever existed on Earth have become extinct is valid, then the table of extinction is no less important than that of emergence. Figure 1.1 is a scheme of the evolution of vertebrates according to Schmalhausen [1.8], which indicates the epochs of emergence, prospering and extinction of a number of groups of these animals.

Among the forerunners of Darwin, mention should first of all be made of Lamarck. An idea of evolution, a clear-cut notion of the historical development of the animal kingdom were proposed in his book *Zoological Philosophy* [1.9] with remarkable clarity and vigour. The preface to this great work contains a very peculiar reasoning. Lamarck writes: "Could I have really been a witness of

Table 1.1. Chronology of events in the universe (with time reduced by a factor of 1.5×10^{10}) [1.7]

Beginning of year	Big Bang
June	Formation of the galaxies
September	Emergence of the solar system and formation of the Earth
October	First living creatures, ancient sedimentary rocks and fossilized imprints of micro-organisms
November	Oxygen-producing microbes are widely developing. Appearance of sexual reproduction. Emergence of synthesizing plants, first eukaryotes
Beginning of December	Formation of the oxygen atmosphere. Intensive eruption of volcanoes. Development of meiosis and sexual reproduction
Middle of December	Development of heterotrophic unicellular organisms, first multicellular organisms. Beginning of macroscopic life
20, XII	Emergence of invertebrates
21, XII	First oceanic plankton. Prospering of trilobites
22, XII	The Ordovician period; first vertebrates (fishes)
23, XII	The Silurian period; sporophytes conquer land
24, XII	The Devonian period; first insects. Animals conquer land, first amphibians, flying insects
25, XII	The Carboniferous period; first conifers, first reptiles
26, XII	The Permian period; first dinosaurs
27, XII	The Triassic period; first mammals
28, XII	The Jurassic period; first birds
29, XII	The Cretaceous period; first flowering plants, extinction of dinosaurs
30, XII	The Tertiary period; first primates, prospering of mammals, first hominids
31, XII	
About 14.00.00 hr	Appearance of Proconsul and Ramapithecus
About 22.30.00	First human beings
About 23.00.00	Stone tools
About 23.59.00	Appearance of agriculture
About 23.59.30	First urban centres
About 23.59.54	Writing
About 23.59.56	Bronze metallurgy
About 23.59.57	Iron metallurgy
About 23.59.59	Euclidean geometry. Archimedean physics
24.00.00	Time recording
1.1 (New Year)	
About 00.00.01	Introduction of zero and decimal system
About 00.00.02	Renaissance and modern science
About 00.00.03	Modern times

the extremely interesting degradation observed in the organization of animals, in a consideration of the series formed by them, from the most perfect to the most primitive of them, without attempting to explore the causes of this indubitable and remarkable fact, the obviousness of which is proved by so much evidence?".

Lamarck writes about the descent rather than about the ascent up the evolutionary ladder; he reversed, so to say, the motion of the camera film. The

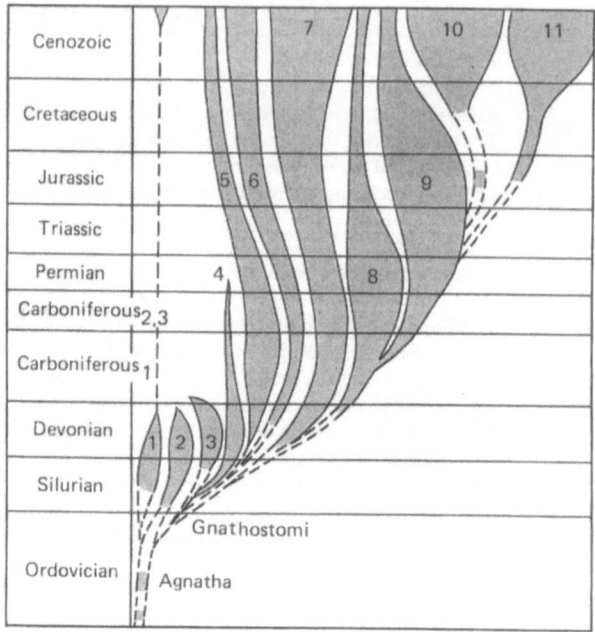

Fig. 1.1. Evolution of vertebrates: (*1*) Monorhini; (*2*) Diplorhini; (*3*) Placodermi; (*4*) Acanthodei; (*5*) Holocephali; (*6*) Elasmobranchii; (*7*) Osteichthyes; (*8*) Amphibia; (*9*) Reptilia; (*10*) Aves; (*11*) Mammalia

course of thinking is quite unusual. Perhaps the excellent verse of Mandelstam is associated with Lamarck's reasoning*:

Lamarck

> *An old man, once shy as a boy,*
> *a timid, clumsy patriarch ...*
> *dueling for Nature's honor?*
> *Fiery Lamarck, of course!*
>
> *If everything living is only a brief mark*
> *for a day time takes back,*
> *I'll sit on the last step*
> *of Lamarck's mobile ladder.*
>
> *I'll lower myself to snails and mollusks and crabs,*
> *I'll rustle by lizards and snakes,*
> *I'll go down those firm gangways, over wide ravines,*
> *and shrink, and vanish like Proteus.*

---

*Translated by Clarence Brown and W.S. Merwin.

I'll wear a horny robe,
I'll say no to warm blood,
I'll be over-grown with suckers and I'll sting
deep in the ocean's foam, like a tendril.

We passed ranks of insects
with ripe liqueur-glass eyes.
He said: "Nature is all ruptures,
nothing can see – you're seeing for the last time."

He said: "Enough harmony –
you loved Mozart in vain:
a spider's deafness is setting in, and here
the trapdoor is stronger than we are."

And Nature stepped back away from us –
as if we were superfluous,
and the longitudinal brain she inserted
in the dark sheath was like a sword.

And she forgot the drawbridge,
it came down too late
for those with green graves,
red breathing, and supple laughter ...

In a remarkable book by Yu.F. Karyakin about Dostoevsky [1.10] there is a chapter called "Lamarck's footnote and the dreams of Dostoevsky". He meant the footnote to the work *The Analytical System of Positive Knowledge of Man* written by Lamarck at the end of his life [1.11]. The footnote reads:

"A human being infatuated with egotism becomes insufficiently prudent even when his own interests are concerned: because of his inclination to extract pleasure from everything at his disposal, in short, because of his careless attitude toward the future and his indifference toward his own kind, he contributes, as it were, to the destruction of the means of self-preservation and, thereby, to the extermination of the human race. To satisfy his passing whims he destroys useful plants that protect the soil as a result of which it becomes barren and infertile and the springs dry out; the animals living in the area are forced by man to leave their habitat, where they could find means of existence. The result is that vast territories of land that were once very fertile, and thickly populated by various living creatures, are converted to barren and irrecoverable deserts. Falling victim to his fleeting passions and ignoring all lessons from past experience, he is in a state of permanent war against his own kind, destroying and killing everywhere for one reason or another as a result of which, nations that were once large in number are gradually disappearing from the face of the earth. One would say that man is destined to exterminate himself after having rendered the globe uninhabitable."

In the same work Lamarck wrote that man is the culmination of the most perfect that Nature could have created on our planet. But, as follows from the words cited above, the human species is not only a dead-end species but also one that destroys all living creatures. Lamarck died in 1829 and could have known nothing about the atomic weapon and about the fate of the Aral sea. But he really predicted the events of the 20th century.

Yu.F. Karyakin treats Lamarck's words given above and Mandelstam's verse as describing the gradual simplification and extinction of life. The only creatures that have a chance of surviving on the devastated Earth are the *Arthropoda* and more primitive organisms. In the light of Lamarck's gloomy prophecy it may be true. However, Lamarck's evolutionary theory leads in another direction. Of course, the irreversibility of evolution may be demonstrated on both the descent and the ascent. However, there is an "arrow of time" directed only upwards.

In his *Zoological Philosophy* Lamarck wrote: "Should not I come to the thought that Nature created various living bodies, making a gradual transition from the simplest to the most complex, since, if one climbs the ladder of animals from the most imperfect to the most perfect, we shall see that their organization becomes increasingly more complex and perfect."

This problem is not so simple, however. Of course, man is more complex than the amoeba and even the ant. But is he better? The notions of complexity and perfection require a special inquiry (see Chap. 10).

If we combine *Zoological Philosophy* and *The Analytical System*, we will see that the "arrow of time" behaves in a peculiar manner. Having reached its peak by creating the species *Homo sapiens*, evolution could now go into reverse due to the activities of the species produced by it; and very rapidly at that. Whether this process will happen or not depends, however, on man, who still has time to come to his senses and acquire wisdom.

Lamarck was the first to have truly understood the evolutionary character of the development of the biosphere and the ability of organisms to adjust themselves to their environment. At the same time he was guided by the non-constructive idea of the innate drive of organisms toward perfection; he presupposed a multitude of acts of spontaneous generation and presumed that the acquired characters were inherited. These ideas have been fully rejected today. The inheritance of acquired characters would have meant the transfer to sexual cells of adequate information about the events that have taken place in somatic, asexual cells of an adult organism. Nobody has ever sought mechanisms of such transfer and one cannot even visualize their existence. An enormous number of experiments have unambiguously rejected the idea of the inheritance of acquired characters [1.12]. These conceptions have long been left outside the realms of science. It is exactly the belief in the inheritance of acquired characters that was the key idea of Lysenko in the Soviet Union. However surprising it is, attempts to revive the concept of the inheritance of acquired characters are occasionally undertaken. It should be noted in passing that Darwin himself did not deny this idea. However, this proposition was not important for his basic concept of driving natural selection.

The great merit of Darwin lies not in the fact that he was an evolution-ist but in that he disclosed the mechanism of evolution, i.e. natural selection. Darwinism is based on the following basic suppositions [1.13]:

1. The world around us is constantly changing. Some species become extinct and other species appear. This position is undisputed, and it was also ad-vanced by the forerunners of Darwin.
2. The evolutionary process is continuous and gradual and occurs without leaps. In the light of modern conceptions, this position appears to be con-troversial (see Chap. 4).
3. Related species emerge from a common ancestor, which is also valid for higher taxa. In the long run, all multicellular organisms originated from unicellular ones. These positions may be considered to have been proved.
4. Evolution is realized by way of natural selection. There is heritable variabil-ity, which is considerable in each generation. As a result of the struggle for existence, the selection of organisms that are best adapted to environmental conditions occurs. This is the key position of Darwin's theory, which dis-closes the mechanism of evolution.

Darwin's theory originally caused a number of objections. The first objec-tion was the "Jenkin nightmare". As was pointed out by Jenkin immediately after the publication of *The Origin of Species*, new characters selected during evolution cannot be fixed but must be dissipated, lost upon crossing. Crossing does not bring out "rational characters", it absorbs them. This difficulty was overcome only by modern evolutionary biology, the synthetic theory of evolu-tion, which united Darwinism with genetics. The selected characters do not disappear since they are determined by discrete genes.

The second objection, which is as old as the first, is a tautology which is supposedly built into Darwin's theory. In the struggle for existence, the winners are those best adapted to environmental conditions. The criterion of fitness is the survival of the larger fraction of offspring. It is exactly the victory in a competitive struggle that implies survival. Hence, we are speaking here of the "survival of the survivor." In fact, the theory of evolution makes use of the independent definitions of adaptation. One of the best known examples is the "industrial melanism" of the black moth *Biston betularia*. In industrial regions of England there once occurred the darkening of the bark of oaks, as the smoke hindered the growth of light-green lichens. As a result, the mutant form of the moth, with dark wings, prospered (Fig. 1.2). Such mutants are less conspicuous on the bark of the tree than the light forms and are less preyed upon by birds. The reverse process has recently occurred due to measures taken to prevent the contamination of the environment; moths with light-coloured wings have again gained the advantage. Thus, adaptation has a clear-cut and independent mean-ing. We shall return to these problems in Chap. 4.

It is appropriate at this point to dwell on the ideas expressed by the philoso-pher K. Popper, which are often cited in modern scientific literature [1.14]. Popper does not consider the theory of evolution to be a true theory since it

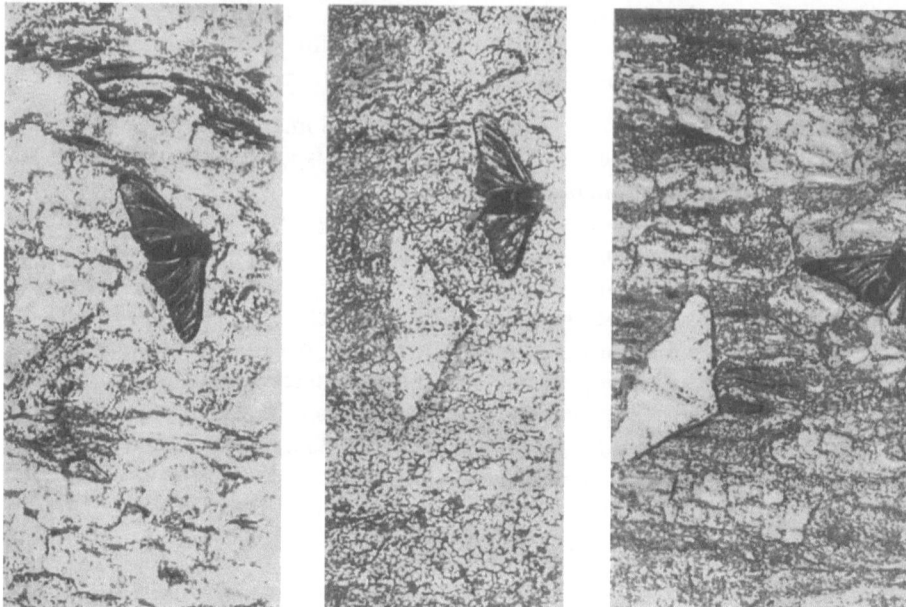

Fig. 1.2. The moths *Biston betularia* on an oak tree. The dark form is inconspicuous on the bark devoid of lichen, but can be distinguished on the lichen-covered bark

cannot be "falsified", i.e., no experiments can be conducted that would refute it. Popper repeats the above-given statement that the entire theory of natural selection rests on a tautology and proposes 12 positions of his evolutionary theory.

All this does not stand criticism. The reasoning as to "falsifiability" is not valid here. The criteria of historical theories in cosmology, geology, and biology are their self-consistency, the non-contradictory explanation of a large collection of facts based on a few basic principles, and the predictions, even though they are directed "back". Thus, scientific cosmology is supported by the discovery of relict radiation, the theory of evolution by the discoveries of relict species (for example, the gingko tree or the fish *Latimeria*).

The universe and the biosphere each exist as a single copy. This, however, does not interfere with their study.

The theses of Popper are formal. Organisms, rather than populations, are erroneously believed to be evolving systems, and survival, rather than reproduction, is believed to be important. This is bad natural philosophy and does not have anything to do with biology.

These views of Popper are well known. Less known, however, is the fact that he later changed his views. In 1980 Popper wrote [1.15]: "Some scientists think that I refuted the scientific character of historical sciences, such as palae-ontology or the history of evolution of life on Earth; or, say, the history of literature, technology or science ....

This is not correct and I wish here to confirm that these and other historical sciences are scientific in my opinion: their hypotheses can be tested in many cases.

It seems that some scientists believe that historical sciences cannot be tested since they describe isolated events. However, the description of individual events can be checked very often by their deduction from predictions or retrospection that can be tested."

Monod gives two examples of important inferences from Darwin's theory, which were confirmed by the subsequent development of science [1.16]. Being unaware of thermonuclear processes, Kelvin estimated the age of the solar system at 25×10^6 years. Darwin showed that the sun and the Earth are much older, which was confirmed in modern cosmology. The second example is the refutation of the "Jenkin nightmare" by the subsequent discovery of the discreteness of genes.

The present-day disagreements with the theory of evolution partly repeat those given above but are also partly different.

Berg [1.17] and Lyubishchev [1.18] erroneously refute "selectionism", i.e. natural selection, as a driving mechanism. Berg opposes Darwinism by proposing his theory of nomogenesis. Nomogenesis is based on the principle that evolution is determined by the established structure of an organism and by constraints imposed on its possible means of change. As we shall see, this idea is in itself valid, but it does not contradict Darwinism (see Sect. 2.5). The works of Berg and Lyubishchev were for a long time believed to be heretical and were not available to the general public. Fortunately, they have now been published and are deservedly popular among readers since, apart from the erroneous proposition indicated above, they contain many valuable ideas. The appeal of these works was, however, such that many of their readers, being unable to separate the kernel from the husk, refuted Darwinism.

Another psychological source of anti-Darwinism, which was popular in this country, was that not very long ago Lysenko and his followers distorted Darwin's ideas, proposing the so-called "creative Darwinism", which has nothing to do with science. Even today true Darwinism is mistaken for its caricature. The natural reaction to these tendencies is superorthodox Darwinism, which rejects new ideas in the theory of evolution, particularly those expounded in this book. However, Darwinism (the theory of evolution) is far from being a fixed, frozen system; it is a vigorously developing and, hence, changing branch of science.

Among people who have nothing to do with biology, a belief has lately become popular that evolutionary theory has failed to provide answers to three principal questions, namely:

1. How could the material have been enough for evolution by natural selection, which led to the diversity of the biosphere, if selection occurred among point mutations which have very low probabilities?

2. How could there have been enough time for the emergence of the modern biosphere if it occurred only by way of selecting rare mutants?
3. How could such complex and perfect systems as (e.g.) the brain or the eye of vertebrates have arisen by way of this reshuffling process?

As we shall see, modern evolutionary doctrine does provide answers to all these questions. The answer to the first question is simple.

Genetic investigations have shown that selection occurs not among individuals, which differ in rare point mutations (local changes in the structure of DNA), but rather in natural populations which possess enormous reserves of variability. Genetic variability is predominantly determined by the high degree of heterozygosity which is inherent in natural populations of any organisms. Recall that a heterozygote is an individual that has inherited different genes from the egg and the sperm. The average degree of heterozygosity is 13.4 percent for invertebrates and 6.6 percent for vertebrates and 18 percent for 8 species of plants. The percentage of mutant forms, i.e., the percentage of gametes (sex cells) containing mutant genes is quite enormous: it reaches 25 percent in the fruitfly *Drosophila*. The diversity of characters selected for by natural selection is determined not by improbable mutations, but by variability measured in percent and in tens of percent.

The heterozygosity in diploid organisms (i.e. organisms whose somatic cells contain two sets of chromosomes) implies the simultaneous presence in these cells of two allele genes, in particular a dominant and a recessive gene. Upon subsequent crossings the hidden characters are revealed.

The diallele nature of sexually reproducing diploid organisms allows one to test new alleles in the presence of the old ones that have already been tested.

An evolving system is not an individual organism but a population, i.e. a group of organisms belonging to a single species living under similar conditions.

The genome, a set of genes of each new individual, is a recombination of parental genomes. The recombination is the mechanism of assembly of gene combinations. There are no two identical individuals (except for monozygotic twins). The birth of an individual implies the generation of new information (see Chap. 10), i.e. the "accidental choice remembered", since no law of nature presupposes the emergence of a particular offspring from a given pair. The appearance of sexual reproduction provided a sharp increase in variability and the acceleration of evolution. Incidentally, the phenomena of chromosome transfer, similar to sexual reproduction, have also been detected and studied in bacteria [1.19].

Thus, the material for evolution is practically unlimited. For any change in the properties of the environment there may be a corresponding selection of optimally adapted organisms.

The second and third questions require more detailed consideration. The answers to them are associated with the directionality of evolution (Chaps. 8 and 9).

Figure 1.3 shows schematically the action of various types of selection on

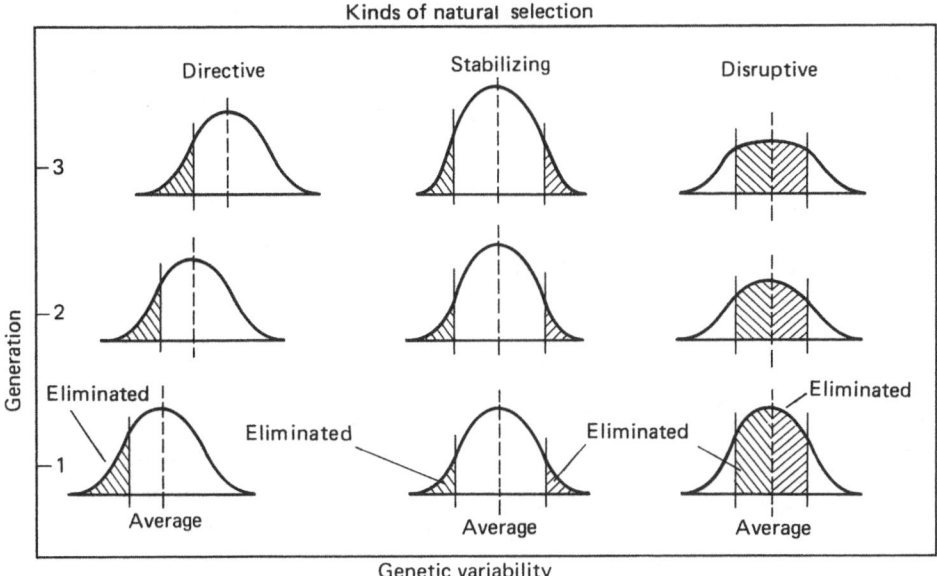

Fig. 1.3. Kinds of natural selection: directive, stabilizing and disruptive

genetic variability in a population [1.20]. Directive selection is a driving force and leads to a change in the genetic constitution of the population, as, for example, in the case of the industrial melanism of the birch geometrid. Disruptive (or diversifying) selection favours the preservation of extreme forms and the elimination of intermediate forms, thereby contributing to speciation. Finally, stabilizing selection performs a protective function, discarding extreme forms. This type of selection has been studied in detail by Schmalhausen [1.21]. It is more widespread in nature. The occurrence of stabilizing selection was not rejected by Berg either [1.17]. The meaning of this kind of selection can be reduced to the following: "One must run in order to be able to stay in one's place." It is a sort of "Red Queen Principle" from *Through the Looking Glass* by Lewis Carroll: "Well, in our country," said Alice, still panting a little, "you'd generally get to somewhere else if you ran very fast for a long time, as we've been doing."

"A slow sort of country!" said the Queen. "Now, here, you see, it takes all the running you can do, to keep in the same place. If you want to get somewhere else, you must run at least twice as fast as that."

The Origin of Species contains only one figure – the diagram of the evolutionary tree (see Fig. 1.4). Darwin emphasized that this representation in the plane is too simple. The branches diverge in different directions; the true evolutionary tree representing the relationships between species requires a three-dimensional model (see Fig. 1.5). The classification of organisms is directly associated with this. Darwin explains the evolutionary meaning of systematics using,

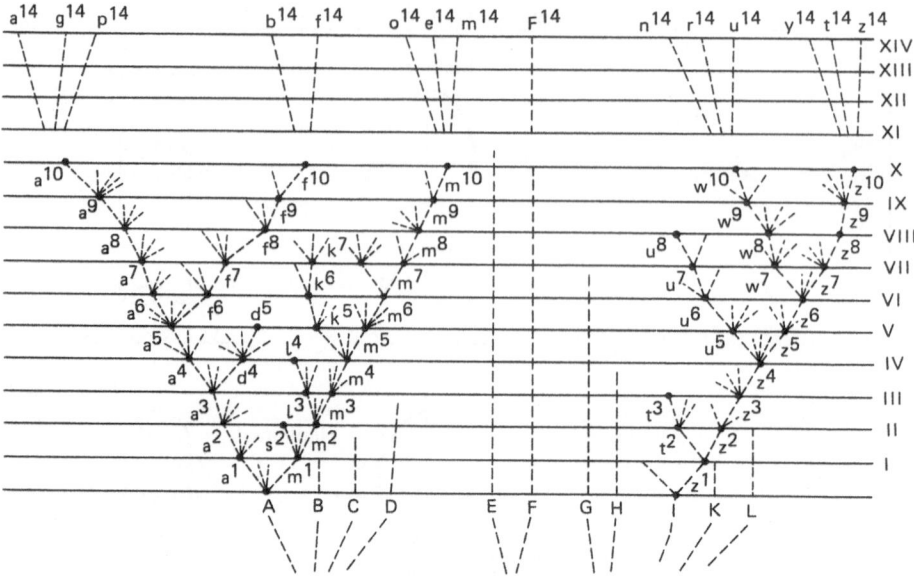

Fig. 1.4. Evolution according to Darwin (The Origin of Species, p. 115): $A, B, \ldots, L$ represent the species of a genus large in its own country; the dashed lines emerging from A and I represent the varying offspring of these species; the intervals between the horizontal lines each represent a thousand or more generations; a^1, m^1, etc., are fairly well-marked varieties; a^{10}, f^{10}, m^{10} are new species; F^{14} is a species which has been preserved without changes

as an example, linguistics. "If we possessed a perfect pedigree of mankind, a genealogical arrangement of the races of man would afford the best classification of the various languages now spoken throughout the world; and if all extinct languages, and all intermediate and slowly changing dialects, were to be included, such an arrangement would be the only possible one" [1.22].

These deep thoughts have only recently found their implementation. Cavalli-Sforza and his associates [1.23] (see also [1.24]) have attempted to reconstruct the evolution of the species *Homo sapiens sapiens* by comparing genetic, archeological and linguistic data. Genetic information was derived from a very large set of gene frequencies that determine the polymorphism of proteins of various peoples of the world. The data were grouped into 42 populations and the number of alleles studied was 120. A genealogical tree was constructed and its correlation with the tree of language families was established (Fig. 1.6). The first divergence of the phylogenetic tree separated Africans from non-Africans, the second gave rise to two main groups: a group including Caucasoids, Eastern Asians, Arctic peoples and native Americans and a group including South-Eastern Asians, the islanders of the Pacific Ocean, the aborigenes of New Guinea and Australia. It has been found that the average genetic distance between the most important groups is proportional to the archeological divergence time. Linguistic families correspond to groups of populations with a number of over-

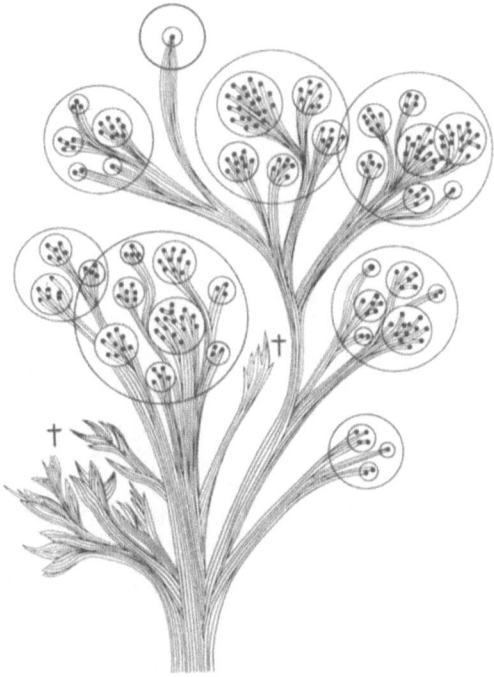

Fig. 1.5. A more detailed schematic representation of evolution according to Schmalhausen. The species are grouped into genera and the genera into families. The crosses represent extinct branches

laps. Linguistic superfamilies are direct evidence of the parallelism between genetic and language evolution. The nostratic theory proposed by Illitch-Switych [1.25] has been confirmed independently. According to this theory, the Afro-Asian, Indo-European, Altai, Arctic and other languages are lumped together into a single superfamily. The Russian word "kora" (bark) sounds like "kereg" in Hungarian, "kerek" in Tamil, and as "khor-akta" in Nanaian.

Cavalli-Sforza and his co-workers have emphasized that during evolution, the populations diverge into two subpopulations, which develop further independently. There is a geographical expansion and mixing of languages, their "horizontal transfer." This process presumably took place after the appearance of modern man and the extinction of Neanderthals. According to available data, *Homo sapiens sapiens* appeared about 100,000 years ago, the subspecies *Homo sapiens neanderthalensis* disappeared 30 or 35 thousand years ago.

A more detailed genetic classification of the European language families has been made by Harding and Sokal [1.26]. The genetic distances between the families were calculated using data on gene frequencies for the antigens of blood, enzymes and other proteins. The classification obtained reflects not so much the origin of the language as the geographical closeness. This is evidence of the importance of "short-range order"; the modern gene pools in Europe

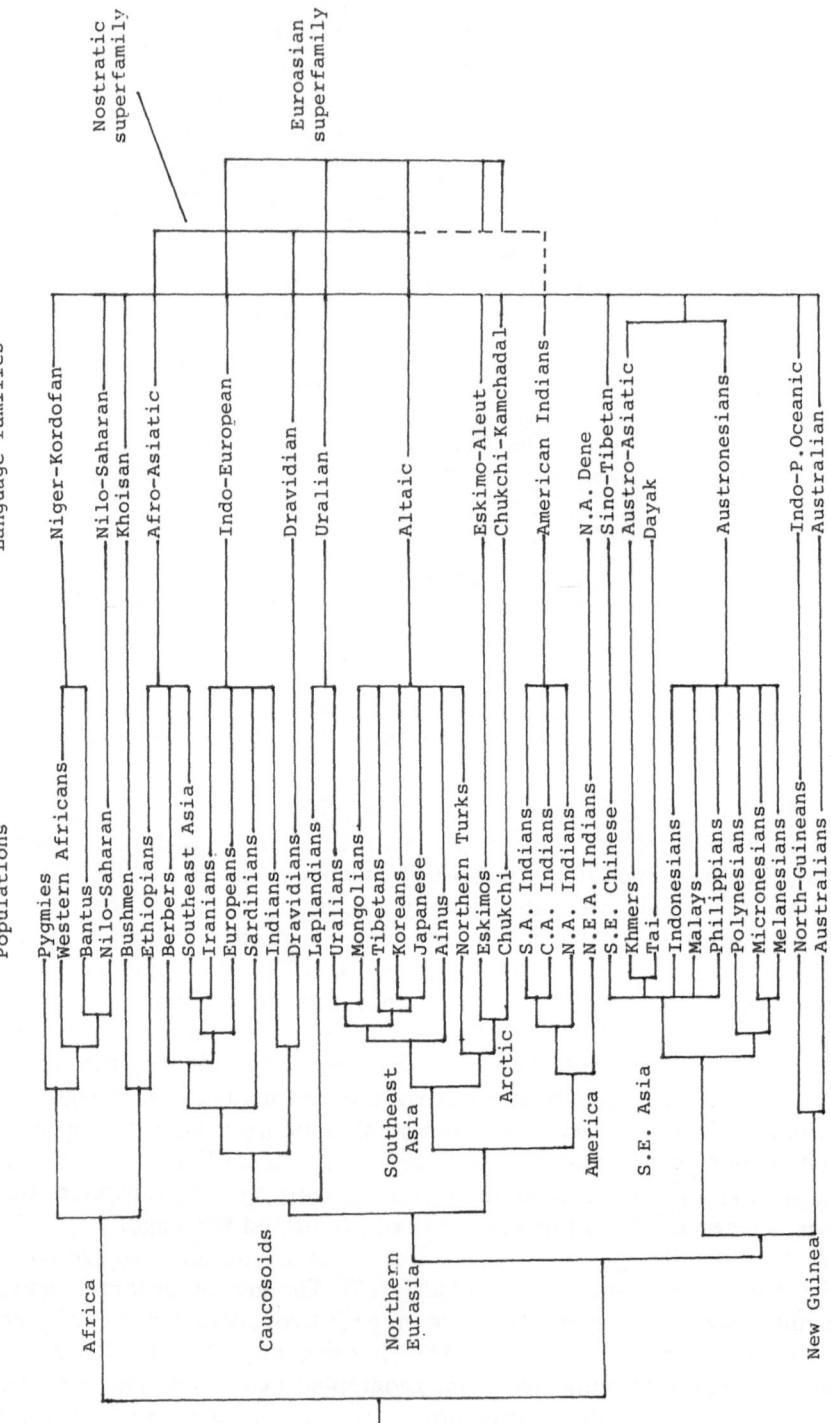

Fig. 1.6. The genealogical tree of populations of *Homo sapiens* and the linguistic families

apparently arose as a result of gene flows, i.e., intraspecies hybridization. However, members of certain language families, say Basques, Finns, Laplanders, and Maltese are not linked with their geographical neighbours (see also [1.26a]).

Of course, the correspondence between genes and languages must be rather crude. This is because man is mentally active and enterprising, capable of migrating to any remotely located regions (Mongoloids migrated to America via the Bering Strait and inhabited the two mainlands). Languages belong to later, post-evolutionary events but they are associated with the genetic features of populations.

As a result of directive and disruptive natural selection, speciation and macroevolution (i.e. the formation of higher taxa) occur. Evolutionary theory starts with the possible emergence of an ordered biosphere from the random variation of original organisms. All multicellular organisms are the descendant of single-celled organisms. Order emerged from chaos. How and why did this happen?

In physics we encounter two types of ordering. First, equilibrium ordering, which arises as a result of a first-order or second-order phase transition with the free energies of the two phases being equal. (Examples are, respectively, the crystallization of a liquid, the paramagnetic-ferromagnetic transition or the ordering of a binary alloy.) Second, the ordering of an open far-from-equilibrium system. Examples are: the transition from ordinary luminescence to laser radiation at the critical value of pumping; the appearance of the Bénard effect; and, as we shall see, a number of phenomena of life activity (see [1.7,27,28]). Ordering *far from equilibrium* occurs as a result of instabilities and an increase in small fluctuations to the macroscopic level and also has the character of a phase transition. These topics will be discussed in Chap. 9.

In the thermodynamic sense, both kinds of ordering imply a decrease in entropy. The entropy of a crystal is lower than that of a liquid. Coherent laser radiation arises due to a decrease in entropy.

More often than not we have to deal with the obscure statement of the "anti-entropy nature" of life. What is meant here is a decrease in entropy during the ontogenetic development of any organism from the fertilized egg cell (a zygote), and also the entire biological evolution which progresses in the direction of increasing complexity and ordering. On this basis, a number of works that have nothing to do with science state that living organisms do not obey the laws of thermodynamics. In fact, there is no problem here. A living organism, a population, and the biosphere are all open systems, whose mass, energy and entropy may increase and decrease (see [1.5,7,27,28]). A decrease in entropy in an open system, i.e. non-equilibrium ordering, occurs due to the export of entropy to the surroundings. This proposition is the basic one in a phenomenological, thermodynamic treatment of the events that occur in living nature. According to Schrödinger, an organism feeds on negative entropy [1.29]. This is a figurative expression; what he really meant is the efflux of entropy.

The entropy balance of the Earth with its biosphere is not diffcult to estimate (see [1.7], page 69). The Earth absorbs on average $J_E = 1.2 \times 10^{17}$ watts of short-wave radiation emitted by the Sun at a temperature of $T = 5770$ K. The same amount of energy is emitted by the Earth, but in this case at an average temperature of 257 K. The emission energy is given by

$$E = \frac{4\sigma}{c} V T^4 \ . \tag{1.1}$$

This is the Stefan-Boltzmann law. Here V is the volume, c is the light velocity, and σ is the Stefan-Boltzmann constant:

$$\sigma = \frac{\pi^2 k^4}{60 \, \hbar^3 c^2} = 5.67 \times 10^{-5} \ \mathrm{g \cdot s^{-3} \ deg^{-4}} \tag{1.2}$$

where k is Boltzmann's constant; $\hbar = h/2\pi$, in which h is Planck's constant.
    The free energy of radiation is given by

$$F = -\frac{4\sigma}{3c} V T^4 \ . \tag{1.3}$$

The entropy of radiation is

$$S = -\left(\frac{\partial F}{\partial T}\right)_V = \frac{16\sigma}{3c} V T^3 = \frac{4E}{3T} \ . \tag{1.4}$$

The density of entropy flux is given by

$$J_S = \frac{S}{V} = \frac{16\sigma}{3c} T^3 = \frac{4E}{3VT} = \frac{4}{3} \frac{J_E}{T} \ . \tag{1.5}$$

The total balance of the entropy flux (J_E is the same for the Sun and the Earth) is

$$\frac{d_e S}{dt} = \frac{4}{3} 1.2 \times 10^{17} \mathrm{W} \left(\frac{1}{5770\mathrm{K}} - \frac{1}{257\mathrm{K}}\right) = -6 \times 10^{14} \ \mathrm{W/K} \ . \tag{1.6}$$

The density of the entropy flux at the Earth's surface, the area of which is approximately equal to 5×10^{14} m^2, is given by

$$\frac{1}{5 \times 10^4} \frac{d_e S}{dt} = \frac{4}{3} 230 \ \mathrm{W/m^2} \left(\frac{1}{5770 \ \mathrm{K}} - \frac{1}{257 \ \mathrm{K}}\right) \cong -1 \ \mathrm{W/m^2 K} \ . \tag{1.7}$$

This export of entropy compensates for its average production of the order of 10^{-3} to 10^{-4} W/m^3 K in the lower layers of the atmosphere. A kind of an entropy pump is in operation [1.7]. The thermodynamic interaction with the Sun provides life on the Earth.
    The export of entropy determines the organized structure of the universe as a whole, i.e. its ordering in the form of the galaxies and stars. In this sense, life and evolution are similar to cosmic processes. What is common to the

evolution of the universe, the Solar System and the Earth, and to biological evolution? In all these cases we are dealing with the creation of new information, be it galaxies or stars, planets or folded mountains. New information is generated as a result of the "accidental choice remembered" (Quastler [1.30]), which arises in the case of the instability of the original state and the multistationarity of the system (see Chap. 10). The creation of new information has the character of a phase transition.

The similarity between cosmic and biological evolution is not limited to this feature. The appearance of inhomogeneities, i.e. order, in the form of stars and galaxies due to gravitational instability implies competition and natural selection. The gravitational inhomogeneities compete with one another for the material to be condensed. The condensation is accompanied by the efflux of entropy into the outer space filled with relict radiation at a temperature of 3 K.

Thermodynamics tells us not about what actually exists but rather about what might exist in principle. The development of life is possible because of the efflux of entropy, but this is a phenomenonological explanation, which is rather general and does nothing for the understanding of the course of evolution on Earth, the course of phylogeny or ontogeny.

Nonetheless, the fact that there is no contradiction between thermodynamics and theoretical biology is promising. Sooner or later the most important life phenomena will find their correct physical (physico-chemical) interpretation.

1.3 Physics and Biology

As was pointed out in the preface, the objective of this book is to describe physical approaches to biological evolution. But this task requires discussion. First of all, it is necessary to establish the relationship between physics and biology, and also the interplay between these sciences. I have already discussed these problems in the past [1.27,28,31]. Here they are considered in more detail.

The problems of these relationships are usually discussed without any definitions of the basic concepts. What are physics, chemistry and biology (the three most important areas of natural science)?

Physics is a science dealing with the structure and properties of all kinds of matter, substances and fields, and the forms of existence of matter: space and time. Physics is thus the foundation of all natural science.

Chemistry is a science concerned with the study of the structure of atoms and molecules and of the changes in structure as a result of their interactions.

Biology is the science of life, of living nature.

We see that these three definitions do not form a logical series. These definitions are incompatible. Does this mean that physics coincides with natural science on the whole, that we return to Aristotle? By no means. While being the deep-seated theoretical foundation of any branch of natural science, physics does not abolish the independence of chemistry or biology, but it discloses their content. The notion of the "reducibility" of these sciences to physics is absolutely pointless.

Any branch of natural science is a multilevel system. Physics is the lowest, most basic level.

Biology is concerned with the study of life. Life is the way of existence of macroscopic, heterogeneous, dissipative systems, i.e. open far-from-equilibrium systems, capable of self-organization and self-reproduction. The most important functional substances of these systems are proteins and nucleic acids.

Macroscopicity implies that any living organism, beginning with a bacterium, must contain a large number of atoms. If it were not for this, the order which is necessary for life would have been destroyed by fluctuations (see the book by Schrödinger [1.29]). Heterogeneity implies that a cell or organism is made up of a multitude of various substances. For example, erythrocyte contains more than 100 different proteins. Taken separately, molecules are not alive. Life arises only in a heterogeneous supramolecular system. Dissipative systems have already been discussed.

Proteins and nucleic acids are macromolecular substances of prime importance for life, but there are other necessary compounds: carbohydrates and lipids, numerous low-molecular organic compounds, water, acidic anions and metal cations. In its interaction with biology, physics is inextricably tied up with chemistry. The problem of establishment of theoretical, i.e. physical, foundations has on the whole already been solved in chemistry. There is no theoretical chemistry without physics: quantum mechanics, statistical mechanics, thermodynamics, and physical kinetics. Not all concrete problems have been solved in this area, but the physical content of the basic principles and conceptions of chemistry has been established. It does not mean at all that chemistry is reduced to physics. The techniques and methodology of chemistry are quite independent and indispensable. The establishment of physical foundations implies the realization of principles rather than their replacement. Chemistry is derived from physics rather than reduced to it.

In principle, the results of any chemical reaction can be calculated *ab initio* with the aid of theoretical physics, i.e. quantum and statistical mechanics. However, even in simple cases such calculations require a very large amount of computer time. Chemistry solves these tasks immeasurably faster and better. The situation resembles, to a certain extent, electrical and radio engineering. The physicist and the engineer know that calculations of the corresponding circuits are ultimately based on the Maxwell equations. But nobody would think of using these equations; there exist special, simplified methods for calculations of circuits.

The amazing feature of chemistry is that long before the advent of quantum mechanics, chemistry had, in fact, discovered its most important laws. When physicists were still not sure of the existence of molecules (Ernst Mach, Friedrich Ostwald), chemists already understood their structure. The point is that in chemical reactions the laws of quantum mechanics manifest themselves at the macroscopic level. The basic notions we are taught at school (valency, the position of an element in the periodic system etc.) can be understood only on the basis of quantum mechanics. In contrast to electrical and magnetic forces and the gravitational force, valence forces are discrete and saturable. We are used to thinking that quantum mechanics operates only in the microworld, that in order to encounter its laws we have to delve into the interior of the atom. But in chemistry we deal with a very large number of atoms at a time and, accordingly, with indefinitely mutiplied electronic events. Mendeleev's periodic law, the cornerstone of chemistry, was achieved during the time of absolute dominance of classical physics. Based on its conceptions, Mendeleev arranged all the elements known at that time in order of increasing atomic weight and established the periodicity of their chemical and physical properties. In doing this, Mendeleev discovered the most important characteristic of the elements: the atomic number. Later quantum mechanics demonstrated that the atomic number signifies the number of protons in the nucleus and the equivalent number of electrons in the atom.

The problem of the relationship between physics and biology is much more sophisticated. We are still far from discovering the physical basis of life phenomena in all their diversity. In the first place, this refers to the key problems of theoretical biology: biological evolution and individual development. Theoretical chemistry has, on the whole, been constructed, but theoretical biology is still in its infant stage.

Until recently even very eminent physicists did not understand the essence of biology. The physicist Ernest Rutherford compared biology with "postage-stamp collecting", i.e. he regarded the science of life as purely descriptive. Much later Wigner stated that the reduplication of DNA contradicts quantum mechnics and, hence, in order to understand life phenomena, some novel physics would have to be constructed. This meant the introduction of so-called biotonic laws [1.32], which were postulated by Elsasser [1.33]. Wigner's line of thought is erroneous; he ignored the template synthesis of nucleic acids. All debates about "biotonic" laws are vitalistic in character and absolutely non-constructive.

Polanyi rightly noted the exceedingly important role of the concepts of information in biology. However, he believed that information theory is not associated with physics [1.34]. We cannot agree with him (see Chap. 10).

Küppers has published a collection of articles devoted to the relationship between physics and biology, in which he has presented, in particular, the views of Elsasser and Heitler, and also the conceptions of Bohr and Schrödinger, which will be considered at a later time. Küppers expresses his viewpoint, which is diametrically opposite to vitalism [1.35]. I agree with him completely. However, I would not use the word "reductionism" to describe this viewpoint. We cannot state that biology is "reducible" to physics, i.e. that physics is simpler than biology. This is in conflict with the definitions of the two sciences.

Even today, many biologists see biology as being opposed to physics, assuming that these sciences are incompatible. Let us consider the fundamental work of Mayr, one of the founders of modern evolutionism [1.36]. This extraordinary book is highly instructive, but what is written in it about the relationship between physics and biology raises serious objections.

Mayr asserts that "the physical sciences are not an appropriate yardstick of science". There are no laws in biology similar to the laws of physics: "When they say that proteins do not translate information back into the nucleic acids, molecular biologists consider this a fact rather than a law" ([1.36], page 37). This statement is quite arbitrary. From this point of view the law of preservation of energy, for example, is also a fact rather than a law. Both biology and physics are concerned with the study and establishment of the laws of nature. From the impossibility of the transfer of information from proteins to nucleic acids follows another important law of biology: the law of the impossibility of the inheritance of acquired characters.

Mayr writes, "The phenomena of life have a much broader scope than the relatively simple phenomena dealt with by physics and chemistry" (page 52). He then continues: "This is why it is just as impossible to include biology in physics as it is to include physics in geometry" (page 53). "Theory reductionism is a fallacy because it confuses processes and concepts" (page 62). "The last twenty-five years have also seen the final emancipation of biology from the physical sciences" (page 131). Further, Mayr states that there are numerous situations in which physicalism has exerted a destructive influence on the development of biology.

These positions are all erroneous. The complexity of life does not contradict physics at all; physics has all grounds and possibilities for exploring natural phenomena, however complicated they might be. Of course, it is not the inclusion of biology in physics that should be considered; the task is to disclose the physical foundations of life phenomena. It is not physics which is incorporated into geometry, but, as has been shown by the theory of relativity, geometry which is included in physics. Reductionism will be discussed in the next section. The last 25 years have demonstrated the deep penetration of physics into biology: molecular biology and synergetics have been developed during that period. As regards the destructive effect of "physicalism" on biology, no examples of this kind are given by Mayr and they are difficult to find. On the contrary, we can cite numerous examples which demonstrate the great importance of physical ideas for the development of biology, beginning with Harvey's theory of blood circulation and ending up with the problem of the genetic code.

Mayr contrasts Darwinism with physics: "It was Darwin more than anyone else who showed how greatly theory formation in biology differs in many respects from that of classical physics" (page 521). Mayr writes further, "It was only after 1859 that the biological sciences began to get rid of the influence of the physical sciences."

The actual situation is completely the opposite. It was exactly Darwinism that brought biology and physics together for the first time. It was not incidentally that the great physicist Boltzmann called the 19th century "Darwin's century." Another eminent evolutionist, Simpson, approached the problem of the relationship between physics and biology in a somewhat more rational manner.

Simpson writes that the statement that the principles of investigation of life must be different from those adopted in physics does not imply the dualistic or vitalistic view on nature. Life should not be looked upon as something non-physical. But living things have been subjected to historical processes over billions of years. The results of these processes are systems which are radically different from any inanimate systems and which are incomparably more complex. All known material processes and the underlying principles are applicable to organisms, but only a limited number of these can be applied to nonliving systems. Biology is thus a science which stands in the centre of all sciences [1.37].

Noting justly that historical value is the most important feature of life, Simpson does not discuss at all the possibilities of physics in the study of historical processes in nature. The statement concerning all processes and principles in biology is in error. For example, the physics of elementary particles has no relation to life activity. The words about "the centre of all sciences" have no clear-cut meaning.

In an extensive article, the biologist Skvortsov [1.38] also states that physics is incompatible with biology. "Biology is interested in what differentiates living things from inanimate nature. Should not therefore theoretical biology differ from theoretical physics and would not biological theory be more biological the less it resembled the physical theory?... Therefore, presumably, in principle the most general and fundamental biological theories can be formulated not in an abstract mathematical form but only in the form of discussions, in the form of instructive, logically non-conflicting propositions." Skvortsov does not think highly of mathematical models: "They are constructed on the basis of certain basic premises, which, as a rule, strongly simplify the real situation, and the abstract nature of a model more often than not complicates the judgement about its correspondence to the phenomenon being studied." According to Skvortsov, molecular biology has not justified hopes either: "Nobody has detected even hints of any unified regularity from which one would infer that upon protein synthesis, a certain amino acid must be coded for by a certain set of nucleotides, or that some protein must have a certain amino-acid sequence, or that the protein must perform a certain function, etc."

This is far from being true. The problem of the genetic code has been solved. It has been established that there is a degenerate correspondence between the primary structure of protein and its biological function (see Chaps. 5 and 6).

Mayr, Simpson and Skvortsov are biologists who have little to do with physics and are not acquainted with those modern branches which deal with non-equilibrium historical processes. It is true that in the article written by Skvortsov there is a footnote in which he writes: "Views concerning the concepts of the evolutionary process, historicalness, originated in biology have also lately been put forward by physicists, particularly by Prigogine." As a matter of fact, physicists have done much more than that: they have developed an extensive new branch of physics: physics of dissipative systems, or synergetics.

We have cited the viewpoints of biologists based on the ideas of "anti-reductionism" (which will be considered in the next section) as being sufficiently typical. On the whole these ideas do not deserve attention.

Let us consider the views expressed by the great physicists Bohr and Schrödinger. While being concerned with the most general problems of science, Niels Bohr speculated much about the relationship between physics and biology. In doing so, he followed his family tradition. His father, Christain Bohr, was an eminent physiologist, to whom is accredited the discovery of the Bohr effect (the specific effect of the pH of the medium on the affinity of hemoglobin for oxygen).

The central point in the philosophy of Niels Bohr is the principle of complementarity formulated by him as a broad generalization of the discoveries made in quantum mechanics. The complementarity of two ideas, concepts or physical quantities implies the impossibility of their simultaneous cognition. By measuring a certain physical quantity with precision, we completely lose any information on the complementary quantity. A special case of complementarity is provided by the Heisenberg uncertainty relations. According to Bohr, complementarity goes far beyond classical and quantum physics. These ideas were put forth in 1954 in an article entitled "The unity of knowledge" [1.39]. Bohr writes about the complementarity of intuition and logic, of art and science, thought and action. "Any new knowledge comes to us in an envelope of old conceptions, which is fitted for explaining the previous experience, and any such envelope may prove too narrow to encompass new experience." This refers not only to knowledge but also to behaviour. "An especially striking example is the relationship between situations, in which we start thinking over the motivation of our actions, and those in which we experience the feeling of resoluteness." A thought and an action are complementary. This idea was excellently expressed by Shakespeare in the central monologue of Hamlet. The complementarity principle is of great importance for the understanding of art [1.40].

The views of Bohr on the relationship between biology and physics changed with the development of natural sciences, which is why it is interesting to consider his articles chronologically. In his essay "Light and Life" (1933) [1.41], he expounds his basic positions. First, "the recognition of the essential importance of fundamentally atomistic features in the functions of living organisms is by no means sufficient, however, for a comprehensive explanation of biological phenomena." Second, "the conditions holding for biological and physical researches are not directly comparable, since the necessity of keeping the object of investigation alive imposes a restriction on the former, which finds no counterpart in the latter." Third, "in biology we are dealing with material systems, whose complexity is of fundamental nature." From this Bohr arrived at his fundamental inference that "the existence of life must be considered as an elementary fact that cannot be explained, but must be taken as a starting point in biology, in a similar way to the quantum of action, which appears as an irrational element from the point of view of classical mechanical physics."

In his speech "Biology and Atomic Physics" (1937) [1.42] Bohr presented the same ideas in a more clear-cut form: "Biological regularities are, in essence, the laws of nature, which are complementary to those which are applicable to explaining the properties of inanimate objects." At the same time, Bohr emphasizes the incompatibility of his view with vitalism. He rejects "as unreasonable all attempts to introduce any special biological laws incompatible with rigidly established physical and chemical laws ... none of the results of a biological investigation can be described unambiguously in any manner other than on the basis of the concepts of physics and chemistry." As we have seen, this most important proposition has been refuted by many scientists. Before Bohr, the same idea was expressed by Berg [1.17] (see [1.43]): "No wonders happen in the world, nature works exclusively according to the laws of physics and chemistry." The theories advanced by Berg will be considered in Section 3.5.

In his speech "The Unity of Knowledge" (1955) [1.39], Bohr based his views entirely on the principle of complementarity. Complementarity "... exists between the considerations of physico-chemical nature used in biology, and the concepts that are directly associated with the integrity of an organism and go beyond the framework of physics and chemistry .... Only when we stop trying to explain life (explanations in the ordinary sense of the word), do we make it possible to take into account its specific features."

Thus, it was the first time that Bohr had revised his view. In the articles written in the 1930s he wrote, in fact, about the complementarity of biology and physics, and in 1955 he spoke of the complementarity of the atomic-molecular study of life phenomena and of the study of the whole organism.

In the article "Physical Science and the Problem of Life" (1957) [1.44] these ideas were developed further. Bohr wrote about the almost unlimited expansion of the applications of physical and chemical ideas in biology, but he pointed out that "an account of all continuously exchanging atoms of a living organism, exaustive in the context of quantum physics, is not only impossible but, evidently, would have required conditions of observation that are incompatible with the manifestation of life."

Note that such an "account" is hardly necessary for the solution of any scientific problem.

The views of Bohr described above are often regarded as arguments in favour of the so-called irreducibility of biology to physics. One of the eminent physicists of this century, Heitler, treats Bohr's views as follows ([1.45], see also [1.35]): "We may ask whether the laws of physics that are obeyed by inanimate matter could be applied to a living system. If the answer is positive, then a living organism will not be different in any way from non-living matter and there will be no place for the notion of life."

However, Bohr wrote nothing of the kind. He did not set any boundaries on the application of atomic-molecular physics in biology, and emphasized that no biological experiment could be described unambiguously in any way other than in terms of physics and chemistry. Heitler's reasoning is erroneous. From the applicability of the general laws of physics to living organisms, it does not follow at all that a frog is not different from the stone on which it is sitting, or that there is no place for the notion of life. The subsequent development of the natural sciences has shown that the laws of physics are obligatory in biology, but the manifestation and expression of these laws is different from that inherent in living nature.

Eigen has shown that contemporary physics is sufficient for constructing the true theory of molecular evolution (see Chap. 8).

Heitler states further that, according to Bohr, physical measurements conducted on organisms with the aid of X-rays etc. inevitably destroy organisms, and that these measurements are incompatible with life. However, as a matter of fact, such measurements provide a wealth of information about life. The discovery of the double helix of DNA by means of X-ray diffraction has given answers to the questions of the factors responsible for the stability of genes, the nature of chromosome doubling, and mutations. Today, the method of nuclear magnetic resonance is used in investigations of living cells and tissues. The number of such examples is unlimited.

In effect, Heitler repeats the words of Mephistopheles in "Faust" by Goethe [1.46]:

> Wer will was Lebendig's erkennen und beschreiben,
> Sucht erst den Geist herauszutreiben,
> Dann hat er die Teile in seiner Hand,
> Fehlt leider! nur das geistige Band.
> Encheiresin naturae nennt's die Chemie,
> Spottet ihrer selbst, und weiss nicht wie.

Heitler states that Bohr spoke of the strict complementarity of animate and inanimate matter. The very existence of life is incompatible with a detailed knowledge of the atomic-molecular structure of an organism. In other words, there is a complementarity relation which may be written provisionally as follows:

$$\text{Atomic structure} \times \text{Life} \sim \text{Constant}$$

But Bohr's statement presented in a similar form reads:

$$\text{Atomic structure} \times \text{Integrity of organism} \sim \text{Life}$$

The advances of molecular biology in the 1950s made Bohr, who attentively followed the development of natural sciences, revise his views on the relationship between biology and physics.

In the article entitled "Quantum Physics and Biology" (1959) he writes, "Quite different prospects have lately appeared of the gradual interpretation of biological regularities based on rigidly established principles of atomic physics. This was possible due to the discovery of remarkably stable special-purpose structures carrying genetic information and also due to an ever increasing insight into processes by which this information is transferred" [1.47].

What is meant here is DNA, the substance of genes. Bohr again emphasized the absence of any limitations on the application of physics and chemistry in biology. However, because of the extreme complexity of biological systems "the concepts pertaining to the behaviour of organisms, and seemingly contradicting the method of describing the properties of inanimate matter, have found successful application in biology". In conclusion, Bohr indicates that the rigorous quantum-mechanical complementarity has already been taken into account in applications of chemical kinetics in biology. A "complementary" approach in biology is required only because of the practically limitless complexity of organisms.

It is characteristic that Bohr uses the word "seemingly" and that he put in quotes the word "complementarity". In his speech in 1960 [1.48], Bohr speaks of the absence of any limitations or violations of the principles of thermodynamics in living nature, of the similarity between living organisms and automatons. At the same time, Bohr points out that life reveals natural resources much richer than those used in designing machines. Evolution is "the picture of results of testing in nature of the enormous possibilities of atomic interactions."

Bohr states further: "In biology, the foundations for a complementary description are not associated with the problems of control over the interaction between an object and measuring devices ... the necessity of complementarity of the description is associated with the practically inexhaustible complexity of an organism. Thus, as far as the word 'life' is retained (whether it be for practical or gnosiological reasons), the dual approach in biology will undoubtedly be valid."

In 1961 I wrote a letter to Bohr and received an answer, in which he confirmed that he had revised his views. These letters have been published [1.49].

The last of Bohr's speeches devoted to the relationship between physics and biology was made on June 21, 1962 at the Institute of Genetics in Cologne, and was entitled "Light and Life, Once Again" [1.50]. In this speech Bohr did not say anything about complementarity in biology. He emphasized that "the formation of all macromolecular structures ... involves basically irreversible processes, which determine the stability of an organism under conditions specified by feeding and breathing." Bohr abandoned the previously formulated conceptions, noting that "it cannot be the task of the biologist to give an account of the fate of every one of the countless atoms that are present, transitorily or permanently, in a living organism."

Thus, the complementarity of the atomic-molecular and integral description of an organism stated earlier by Bohr no longer exists. Complementarity is mentioned only in connection with psychology: complementary thoughts and feelings. The great Russian poet Tyutchev wrote, "A thought expressed out loud is a lie". But we still know little about this. The transition from fundamental to practical complementarity marks the radical revision of Bohr's views. This change has not been noticed by many scientists; the last two of Bohr's speeches are known only to a few, but his earlier views, which he repeatedly expressed, are widely known. The incompatibility of biology with physics seems to be more convincing at first sight than are the modern conceptions of the physical basis of life, the understanding of which requires special knowledge.

One of the few scientists who noted the evolution of Bohr's views was V.L. Ginzburg [1.51]. He stresses that up to his last days, Bohr could change his

views as a result of new, undisputable facts, such as, for example, the facts that
have determined the outstanding successes of molecular biology.

Niels Bohr had nothing to do with dogmatism. He looked upon the world
with widely opened eyes and clearly understood everything that was happening
in this world: the development of science and the threats, first of fascism and
then of nuclear war.

The conception of practical rather than fundamental complementarity cannot raise any objections. In essence, practical complementarity of atomic-molecular and integral investigations reduces at present to the complementarity of
atomic and phenomenological physics. The life activity of whole integral systems, beginning with the biosphere and ending with the cell, is studied with the
aid of physico-mathematical models based on thermodynamics, cybernetics and
synergetics.

Bohr's views on the relationship between physics and biology seem at present to be of largely historical interest, since they characterize the evolution of
those views of the great scientist and thinker which were directly associated with
the evolution of science. Bohr's views did not have a substantial impact on the
development of biophysics and theoretical biology, in contrast to the ideas of
Schrödinger.

Erwin Schrödinger, one of the founders of quantum mechanics, published
his lectures (which he read in Dublin in 1943) in a book entitled "What is
Life from the Standpoint of Physics?" [1.29]. In this book he said nothing
about complementarity and there were no doubts as to the application of physics to the study of life phenomena. Schrödinger set himself the task of formulat-
ing the basic questions that biology puts to physics, and trying to answer these
questions. He accomplished this task.

The first question is of a phenomenological nature and pertains to integral
systems. What are the thermodynamic bases of life, the bases of the existence
and development of dissipative systems? Schrödinger provided an answer to this
important question. Let us cite his words:

"What then is that precious something contained in our food which keeps
us from death? That is easily answered. Every process, event, happening – call
it what you will; in a word, everything that is going on in Nature means an
increase of the entropy of the part of the world where it is going on. Thus
a living organism continually increases its entropy – or, as you may say, produces positive entropy – and thus tends to approach the dangerous state of
maximum entropy, which is death. It can only keep aloof from it, i.e. alive, by
continually drawing from its environment negative entropy – which is something very positive, as we shall immediately see. What an organism feeds upon
is negative entropy. Or, to put it less paradoxically, the essential thing in metabolism is that the organism succeeds in freeing itself from all the entropy it cannot
help producing while alive."

A similar thought had previously been expressed by the physicist Emden in
1938 (see [1.52]). The maintenance of life by the efflux of entropy has already

been mentioned. This is a very important proposition, which completely refutes the empty resonings about the "anti-entropy nature" of life, and the apparent contradictions between the phenomena of life and the second law of thermodynamics. Note that the foundations of the thermodynamics of living systems were first formulated in 1935 by the biologist Bauer. Bauer wrote that living systems are never at equilibrium and perform work against equilibrium at the expense of their free energy [1.53].

The second question put by Schrödinger was as follows: Why are atoms so small? The question is paradoxical, and scientifically pointless. One should ask: Why are atoms so small with respect to the metre or foot, i.e. to measures of length corresponding in order of magnitude to the size of the human body? The question should, in fact, be reformulated, said Schrödinger. Why must living bodies be larger than atoms? In other words, why do organisms consist of a very large number of atoms? (The smallest bacterium, *Mycoplasma laidlavii*, is composed of 10^9 atoms).

Schrödinger gave an answer to this question. It is because with a small number of atoms an ordered system cannot arise as the order would be destroyed by fluctuations. Life is macroscopic.

The third question is: What is the physical nature of mutations? Based on the classical work of Timofeeff-Ressovsky, Delbrück and Zimmer (1934), Schrödinger pointed out the quantum nature of mutations caused by radiation, considers physical estimates of the size of the gene and explains how the nucleus of the egg cell or the head of the sperm can accommodate "the diversified content pressed into a miniature code." These ideas were completely confirmed during the subsequent development of molecular biology.

The fourth question is: How can molecules that form chromosomes and are built up only of light atoms (C, H, O, N, P, S) be so stable that the inherited characters are preserved over many generations? An exmaple is the drooping lower lip of the members of the Habsburg dynasty. Schrödinger did not know an answer to this question. The answer was provided by the discovery of the double helix of DNA in 1953.

Schrödinger thus threw light on the two practically complementary branches of biophysics – phenomenological and atomic-molecular cognition of living systems. According to a number of scientists who had made significant contributions to the development of biophysics and molecular biology, this development was stimulated by Schrödinger's book.

The relationship between physics and biology in connection with problems of the origin of life and biological evolution has been discussed in the monograph by Küppers [1.54]. The author indicates three possible approaches to the problems of the origin of biological information, namely, (1) the chance hypothesis; (2) the teleological approach; (3) the molecular-Darwinistic approach.

The hypothesis of the origin of life as a singular chance event, the probability of which is negligibly low, was supported by Monod in his well-known book [1.55]. As a matter of fact, if we are speaking of random rather than pre-determined events, their probabilities are vanishingly small. For example, the number of possible polypeptide sequences 100 units long is 10^{130}, and, accordingly, the random synthesis of such a protein has a probability of 10^{-130}. If the appearance of nucleic acids and proteins (the emergence of life) is a purely random event, then physics and chemistry play no role in this process, which may be treated as an absolutely unique act of creation. These ideas go beyond the framework of science.

The teleological approach recognizes the presence of a purpose in a biological system (the Greek word "telos" means completion, end). In fact, this approach is the modern modification of vitalism. It has been developed in the works of Elsasser, Wigner, and Polanyi mentioned earlier. According to Polanyi, the working of an elaborate biological machine of a living system cannot be explained on the basis of physics and chemistry. Special boundary conditions are imposed on the laws of physics and chemistry. The special nucleotide sequence in the genome, which carries biological information, is similar to the construction principle of a machine in the sense that it produces boundary conditions specified by biology on the laws of physics and chemistry. These boundary conditions govern the action of the laws of physics [1.34,35].

A number of scientists proceed from anti-reductionism and anti-physicalism on the grounds that life is the property of a complex, integrated, whole system. The whole is more than the sum of its parts, it possesses special properties which require a biological rather than a physical approach. This viewpoint is associated with the philosophy of the so-called holism, which introduced the scientifically uncognizable "wholeness factor". According to holism, the world is governed by the basically mystic process of creative evolution, which produces new "wholenesses". Holism has nothing to do with natural sciences.

We have already seen how Niels Bohr treated the problem of the whole and the parts. Evidently, the necessity of studying a whole system does not contradict physics at all. Let us cite in this connection the words of the eminent mathematician and physicist Weil. Refuting vitalism as a conceptual system based on "a too narrow understanding of the mechanical or physical explanation of nature", Weil writes [1.56]: "The wholeness is not a distinctive feature of the organic world alone. Each atom is already a quite definite structure; its organization serves as a basis of possible organizations and structures of the highest complexity."

The third approach to the origin of biological information is the molecular-Darwinistic approach. It has been worked out most thoroughly in the works of Eigen and his school (see Chap. 8). Eigen and his associates have developed a rigorous physico-mathematical theory of the self-organization and natural selection of macromolecular systems, which has been proved experimentally. It was shown for the first time that natural selection can and must be understood on the basis of modern physics. The Eigen theory is inseparable from the physics of dissipative systems and from information theory.

Küppers, in his monograph [1.54], convincingly disproves so-called anti-reductionism, which will be discussed in the next section. We give here the

Table 1.2. The most important classes of biological theories

	External conditions	General laws	What is explained
1. Ontobiological theories			
Genetic determinism	Physical environment and biological constraints	Laws of physics and chemistry	All life phenomena
Classical vitalism	Physical environment and biological constraints	Laws of physics and chemistry and specific laws of life	All life phenomena
2. Theories of biological development			
Molecular Darwinism	Physical environment	Laws of physics and chemistry	Biological boundary conditions
Scientific vitalism	Physical environment	Laws of physics and chemistry and specific laws of life	Biological boundary conditions

classification of fundamental biological theories composed by Küppers (see Table 1.2).

Science refutes both classical and "scientific" vitalism. Küppers convincingly demonstrates the great importance of molecular Darwinism for the further development of natural science.

The basic problems of theoretical biology, which are still far from being solved, concern phylogeny and ontogeny: biological evolution and the biology of individual development. In the 1930s Dobjansky, Fisher, Haldane, Wright and others developed the synthetic theory of evolution, which united classical Darwinism with genetics, primarily population genetics. Evolutionary theory is a vigorously developing branch of knowledge. In an article under the title "Sturm und Drang and Evolutionary Synthesis", Futuyma writes [1.57] that this area of science has never been so alive, so exciting, and so successful as it is today. The synthetic theory, which radically deepened the understanding of biological evolution, did not remove the considerable inconsistency between historical biology, i.e., paleontology and systematics, and the biology of the present-day biosphere. Nonetheless, evolutionary synthesis undoubtedly counts as one of the biggest intellectual achievements of our century.

Today we are living in another period of development of evolutionary theory and ontogeny. In the second half of the 20th century new scientific disciplines have emerged, whose importance cannot be overestimated. These are molecular biology, which is inseparable from physics; the physics of dissipative systems or synergetics; and cybernetics, which includes information theory. The current task of evolutionary theory is to unite with these branches of natural science, i.e. with physics, since molecular biology is inseparable from physics. Of great importance for the further development of evolutionary theory on the basis of this new union are the Eigen theory (Chap. 8) and the Kimura theory (Chap. 6). What follows in this book concerns the approaches to such a union.

1.4 Concerning Anti-Reductionism

What has been expounded in the previous section is, I think, unacceptable to many biologists and physicists (not to mention philosophers), beginning with such an outstanding scientist as Ernst Mayr. Mayr is an outspoken and convinced anti-reductionist; he thinks that "physicalism", i.e. a physical interpretation of biological phenomena, is unacceptable and even harmful. His words have been cited above.

Mayr and his numerous followers are very far from physics and, in fact, do not understand its importance in the natural sciences. They would perhaps be greatly surprised and distressed if they learned that anti-reductionism coincides with the positions of official philosophers in the Soviet Union, who were fully supported by the authorities over many years. Lysenko was also a convinced anti-reductionist.

The combat against "reductionism" and "physicalism" is harmful and pointless. These notions as such are devoid of meaning, at least with respect to biology. In the Soviet Union they were bugaboos invented to scare naturalists. Today the times of fear are ended and we can treat the problem on its merits. We have already given arguments in favour of the union of biology with physics.

We read in the Philosophical Encyclopedic Dictionary ([1.58], page 551): "Reductionism ... is a methodological principle, according to which the higher forms of matter can be fully explained on the basis of the regularities inherent in the lower forms, i.e., can be reduced to the lower forms (for example, biological phenomena with the aid of physical and chemical laws ...)." Here the words "fully ... reduced" are meaningless. Today there is no scientist who might think that biology is fully reducible to physics; the very notion of "reduction" is devoid of any meaning. From the definitions given above it does not follow that biology is more complicated than physics. What is written in the dictionary is based on the ideas of Engels, which contain two principles: first, every branch of science created by man is concerned with the study of a definite form of motion of matter, which exists independently of our consciousness; second, there is the hierarchy of sciences proceeding from the simple to the complex:

$$\text{Mechanics} \rightarrow \text{Physics} \rightarrow \text{Chemistry} \rightarrow \text{Biology} \rightarrow \text{Sociology} \ .$$

In the past Engels' ideas seemed attractive, since they refuted the positivistic classification of sciences, which did not explain phenomena but only described them. However, today we understand that nature is not constructed that simply, and that statements such as "Physics is concerned with the study of the physical form of motion of matter, and biology deals with the biological form of motion of matter" have long become a meaningless tautology. Each science is a multi-level system of knowledge which is developing and changing with time. Both physics and biology are concerned with the study of numerous diversified "forms of motion". As regards the hierarchy of sciences, we first of all need the definition of complexity, which is far from being simple (see Chap. 10).

Of course, physics will never replace biology. We are not speaking here of biology being reduced to physics. We should rather be concerned with disclosing the deep-seated physical basis of life phenomena.

Reductionism in natural sciences is the necessary and most constructive method of cognition. At all times the task of science is to reveal simplicity in the observed complexity. Occam's principle that the hypotheses should not be multiplied without necessity, must work. Harvey's theory of blood circulation was "reductionistic" and "physicalistic"; it was even mechanical. But it was a big step forward in physiology and the further development of our knowledge of blood circulation would have been impossible without this first step. The same can be said about the elucidation of the similarity between breathing and burning (Lavoisier), about the discovery of the genetic code, and about numerous other events in the history of science.

In application to biology it is rather difficult to separate anti-reductionism from vitalism. These trends of thought are interrelated. We shall not speak of vitalism, which has long been abandoned. It should only be noted that anti-reductionism has always been absolutely non-constructive; it has merely created obstacles to the development of science.

The anti-reductionism of a number of biologists can evidently be explained in a simple way. These scientists, among whom there are outstanding individuals such as Mayr, are not acquainted with modern physics. Of course, it is not easy for a scientist to step out of the framework of his science. Great achievements in biology are not yet the guarantee of the breadth and depth of the general scientific approach.

2. Evolutionary Methods

2.1 Palaeontology

"In the Pampean deposit at the Bajada I found the osseous armour of a gigantic armadillo-like animal, the inside of which, when the earth was removed, was like a great cauldron; I found also teeth of the Toxodon and Mastodon, and one tooth of a Horse in the same stained and decayed state. This latter tooth greatly interested me."

These are lines from Darwin's book on his *Beagle* voyage [2.1], a book which he devoted to Charles Lyell, one of the founders of geology. Darwin made a number of palaeontological discoveries during his voyage on the Beagle; in particular, in 1833 he discovered in Uruguay a remarkable extinct ungulate animal – Toxodon, which is known at present as *Toxodon darwini* (Fig. 2.1). The Toxodon is "perhaps one of the strangest animals ever discovered: in size it equalled an elephant or megatherium, but the structure of its teeth, as Mr. Owen states, proves indisputably that it was intimately related to the Gnawers ..." [2.1].

Darwin not only collected and described, he also speculated. The fossil remnants found by Darwin in his youth became one of the real foundations of his great theory.

For the majority of people, including those engaged in various branches of natural sciences, palaeontology is a rather strange science. Many people believe that it boils down to the collection of interesting, more or less spectacular fossilized remains and imprints. Something like "postage-stamp collecting", according to Rutherford.

In fact, palaeontology, which is concerned with the study of animal history, is of fundamental importance to biology as a whole. As has already been said, biology is also a historical science, the objects of which undergo constant change with time and retain the evidence of their evolutionary development. The palaeontological record contains direct evidence of evolutionary history. There is no other direct proof of speciation and macroevolution. Neontology, i.e. a science about currently existing taxa, provides only indirect information about these processes, which are too long for direct observation.

Palaeontology is a rapidly developing science. Firstly, new material is constantly being accumulated. Secondly, methods of its treatment are being refined such that the time during which extinct plants and animals lived can be final-

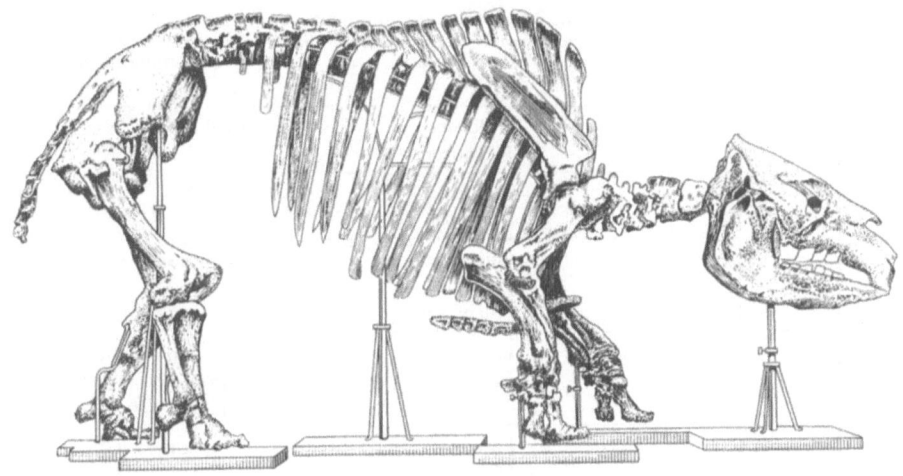

Fig. 2.1. The skeleton of *Toxodon darwini*

ized. Thirdly, (and this is of prime importance) the results obtained assume a deeper significance on the basis of evolutionary theory.

The dating of fossils is predominantly based on the radioactivity of a number of atoms; this is a scientific achievement of our century. The half-life of the carbon isotope ^{14}C is 5570 years, which is why its use in palaeontology is limited. The carbon isotope ^{14}C is important for archaeology, i.e. for human history. In palaeontology, efficient use is made of other elements: the uranium–lead pair ^{235}U–^{207}Pb with the half-life of uranium 7.1×10^8 years; the pair ^{238}U–^{206}Pb with the half-life of uranium 1×10^{16} years; and, most importantly, the potassium–argon pair ^{40}K–^{40}Ar with the half-life of potassium 1.32×10^9 years. What is determined is the ratio of the amount of a radioactive element to that of its decay end-product in a given mineral.

Another method is based on palaeomagnetism. The magnetic poles of the Earth have exchanged their positions from time to time. Magnetized minerals retain an accurate memory of these events. In conjunction with radioactivity dating and the study of the sequences of fossilized organisms, the magnetic method provides useful results.

Table 2.1 presents the modern geochronological scale [2.2,3]. The dating of the periods is based on the radioactive technique, whose accuracy diminishes with increasing remoteness from present time. Each period and each epoch is characterized by certain fossil organisms. Moreover, in a number of cases the epoch is subdivided into so-called centuries. This has proved useful and necessary, particularly for the abundant palaeontological record of the land mammals of South America [2.4].

The established chronology allows one to trace the history of taxa, i.e. classes, orders and so on. Figure 2.2 shows, as an example, the history of the class *Reptilia*. The line thickness or band width indicates the relative size of a

Table 2.1. Geological time scale

Eon	Era	Period	Epoch		Events, millions of years ago
Phanerozoic	Cenozoic	Quaternary	Holocene Pleistocene		Evolution of man, 1.8
		Neogene	Pliocene Miocene		25 ± 2
		Palaeogene	Oligocene Eocene Palaeocene		Radiation of mammals, 66 ± 3
	Mesozoic	Cretaceous	Late Early		Last of dinosaurs. First primates. First flowering plants, 136 ± 5
		Jurassic	Late Middle Early		Dinosaurs. First birds, 190 − 195 ± 5
		Triassic	Late Middle Early		First mammals. Domination of therapsids, 230 ± 10
	Palaeozoic	Permian	Late Early		Mass extinction of marine organisms. Domination of pelicosaurs, 280 ± 10
		Carboniferous	Late Middle Early		First reptiles. Lepido-dendrons. Seed ferns, 345 ± 10
		Devonian	Late Middle Early		First amphibians. Increase of diversity of fishes, 400 ± 10
		Silurian	Late Early		First vascular land plants, 435 ± 10
		Ordovician	Late Middle Early		Burst of diversity in Metazoa families, 490 ± 10
		Cambrian	Late Middle Early		First fishes. First Chordata, 570
Cryptozoic (Precambrian)					First elements of skeleton. First molluscs. First traces of animals, 650 − 690 ± 20
		Proterozoic	Upper (Riphean)	Upper Middle Lower	1050 ± 30 1350 30 1650 50
			Lower (Karelian)		2500 ± 100
		Archean			3500

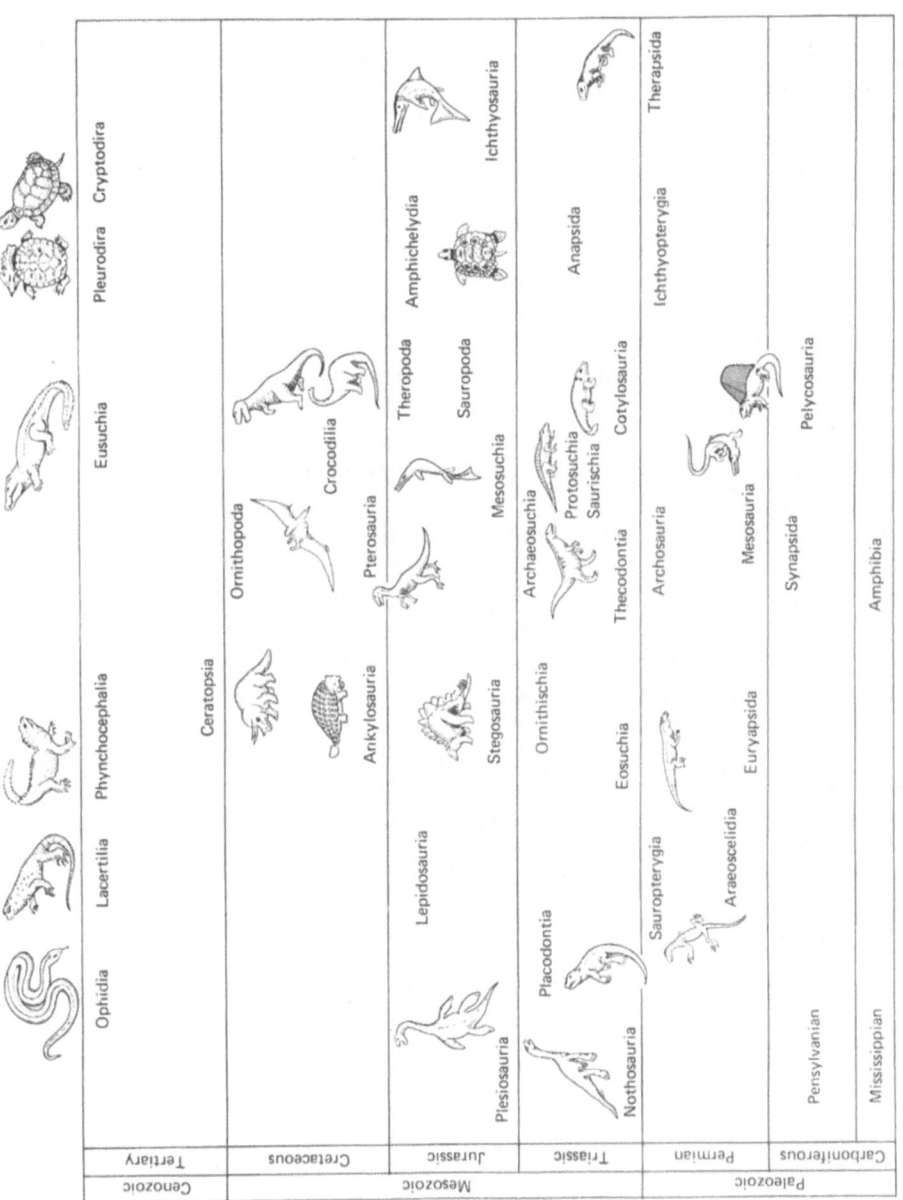

Fig. 2.2. The history of the class *Reptilia*

given family, i.e., the approximate number of species in a given geological period
([2.3], page 164). We see that dinosaurs flourished in the Jurassic period
(plesiosaurs, stegosaurs, ichthyosaurs) and in the Cretaceous period (ceratop-
sians, pterosaurs, iguanodons, etc.). Only a few ichthyosaurs survived up to the
Cenozoic era.

The classification of fossil organisms is based exclusively on the morphology of their fossilized remnants. There are characteristic and best-preserved
parts. Thus, especially remarkable for the history of vertebrates are the remnants of teeth; it is not by chance that Darwin became especially interested in a
certain tooth. The structure of teeth shows the feeding pattern of an animal and
what food it ate, i.e. whether it was a predator or a herbivore and, in the latter
case, whether its food was hard grass or soft fruits.

The systematics of extinct organisms thus encounters considerable difficulties. Is a given organism a new species or only a variety of it? Palaeontology,
of course, does not provide any information about reproductive isolation, or
indeed about the differences at the molecular level. The higher the taxon, the
more reliable its determination, but to trace out directly the emergence of a new
species is not that simple. Incidentally, the important law of speciation was in
fact discovered through palaeontology. Evolution proceeds irregularly: relatively short periods of speciation alternate with very prolonged periods of morphological invariability, periods of stasis. This will be discussed at a later time.
This regularity was known to Darwin.

Palaeontology is concerned with the similarities and differences between
organisms (as is neontology). Substantial differences arise in the course of divergent evolution. According to Schmalhausen [2.5], divergence implies the independent acquisition by closely related organisms of different characters. In Fig.
2.3, which was taken from Schmalhausen [2.5], the letters a, b, c, d, f, g signify
these characters. The scheme proposed by Schmalhausen coincides with the
Darwinian scheme: the only figure in *The Origin of Species* (page 115) shows
divergent evolution. The similarity of organisms belonging to different taxa is of
dual nature. In the first place, there is parallel evolution, i.e. the "independent

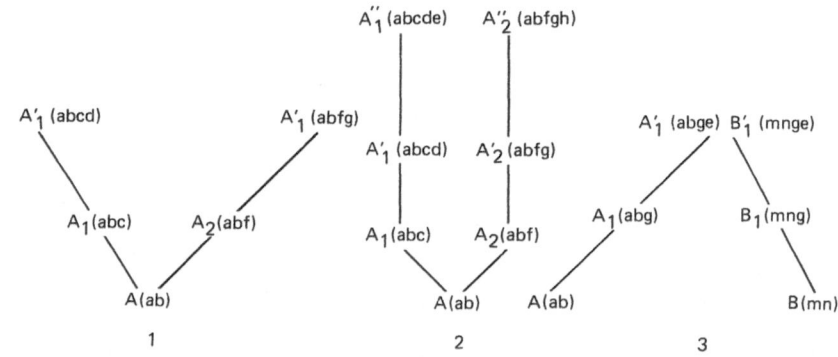

Fig. 2.3. Schematic representation of divergent (*1*), parallel (*2*) and convergent (*3*) evolution

acquisition of similar characters by closely related organisms" [2.5]. Closely related organisms are those which emerged from a common ancestor. In the second place, the similarity arises in a number of cases in unrelated organisms. This is convergent evolution, which implies the "independent acquisition by unrelated organisms of similar characters". Such is, for example, the similar structure of the eye in the octopus and a vertebrate. Both ways of acquiring similar characters during evolution are shown in Fig. 2.3 [2.5]. The intrinsic meaning of these phenomena will be considered in Chap. 3, which is devoted to the directedness of evolution.

The rate of evolution of characters is measured in a number of cases in darwins. One darwin is equal to a change of some character by one percent of the original character during 1000 years. It is obvious that such estimates are possible only for those characters that are described by a quantitative measure, and different characters of the same organisms may evolve at different rates.

2.2 Species

As has already been pointed out, a species is the real unit of evolution. Mayr defines a species as "a reproductive community of populations (reproductively isolated from others) that occupies a specific niche in nature" ([2.6], page 273). This definition, in which the emphasis is on reproductive isolation, is directly applicable to bi-parental organisms reproducing sexually. A species is thus characterized by a protected gene pool. Certain limits of genetic variability exist, which can accumulate in a gene pool, without leading to the appearance of an extremely large fraction of non-viable recombinations. The presence of protected gene pools, i.e. species that introduce order into genetic diversity, is a guarantee that these limits will not be surpassed [2.7]. Isolating mechanisms are varied – they may be defined as the biological properties of individuals which prevent the interbreeding of sympatric populations (those living in the same geographical region). Isolation may occur either before or after matings, and may have the character of ethological barriers, may be seasonal, etc.

The definition proposed by another well-known evolutionist, Simpson [2.8], is less clear-cut: an evolutionary species is a lineage which develops separately from others and which plays a unified evolutionary role of its own and has tendencies of its own.

One of the founders of the synthetic theory of evolution, Dobzhansky [2.9], regards species as a step of the evolutionary process "... at which the once actually or potentially interbreeding array of forms becomes segregated into two or more separate arrays which are physiologically incapable of interbreeding."

Since it is reproductive isolation that is the exact criterion of a bi-parental species, there arises a natural question as to the species of asexual prokaryotes and Protozoa. A number of authors maintain that the species definition itself loses its meaning if applied to these living things (see, for example, [2.10]).

Poljansky [2.11] maintains that there are other criteria for the protection of the gene pool, apart from reproductive isolation. Such criteria are morphology, geographical position, population structure, biochemical features, relative stability and integrity. Poljansky illustrates these positions by a number of examples referring to Protozoa. He comes to the following conclusions: "The species is a form of life existence common to all living organisms. At various steps of evolution the phenomenon of species bears on its qualitative peculiarity. This is the case with protozoa, especially in connection with their unicellularity, diversity of reproduction forms and complexity of their life cycles."

Takhtajan proposes a species definition, which is as general as the above one and which applies to both bi-parental and uniparental organisms [2.12]. A species is "a system of clones or populations which has segregated during evolution and which is united by common attributes (morphological, ecological, biochemical, genetic, cytological and others), by a common origin and a common geographical area and which is sufficiently distinctly isolated from related species by both a totality of its characters and by different (different for different species) isolating barriers".

One might inquire as to the causes of the discreteness of species, the causes of the absence of continuous transitions between them. This discreteness is eventually determined by the inhomogeneity of the environment in space and time. The interactions within an inhomogeneous population inhabiting an inhomogeneous environment lead to the splitting of the population into clusters, i.e. groups united by larger interactions. Information appears which serves as a source of new information (see Chap. 10). Such "clustering" is inherent in biological evolution, the evolution of human communities and the evolution of languages.

Thus, there are no doubts as to the reality of the biological species concept. It is also clear that the differences between the existing species and the differences in their evolutionary origin eventually have molecular foundations at the level of nucleic acids and proteins.

Species differ. There are fundamental differences not only between Metazoa and Protozoa but also between the different species of a single group. As has been suggested by Elizabeth Vrba, it is reasonable to divide species into "generalists" and "specialists" [2.13,14]. "Specialists" occupy narrow ecological niches and are therefore tolerant with respect to closely related species, and there is no competition between them. However, it is for this reason that minor changes in environmental conditions can easily "push" them from the niche. As a result, "specialists" are easily changed or die; in this sense, their evolutionary development must be faster than the evolution of "generalists". Vrba gives examples referring to the subfamilies *Alcelaphinae* and *Aepyceratinae* of the family *Bovidae* (hollow-horned). The first (cow antelopes) are "specialists", the second (gazelles) are "generalists". "Generalists" adapt to various niches, can exist in diverse conditions and therefore evolve slowly. Over a period of 6×10^6 years 27 species of *Alcelaphinae* have emerged, but only 2–4 species of *Aepyceratinae*. The scheme of the evolution of "specialists" and "generalists" is given in Fig. 2.4.

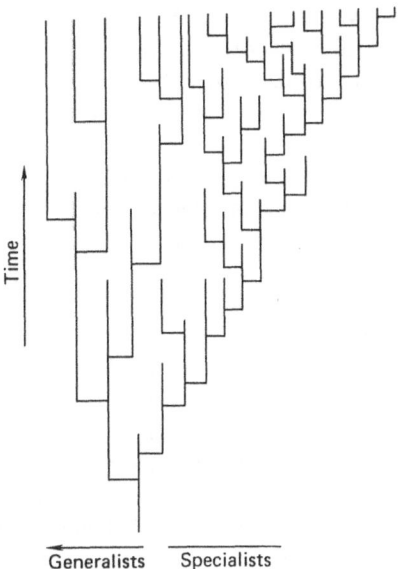

Fig. 2.4. Evolution of specialists and generalists

Accordingly, the number of species of "specialists" must considerably exceed the number of species of "generalists". At present the number of insect species is estimated at 750,000 but it may reach three million since not all species have been described so far, even in moderate regions. Timofeeff-Ressovsky told me once that one entomologist, who visited the Ilmen reservation in the Urals, discovered during one summer 20 new species of ichneumon flies (*Apocryta*). (Note that the total number of vertebrate species is 47,000.) The overwhelming majority of herbivorous insects are "specialists" adapted to feeding upon certain species and even certain plant tissues. It is this fact that accounts for such a large number of insect species. Predatory insects, such as the praying mantis (*Mantis religiosa*) and the diving beetle (*Dytiscidae*), are, on the contrary, "generalists". This is also valid for other classes and phyla of animals: the number of herbivores that are "specialists" by far exceeds the number of "specialist" carnivores.

The number of plant species is estimated at 250,000 and the number of fungus species at 100,000. Both groups consist mostly of "specialists". The coevolution of plants and the insects that pollinate them is evident. The smaller number of plant species (as compared to insects) is probably accounted for by a different character of the evolution of fungi and plants.

Grant writes that in all large groups of higher plants we have to deal not with phylogenetic trees but with phylogenetic networks [2.15]. Speciation and macroevolution with regard to plants and fungi are probably connected with their simpler organization and with their natural hybridization. Parasitic organisms are narrow specialists. An example is the nematode, which parasitizes two hosts, an ant and a cow. An ant infected by this nematode climbs to the top of a blade of grass, where it dies. A cow eats grass and thus provides new conditions suitable for the nematode. Other herbivores also eat grass with ants infected by nematodes, but these nematodes cannot live in horses, sheep, goats, etc.

Among mammals, apart from those described by Vrba, typical specialists are the giant panda *Ailuropoda melanoleuca*, whose diet is limited to bamboo roots and shoots, and the koala *Phascolarctos cinereos*, which eats only certain types of eucalyptus leaves. It is for this reason that these animals are difficult to keep in zoos.

Among the herbivorous species belonging to a single order, larger organisms are presumably more often specialists than smaller ones. An example would be the beaver *Castor fiber*, a specialist; the rat *Rattus norvegicus* is the champion among generalists (alongside *Homo sapiens*).

However, the question of the rate of evolution of specialists and generalists is not simple. Among generalists there are species with a relatively large brain mass, e.g. primates and songbirds. They are "cleverer" than their competitors. Some of them can evolve rapidly since they are capable of adapting to diversified ecological niches and exhibiting behavioural ingenuity. This has not happened to gorillas, but the tits in England quickly learned to peck through the foil caps of milk bottles left at doorways (Fig. 2.5) [2.16].

An excellent example of allopatric speciation among birds with a relatively large brain is Darwin's finches, which have already been mentioned. There are currently 14 species of finches on the Galapagos Islands.

If we measure the rate of evolution by the number of species that emerge per unit of time, I think it will be faster for specialists, although a large proportion of these species will eventually vanish. However, if we measure the rate of evolution by the scale of evolutionary morphological changes, the result may by

Fig. 2.5. A titmouse can peck through and remove the cap of a milk bottle

quite different. The contiguous species of insects are very similar to one another. This refers, in particular, to 800 species of *Drosophila* on the Hawaiian islands, which appeared in a relatively short period of time.

In any case, the division of species into these two groups, specialists and generalists, is quite meaningful. Of course, intermediate cases also exist.

Naturally a question arises as to the molecular basis of this classification. The enzymes of specialists must be especially selective regarding the choice of substrates. Situations are known where certain enzymes necessary for the assimilation of various kinds of food are absent in some populations of the same species. Some good people in Australia, wishing to help the natives, organized kindergartens for their children. The children were well fed and taken care of. However, the children began to fall ill and some nearly died. It turned out that in the bodies of Australian aborigenes, and also of some other peoples, lactose is not synthesized and so these people cannot drink cow's milk.

Specialization at the cellular and molecular levels can be reduced to structural recognition. Lymphocytes, which produce antibodies to a particular antigen, and these antibodies themselves are both narrow specialists. Molecular recognition by specialized cells is the basis of odour perception (see [2.17]).

In the course of normal ontogeny cells differentiate and become specialized. This also refers to tissues – totipotency is replaced by unipotency. In carcinogenesis, on the contrary, specialists are replaced by generalists. Cancer cells are generalists, and this is what determines their oncogenous action. The transformation of a normal cell to a cancer cell is equivalent to a change in symmetry, which increases in this case. Such a process is similar to a second-order phase transition, say the ferromagnetic-paramagnetic transition. Evolutionary events that are similar to phase transitions will be considered in Chaps. 8 and 9.

Contemporary molecular theory of evolution has not yet been developed to such an extent as to account for the differences between generalists and specialists. This is a matter for the, perhaps not too remote, future.

Depending on the type of habitat, even closely related species can significantly differ demographically, undergoing K- or r-selection (organisms are, respectively called, K- and r-strategists) (cf [2.18]). The r-selection is inherent in organisms inhabiting unstable environments and producing numerous offspring. The K-selection is inherent in organisms living in stable habitats and producing a relatively small number of offspring. Thus, frogs undergo r-selection, and higher mammals, in particular *Homo sapiens*, are subject to K-selection. It would seem that natural selection is optimal for r-strategists since it favours individuals that produce the largest number of offspring. However, r-selection does not imply maximal survival: the overwhelming majority of the numerous offspring die before reaching maturity. On the other hand, the conditions of existence exert a direct influence on the development of progeny. A remarkable example is provided by the closely related families of hares (*Leporini*) and rabbits (*Oryctologini*) of the order *Lagomorpha*. The number of offspring in a single brood differs only slightly, but rabbits give birth to blind, naked offspring which are not adapted to independent existence, while hares

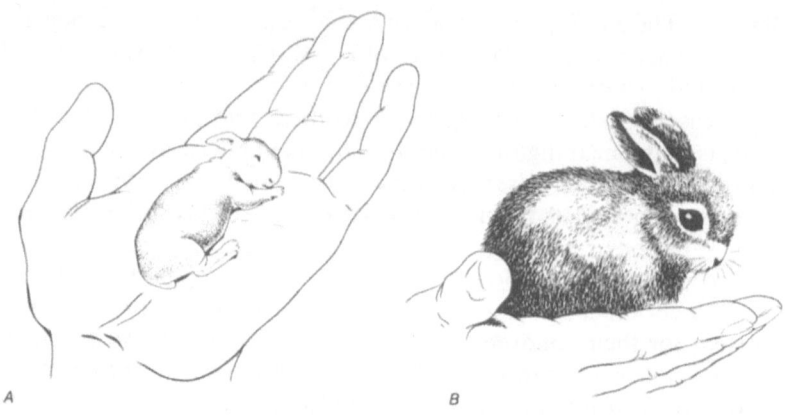

Fig. 2.6. The one-day-old young of a rabbit (*A*) and of a hare (*B*)

bear young with vision and a hairy coat (Fig. 2.6). The causes of such differences should be investigated in ontogeny.

2.3 Gradualism and Punctualism

One of the key positions of classical Darwinism is the graduality of evolutionary changes, as a result of which new species and higher taxa appear. This position was given great significance by Darwin. In the *Origin* Darwin writes: "If it could be demonstrated that any complex organ existed, which could not possibly have been formed by numerous, successive, slight modifications, my theory would absolutely break down."

Natura non facit saltum – this was Leibniz's idea, which turned out to be one of the principal claims of Darwinism. The main arguments in its favour were taken by Darwin from a study of artificial selection [2.19]. The gradual morphological and behavioural changes of doves, dogs and cats have led (and still lead) to the creation of new breeds which differ radically from their ancestors. The activities undertaken by man have enabled these processes to be rapid by human rather than geological standards. Belyaev has recently carried out experiments on domestication of foxes, choosing them according to their contact with man. He observed the rapid appearance of foxes with bent tails, drooping ears, etc.; these breeds got on well with man, in contrast to their wild progenitors [2.20]. However, in all cases of artificial selection of animals, new breeds appeared, but not new species. The karyotype (a set of chromosomes) of dogs is the same for all breeds and coincides with the karyotype of their ancestor, the wolf. At the same time, for many breeds of dogs reproductive isolation has occurred; the crossing of the wolf with the toy-terrier is impossible for physiological reasons. The emergence of a large diversity of artificially bred stocks is, in general, understandable and is well demonstrated by Belyaev foxes. Their selection was realized according to behavioral patterns; the diversity of forms produced was associated, according to Belyaev, with hormonal factors that interact with the brain. It is more natural, however, to think that there was normal variability and under selection conditions only those forms survived which should have vanished in nature. By the way, the emergence of domestic dogs was also originally determined by behaviour. The Palaeolithic man found young wolves. He killed the young that tried to bite or retreated into the corner, and carried those which licked his hand back to his cave for his children to play with.

Darwin regarded artificial selection as a model of natural selection. The palaeontological record presents species that have changed over millions of years. Natural selection based on the adaptation of characters to environmental conditions might, in principle, have produced any results. Let us cite Darwin's words concerning the emergence of the eye, an extremely complex organ [2.21]:

"Reason tells me, that if numerous gradations from a simple and imperfect eye to one complex and perfect can be shown to exist, each grade being useful to its possessor, as is certainly the case; if further, the eye ever varies and the variations be inherited, as is likewise certainly the case; and if such variations should be useful to any animal under changing conditions of life, then the difficulty of believing that a perfect and complex eye could be formed by natural selection, though insuperable by our imagination, should not be considered as subversive of the theory."

This is, no doubt, very strong reasoning. It cannot be refuted by considerations regarding the uselessness of the bad eye. Creatures that possess an organ which differentiates light from darkness have advantages over creatures devoid of such an organ. There is an even greater advantage for creatures capable of seeing a predator or prey. Analogous considerations are valid for other sensory organs as well.

Thus, gradualism is directly associated with adaptationism, i.e. the graduality of evolutionary changes is not realizable if at each stage these changes are not adaptive. Adaptationism or selectionism, no matter which of the terms is used, is very important to Darwinism. At the same time, Darwin wrote "I am convinced that natural selection [and, hence, adaptationism (M.V.)] has been the main but not the exclusive means of modification."

On the other hand, Wallace, who, quite independently of Darwin, formulated the idea of natural selection of the fittest, was a hyperselectionist and refuted any other way of evolutionary changes, including sexual selection. This did not prevent him from maintaining that the human brain had been created by God (see [2.22]).

The gradualness of evolutionary development is usually demonstrated by classical examples of the transformation of the myriapod *Hyracotherium* to the present-day horse (*Equus*) with its hoofs. Another example is the evolution of elephants. However, even in these cases many intermediate forms are missing, something that was accounted for by Darwin by the incompleteness of the biological record. It is really far from being complete in the majority of cases, though, as we shall see, there are exceptions.

Even during Darwin's time, gradualism was not considered to be mandatory, even by the most ardent evolutionists. Thomas Henry Huxley, a friend of Darwin and a defender of his theory, wrote to Darwin in November 23, 1859: "You have loaded yourself with an unnecessary difficulty in adopting *Natura non facit saltum* so unreservedly".

Thirty-four years later Huxley wrote to Bateson: "I see that you are inclined to defend the possibility of appreciable "saltus" on the part of Nature in its variations. I always adhered to the same view, much to the displeasure of Darwin". However, Darwin was right in considering gradualism to be a necessary part of his theory. Discontinuity or saltationism does not really agree with classical Darwinism; nor does it agree with the synthetic theory of evolution, which united Darwinism with population genetics. This "new synthesis" or "neo-Darwinism" is based on the following conceptions (cf. [2.23]):

1. The ultimate source of variability is point mutations. Evolutionary change is the process of gradual replacement of alleles in a population. Mendel's laws are fulfilled everywhere and at all times.
2. Genetic variation provides only raw material for evolution. Evolutionary changes are determined by natural selection. The rate and direction of changes are controlled by selection with some minor constraints that are imposed by the raw material itself.
3. All the genetic changes observed are adaptive, since it is only these changes that are subject to natural selection.

Gould and Eldredge, while citing the words of Huxley given above, state that "gradualism was always based on an adopted prejudice rather than on palaeontological findings... When Darwin was strongly inclined to accept gradualism ... he translated the Victorian society into biology .... Gradualism was part of the cultural context rather than of Nature" [2.24].

One cannot agree to this, however. As has just been said, gradualism is directly connected with selectionism, i.e. gradualness has a deeper meaning.

Marxist philosophy greeted Darwinian gradualism as a striking illustration of one of the commonplace truths of dialectical materialism – the "transformation of quantity into quality". A few words should be said in this respect.

What does the principle of the transformation of quantity into quality mean in natural phenomena? In the space of certain variables that have a quantitative measure, a process of gradual change of these variables occurs, which ends up with the appearance of a new essence, different from the one that has changed. Two points are involved here. Firstly, the degree of difference of the new essence from the old one is itself subject to quantitative estimation. It may be either large or small. Second, any natural process occurs in time. The transition from quantity to quality may be rapid or slow, gradual or discontinuous (saltational). In the language of physics, both thermodynamics and kinetics of the transition are important.

The melting of a crystal as a result of gradual heating is a first-order phase transition, which may be treated as the transition of quantity (the amplitudes of the anharmonic vibrations of atoms of molecules of the crystalline lattice) to a new quality of a melt, a liquid, in which the atoms have acquired degrees of freedom of translational motion. The degree of the transition may be characterized by means of thermodynamic quantities, i.e. the differences between the enthalpies and entropies of the liquid and crystal.

The kinetic characteristic is the rate of melting determined by its mechanism, which depends on the particular nature of the substance. The thermodynamics of phase transitions has been studied sufficiently thoroughly, but in kinetics there are still many unsolved problems.

In contrast to crystals, glass melts gradually. Thermodynamics does not provide an explanation of this process – the very existence of glasses is determined by kinetic factors.

As we shall see, the basic propositions of the synthetic theory have no general, absolute value. They are re-examined at a new stage of development of evolutionary theory.

An attempt to refute gradualism was undertaken by Goldschmidt in 1940 [2.25]. He advanced the idea of "hopeful monsters" which appear as a result of strong mutational changes in a new generation. It remained obscure, however, how a number of "monsters" sufficient for their multiplication and fixation in a population could have appeared simultaneously. For these reasons, Goldschmidt's idea was considered as purely speculative at first.

In 1972 Gould and Eldredge formulated the concept of punctualism (as opposed to gradualism), i.e. "punctuated equilibrium" [2.26]. This concept has been presented in more detail in the journal *Paleobiology* [2.24]. According to punctualism, evolution is considered to be a rapid (in the geological sense) phenomenon of speciation which follows the prolonged stasis. Stasis may continue for millions of years, during which mutational fluctuations of morphology with no definite direction occur. Speciation is realized during tens of thousands of years.

It is necessary to note that this was known to Darwin. In the fourth edition of the *Origin* [2.27] we read: "... the periods, during which species have undergone modification, though long as measured by years, have probably been short in comparison with the periods during which they retain the same form."

These words constitute one of the most important propositions of palaeontology and evolutionary biology in general. However, Darwin did not draw any punctualistic conclusions from what he said.

Gould and Eldredge reject gradualism, assuming that gradual evolutionary changes governed by point mutations occur too seldom and too slowly. Speciation is an independent process, incomparably more important than phyletic evolution resulting from the accumulation of such mutations. Gradualism is incapable of providing an explanation for gaps in the palaeontological record, which Darwin ascribed to its incompleteness. Gradualism does not explain the phenomenon of stasis. According to punctualism, the morphological gaps in the record may be real and each case of stasis is no less important to evolutionary theory than each example of the changes.

The greater part of evolutionary changes is concentrated in small, peripherally isolated populations – allopatric speciation takes place. Allopatric implies geographically separated. An example is provided by Darwin's finches on the Galapagos Islands. But the punctuational model is also applicable to sympatric speciation (when two groups of animals occur together in the same geographical region) if daughter species arise in a small part of a population. The normal state of a species is stasis with fluctuations in a large population. Prolonged stasis is maintained by stabilizing selection, which discards considerable deviations from the average norm.

Punctualism does not refute gradualism completely. As Gould and Eldredge have written, the main point here is not whether gradual evolution exists or not but how often it is observed. Gradual evolution is contrasted to speciation. Here it must be kept in mind that the presence of intermediate forms between two species may result from hybridization rather than from divergence. However, this violates the principle of reproductive isolation, although palaeontology provides no information about karyotypes and intermediate forms may only be varieties. Gould and Eldredge state that the majority of species do not generally serve as evidence of appreciable evolutionary changes; they disappear without leaving offspring. Evolution is not uniform or, in this sense, gradual. During rapid speciation there cannot be complete adaptation of any characters (non-adaptationism). During further, practically invariable existence, a viable species finds and forms its own ecological niche and becomes sufficiently adapted. Thus, punctualism follows from non-adaptationism and vice versa. Adaptation will be discussed in Chap. 4.

Palaeontological argumentation in favour of punctualism is extensive and varied. The replacement of stasis by rapid speciation may be considered to have been established for a number of evolutionary lineages. As Stanley pointed out in his fundamental monograph devoted to macroevolution [2.28], not a single example of phyletic, i.e. gradual, transition from one genus to another is known. Stanley maintains that macroevolution and speciation are punctuational events. Macroevolution will be discussed in the next section.

Williamson has studied successive deposits of fossil molluscs in the Turkana Basin in Africa and has come to the conclusion that their evolution is punctuational [2.29]. In these cases it has been found that fundamental phenotypic changes in relatively large populations are accompanied by appreciable increases in phenotypic variability. Williamson writes that this increase indicates the extreme instability in the development of a population which appears to be in a transient state. Special emphasis should be put on the word "instability" (see Chaps. 8 and 9; see also [2.30]).

In the literature the question of the evolution of hominids which led to *Homo sapiens* has been repeatedly discussed. Gould and Eldredge thought it to be punctuational; arguments in favour of gradualism are given by Cronin et al. [2.31]. The question cannot really be considered to have been solved. Avise

[2.32] has made a comparison of gradualism and punctualism for North American fishes, namely 69 species of *Cyprinidae* and 19 species of *Centrarchidae*. He used as a critical test the electrophoretic analysis of 13 proteins. It turned out that the rate of evolution of the proteins is somewhat lower for the more rapidly speciating *Cyprinidae*. In Avise's opinion these results are not consistent with punctualism and they do not explicitly contradict gradualism. It should, however, be stressed that the comparison of proteins is not a test for the character of evolution; not only the protein structure is essential but also the exact answers to three questions – how much protein is synthesized, when and where.

Sheldon [2.33] points out the gradualness of the evolution of trilobites in the Ordovician. Based on this work of Sheldon, Maynard Smith published an article under the title "Darwinism stays unpunctured" [2.34]. It is worth dwelling on this article because it is a typical case of misunderstanding of punctualism. Maynard Smith writes that punctualism cannot be taken as a universal phenomenon and that adaptation cannot be dismissed completely. Both these statements are correct, but the concept of punctualism does not claim universal applicability and does not refute the existence of adaptations. Accepting long periods of stasis punctuated by brief bursts of rapid change, Maynard Smith discards at the same time the ideas of non-adaptationism, the selection of species and the decoupling of microevolution from macroevolution.

The answer of Eldredge and Gould bears the title "Punctuated equilibrium prevails" [2.35]. They emphasize that variations and variability exist at any time in the history of a species. However, it is precisely the impossibility of transforming these variations to speciation and macroevolution that is the pivotal proposition of punctualism. Sheldon and Maynard Smith maintain that the evidence of "punctuation" must consist of instantaneous, quantum morphological shifts within the stratigraphical sequence of fossils in a single phyletic line. However, punctualism requires the overlapping of two separate taxa, which are interpreted as sibling species; their ancestor may simultaneously be preserved. The task is to understand the origin of these species and their differences. It is only in this context that the problems of speciation may be considered. The idea of the coincidence of punctualism with saltationism is incorrect. Saltationism implies sharp morphological inheritable changes in one generation or in a few generations. Punctualism implies only a much shorter time span of speciation as compared to stasis (see also Dawkins [2.36]). Stebbins wrote: "The origin of a new kind of animal in 100,000 years or less is regarded by palaeontologists as 'sudden' or 'instantaneous'" (see [2.36], page 242). Note that the breeding of new stocks of goldfish or dogs does not produce a new species. Both the lap-dog and the Russian wolfhound are wolves. Gradual natural selection does not lead to speciation. Perhaps this is because there is too little time.

Eldredge and Gould have analysed the arguments put forward by Sheldon concerning trilobites [2.33] and show that Sheldon has no grounds for considering evolution as gradual in a number of lineages. The consideration of palaeontological findings on speciation is always beset with formidable difficulties. One observes only morphological differences which do not go beyond the range

of intraspecies variability. Palaeontology provides no information about reproductive isolation, or about the differences between genomes. Therefore, more
often than not, one can find quite arbitrary palaeontological arguments both in
favour of and against canonical gradualism.

The graphical representation of both gradual and punctuational evolution
has already been given in this book. The only figure in the *Origin* presents
gradual speciation (see Fig. 1.4). Punctuated speciation is sketched in Fig. 2.4.
For clarity Fig. 2.7 is added. Punctuational evolution may be called rectangular.
The vertical lines indicate stasis and the horizontal lines represent speciation.

Gould and Eldredge point out that the gradulism-punctualism dilemma
depends on the "magnification" – on the timescale [2.26]. Figure 2.8 shows
the seeming gradualism in a certain evolutionary branch passing through the
species b and c and the punctuated character of this branch revealed by strong
"magnification".

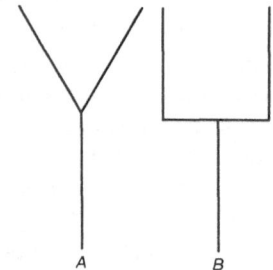

Fig. 2.7. Simplified representation of gradual (*A*) and punctuated (*B*) divergence of species

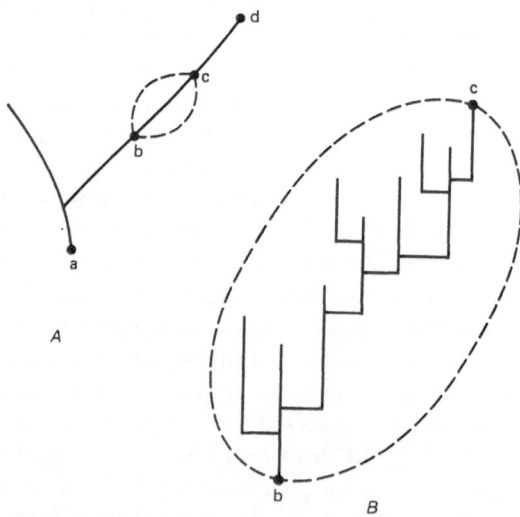

Fig. 2.8. Apparent gradualism (*A*) is actually punctualism (*B*)

Dawkins believes that the difference between the two concepts reduces only to the tempo of evolution [2.36]. Punctualism (or punctuationism) implies not sudden jerks but only episodes of relatively rapid evolution. Gould and Eldredge are actually gradualists; they place special emphasis on the frequently observed rapid speciation. One cannot agree with this. The theory of punctuated equilibrium tells us much more. It explains stasis and, what is especially important, the nonadaptationism of a number of characters (see Chap. 4). The punctuational theory leads to the hierarchial structure of living nature: individuals form populations, in which they compete with one another; populations form species, which are analogues of individuals at the level of macroevolution, i.e. the appearance of higher taxa (genus, family, order, class, phylum).

It is obvious that evolution must be more gradual in Protozoa and especially in bacteria which reproduce asexually. The units of their evolution are not species but clones (strains of bacteria), and variability is determined not by gene recombinations but by the appearance of new clones.

Punctualism is closely associated with directionality, canalization of evolution, which will be considered in Chaps. 3 and 9.

In the light of the concepts of gradualism and punctualism we may consider the scheme of evolution proposed by Severtsov [2.37]. Figure 2.9 shows a three-dimensional evolutionary tree. The branches in the plane alternate with transitions from one plane to another. Branching in a given plane means adaptations (b, b') to various environmental conditions, i.e. the period of idioadaptation. The transition to another plane (aromorphosis) implies the development of new characters of progressive nature $(Q \rightarrow R)$, or the reverse, i.e. simplification $(Q \rightarrow P)$. When speaking of idioadaptations, Severtsov meant microevolution and speciation. Much rarer aromorphoses simulate macroevolution. A modern evolutionist might say that idioadaptations express changes within a species, which are gradual, and that aromorphoses express punctuational speciation and macroevolution.

The ideas of punctualism had been formed by eminent evolutionists long before the works of Gould and Eldredge were published. In 1944 Simpson wrote about quantum evolution, which begins with a random fixation of non-

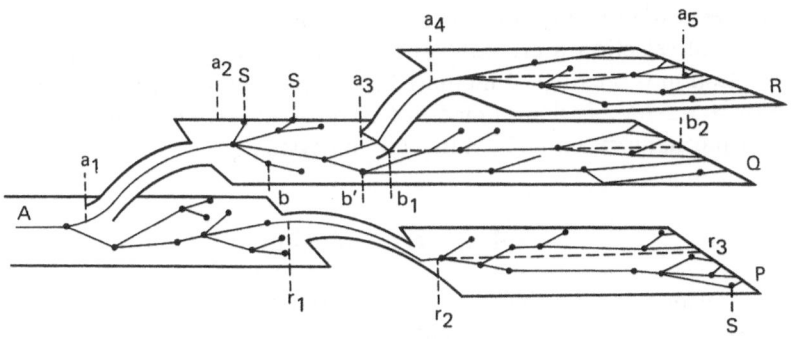

Fig. 2.9. Evolutionary process according to Severtsov

adaptive mutations in small, isolated populations [2.38]. In 1954 Mayr stated that only sharp changes in structure are possible. "The fossil record known to us contains not a single example of phyletic evolution terminating a long morphological transition" [2.39]. Examples of very rapid speciation are known; an especially remarkable example is the emergence of 800 species of the genus *Drosophila* on the Hawaiian Islands.

Of course, the real phyletic line may be more or less punctuational or gradual. The vertical lines in the rectangular scheme of Fig. 2.8 may be inclined.

A detailed and thorough study of gradual and punctuational evolution requires physical approaches based on the general conception described in this book. They include, first, the molecular basis of evolution (which will be discussed in Chaps. 5, 6, 7 and 8) and, secondly, the treatment of biological evolution as specific behaviour of dissipative systems (Chaps. 8, 9 and 10). As we shall see, the gradualist–punctualist controversy is finding a new interpretation.

A question that is often asked concerns the creative role of natural selection. What is creative activity? It invariably implies the creation of new information, i.e. the memorizing and fixation of a random, nondeterministic choice (see Chap. 10). In this sense the outcome of creative work is not only a new species but also a new individual. Indeed, the emergence of offspring from a given couple and the corresponding recombination of parental genomes are not determined by any laws. Creation in speciation implies a change in a number of characters, the preservation of some characters and the rejection of others.

This is what Richard Dawkins writes about constructive evolution [2.36]: "People sometimes think that natural selection is a purely negative force, capable of weeding out freaks and failures, but not capable of building up complexity, beauty and efficiency of design. Does it not merely subtract from what is already there, and shouldn't a truly creative process add something too? One can partially answer this by pointing to a statue. Nothing is added to the block of marble. The sculptor only subtracts, but a beautiful statue emerges nevertheless. But this metaphor can mislead, for some people leap straight to the wrong part of the metaphor – the fact that the sculptor is a conscious designer – and miss the important part: the fact that the sculptor works by subtraction rather than addition. Even this part of the metaphor should not be taken too far. Natural selection may only subtract, but mutation can add."

Is natural selection a creator? There is no doubt that it is, but not only natural selection is creative. Punctuated speciation, which opposes the gradual change of characters, is determined not only by natural selection but also by molecular mechanisms embedded in evolving systems.

2.4 Macroevolution

The gradualistic legacy of Darwin is determined, as we have seen, by three factors: the incompleteness of the palaeontological record, the enormous geological timescale, and the slowness of natural selection.

Macroevolution is the appearance and extinction of higher taxa. If we assume the gradual nature of speciation, then the higher taxa must also arise gradually. We have already pointed out the biological reality of a species. From the gradualness of speciation follows the conditionality of the division of higher taxa, say genera. Indeed, there is no new criterion here, apart from the criterion of reproductive isolation, which already operates at the species level.

Macroevolution is directly consistent with the concept of punctualism, as has been shown in Stanley's fundamental monograph [2.28]. Natural selection may be assumed to be the original cause of evolutionary changes only if its action is episodic and therefore strong enough to work for short time intervals. Palaeontology shows that phyletic, gradual evolution is very slow. Not a single case is known of the gradual transition from one genus to another. The number of genera in the Pleistocene in Europe is estimated at 15–20 and the average time of existence of a genus at 8 million years. Presumably, genera were formed relatively rapidly as a result of divergent speciation. Thus the polar bear, the only representative of the genus *Thalarctos*, probably appeared not more than 20,000 years ago. Stanley has come to the conclusion that the genera of animals appeared by way of "quantum", i.e. punctuated, speciation. The rate of macroevolution and the rate of genus formation depend on the rate of speciation, which is a multiplicative process. Adaptive radiation, i.e. the occupation of various ecological niches, may be represented by an exponential time dependence of the number of species N:

$$N = N_0 e^{Rt} , \qquad (2.1)$$

that is,

$$\frac{dN}{dt} = RN . \qquad (2.2)$$

The rate of increase of the number of species R is equal to the rate of speciation S minus the rate of species extinction E:

$$R = S - E . \qquad (2.3)$$

As a matter of fact, E in this expression is the rate of completion of an evolutionary lineage. There is a second component, which represents the rate of species extinction due to phyletic transitions.

From the concept of punctualism it follows that the taxa characterized by high rates of speciation must also experience rapid large-scale evolution. Figure 2.10 shows the variation of the number of mammalian families with time and Fig. 2.11 presents a similar dependence for bivalve molluscs, which are characterized by a much slower speciation [2.28]. The existence of relict species, "living fossils" (the fish *Latimeria*, or the ginkgo tree) is evidence of the lack of adaptive radiation. These creatures of nature contradict gradualism.

Let us consider, following Stanley, a taxon consisting of N species in a

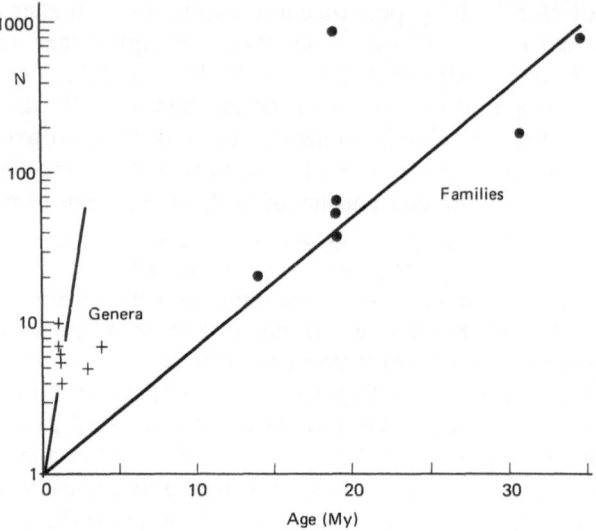

Fig. 2.10. Dependence of the number of mammalian families on geological time [2.28]

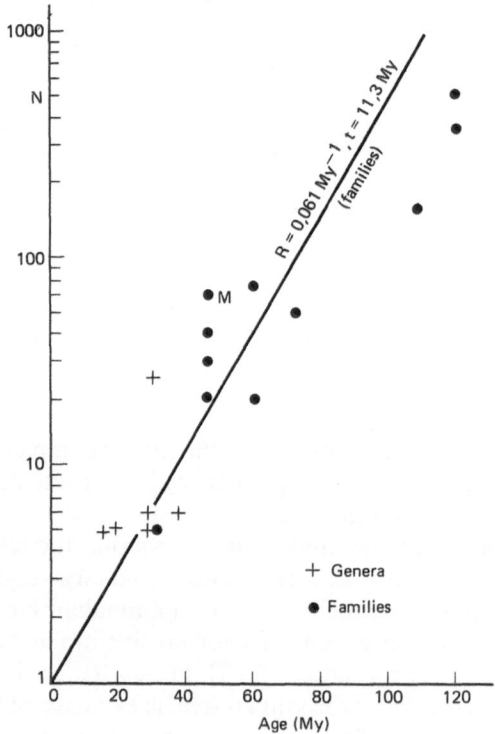

Fig. 2.11. Dependence of the number of families of bivalves on geological time [2.28]

state of adaptive radiation. The number of lineages at a certain previous moment $T = -t$ is given by

$$N_T = Ne^{Rt} . \tag{2.4}$$

The number of lineages that die out at the next moment of time is

$$ENe^{Rt} \tag{2.5}$$

and the total number of lineages that start in monophyletic radiation from a common ancestor is

$$EN \int_{-\infty}^{0} e^{Rt} dt = \frac{EN}{R} . \tag{2.6}$$

The total number of lineages that have ever existed is equal to

$$\hat{N} = \frac{EN}{R} + N = \frac{SN}{R} . \tag{2.7}$$

The period of existence of a species is, as a rule, much shorter than that of higher taxa. Occasionally speciation implies such a dramatic divergence that the resultant forms have to be treated as new species or even as new families. According to Stanley, quantum speciation is the result of chromosomal rearrangements and changes in gene regulation. The molecular basis of these events is considered in Chap. 7. Presumably, most morphological modifications can arise sufficiently rapidly due to changes in growth gradients in ontogeny without a substantial reconstruction of the genotype. Important evolutionary changes occur mostly in small populations during a small number of generations. The founder effect is realized, the role of which may be played by the hopeful monster of Goldschmidt. Such a monster might have been the first bird hatched from the egg of a reptile (Schindewolf [2.40]). No gradual transition from the squama to feathers is possible. Monsters could be fixed if the first one were a female and inbreeding occurred in the offspring. Such events probably occur more frequently in the plant kingdom than in the animal kingdom; speciation in land organisms is realized more rapidly than in marine organisms.

Macroevolution cannot be understood without punctualism. Macroevolutionary events are analogous to microevolutionary events in the sense that species play the role of individuals, and speciation and exctinction replace birth and death. Speciation creates variability at the level of macroevolution as do mutations and recombinations at the population level.

"... philosophically speaking, a species taxon is an individual, the members of a species being "parts" of this individual .... Speciation, thus, is not so much the origin of new types as the origin of effective devices against the inflow of alien genes into gene pools" (Mayr [2.6], pp. 561, 562). However, in actual fact, this is not a philosophical but a biological matter; a species is analogous to an individual in many respects.

Table 2.2. Mechanisms that produce macroevolutionary trends and analogous mechanisms that produce microevolutionary trends

Microevolution	*Macroevolution*
1. Genetic drift	Phylogenetic drift
2. Mutation pressure	Directed speciation
3. Natural selection	Species selection

Components of individual selection and species selection		
Process:	Microevolution	Macroevolution
Unit of selection:	Individual	Species
Source of variability:	Mutation/recombination	Speciation
Type of selection:	Natural selection	Species selection
	A. Survival against death	A. Survival against extinction
	B. Rate of reproduction	B. Rate of speciation

Table 2.2 (borrowed from Stanley) compares the factors and processes of microevolution and macroevolution. Genetic drift will be discussed in Chap. 6. Phylogenetic drift is taken to mean stochastic changes in morphology, which are not associated with the directionality of evolution. The unit of selection in microevolution is a population rather than an individual.

Thus, the modern conceptions of macroevolution are in a natural way consistent with the concept of punctualism. The analogy between a species and an individual (population) leads to an understanding of the great importance of the interspecies struggle and competition. This factor has been underestimated in the past. The ideas expounded here allow one to understand better the role of sex in evolution. Genotypic recombinations in sexual reproduction play a direct role in speciation. Divergent speciation is in fact impossible for asexual organisms, which are characterized by a slow accumulation of valuable mutations. The gradualist model by itself does not account for the appearance of sex. Sex is the creator and keeper of higher levels of life and evolution. It was not without reason that the first woman, Eve, was tempted by an apple.

The dynamics of the turnover of species is determined by the interaction between speciation and species extinction. The latter clears the ecological niche. Here the dual role is played by behaviour, which is different for generalists and specialists. A change in behaviour is an efficient isolating mechanism; at the same time stereotyped behaviour makes a species especially unstable.

The lifetime of an individual and also the longevity of species vary within wide limits. Figure 2.12, which is taken from Stanley ([2.28], page 232) presents the scale of species lifetimes. They are especially long for *Foraminifera* and corals. Figure 2.12 shows that high rates of speciation and extinction of species correlate with the level of biological organization. Even in the first edition of the *Origin* [2.41] Darwin wrote:

"The productions of the land seem to change more rapidly than the productions of the sea."

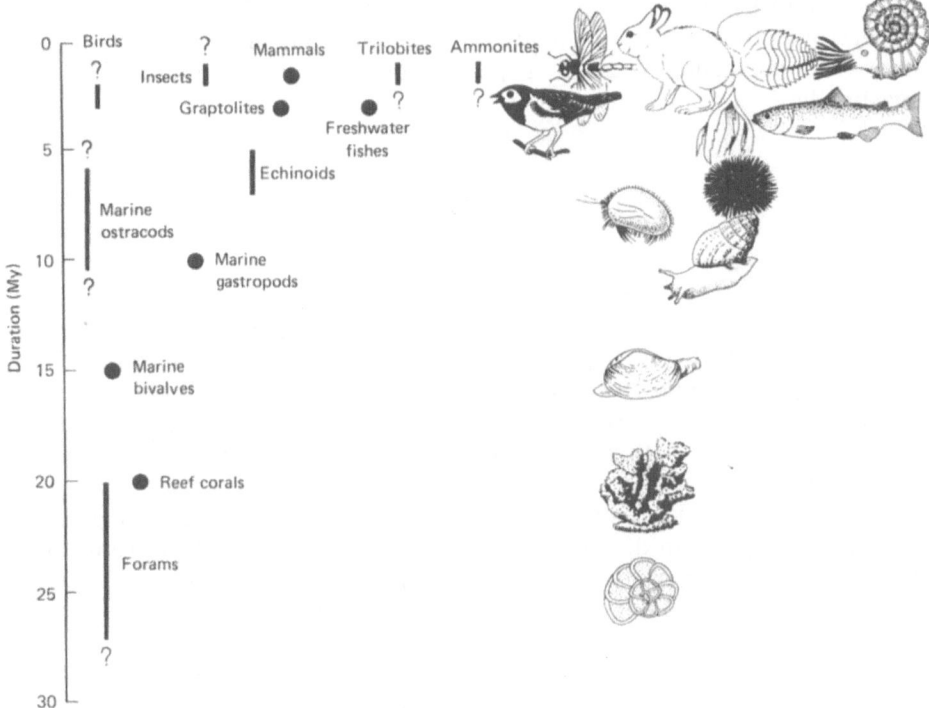

Fig. 2.12. The hierarchy of average species durations estimated for major taxa

"... We have reason to believe that organisms that are considered to be high in the scale of nature become modified more rapidly than those which are low, though there are exceptions to this rule."

To explain this, Darwin proposed a gradualist hypothesis. He maintained that competition is especially severe in more highly developed taxa. Punctualism ties up accelerated evolution primarily with a change in behaviour, which makes possible the penetration into new ecological niches. Developed creatures displaying complex behavior must differ from simpler ones in modes and frequencies of speciation.

The problem is by no means simple, but generalists are, as a rule, more developed than specialists. Generalists are capable of adapting better to new niches, but the evolution of specialists occurs more rapidly if it is estimated by the number of species that emerge and become extinct per unit time.

The selection of species leads to an increase in R; it maximizes the rate of speciation S and minimizes the rate of extinction E. Assuming that fluctuations of these quantities are proportional to their original values, it may be expected that taxa that are characterized by high rates must be less stable. This is valid for specialists rather than for generalists.

No model physico-mathematical theory of macroevolution has yet been worked out. However, it is evidently possible to proceed from the fundamental position of the analogy between a species and an individual. Species are born, age and die.

2.5 The History of the Eye

Let us consider, as an example of the origin and evolution of a complex biological system, the history of the organ of vision. The evolution of the eye has been described in detail by Wolken [2.42]. This example is not chosen at random; while speaking of evolution, the layman is always strongly impressed by the complexity and perfection of the eye and asks how it could have been produced. It is clear that light reception has travelled a long micro- and macro-evolutionary path.

The biosphere, life, exists due to solar radiation. The radiation in the visible region of the spectrum determines the most important biological phenomena, primarily photosynthesis, which is responsible for the existence of green plants and the oxygen-containing atmosphere produced by green plants. Due to this there has appeared and exists the animal kingdom.

Light has various effects on a system if it is absorbed. Visible light is absorbed in plants and animal organisms by coloured organic substances, pigments, which contain conjugated π-bonds. The number of types of such pigments is limited. Of primary importance are the porphyrins and carotinoids. The porphyrin ring shown in Fig. 2.13 is the basis of chlorophylls – the substances in green plants, and a number of bacteria and algae which absorb light. Figure 2.14 shows chlorophyll a. Carotinoids and phycobilins are also of importance. Figure 2.15 shows the structure of β-carotene and retinal, the latter being similar to half of the former. Retinal determines vision, as it enters into the composition of rhodopsin, the visual pigment. Figure 2.16 shows the absorption bands of chlorophylls, β-carotene, and rhodopsin against the background of the spectrum of the sunlight. Photosynthesis and photoreception have been described, for example, in a book by Volkenstein [2.17]. The paths of synthesis of these substances in cells have already been studied in detail. Haem-containing proteins have a haem group, i.e. a porphyrin ring with an iron ion in the centre. The haem is a group which determines the biological functions in myoglobin and haemoglobin, cytochromes, etc. Of course, the haem group does not play a role in photochemistry; its oxygen storage and transport functions, oxidation and reduction, are realized in the dark, within an organism.

A history of the eye should begin with phototaxis phenomena, the orientations of an organism or of its parts with respect to light. An example is phototropism, a differential growth response in which a plant grows towards, or away from, light. The sunflower orients itself towards the sun. With regard to freely moving organisms (animals, certain bacteria) we may speak of photomovement.

Fig. 2.14

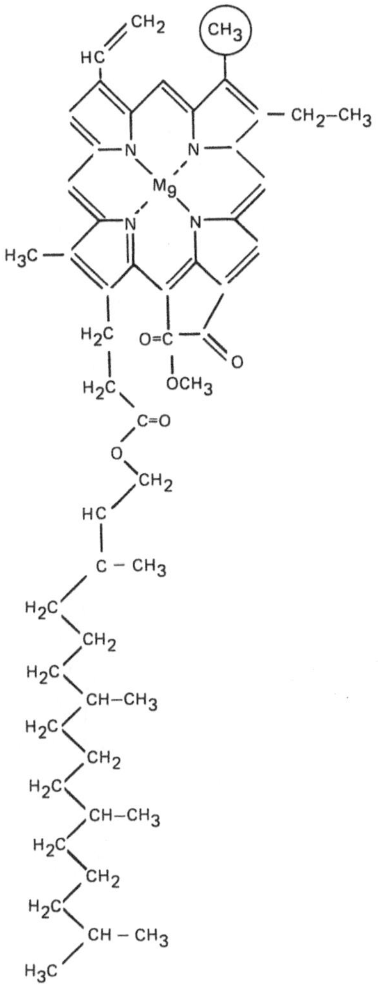

Fig. 2.13

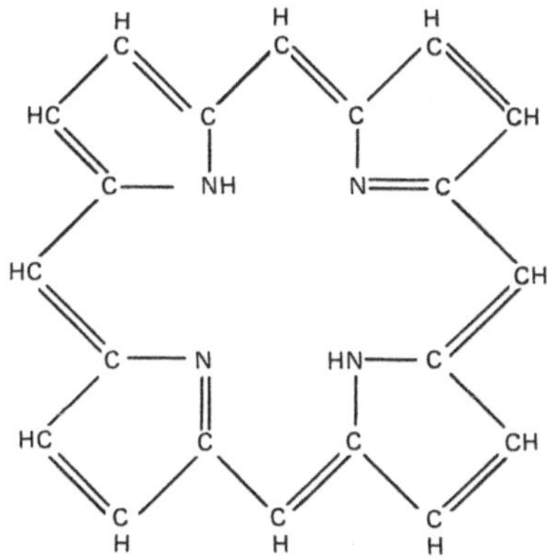

Fig. 2.13. The porphyrin ring

Fig. 2.14. Chlorophyll *a*. In chlorophyll *b* the CH_3 group enclosed in a circle is replaced by $CH=O$

The phenomenon of phototaxis relies on photoreception, the absorption of light by organized pigment molecules. The phototropism of the fungus *Phycomyces* has been studied in detail. Photosensitivity is exhibited by the sporangiophore of this organism, which contains a number of carotinoids. The bending and curvature of the sporangiophore upon illumination are associated with the action of contractile proteins; like muscle proteins, they respond to photostimulation, which causes a change in the electrochemical potential of specialized cells. The phototaxis of the fungus is directly governed by molecular events. Light

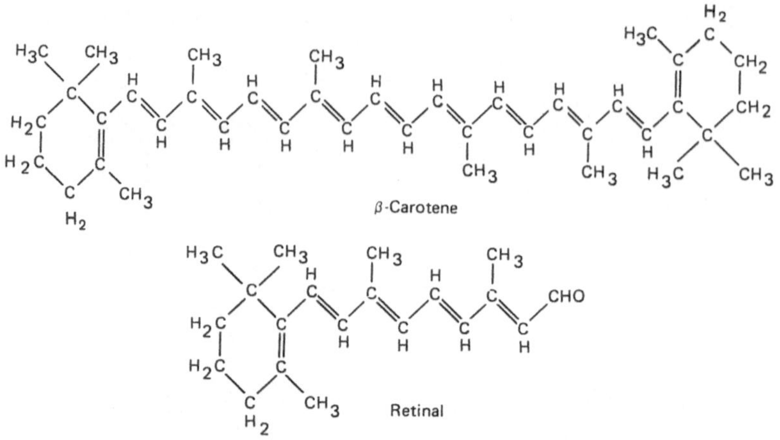

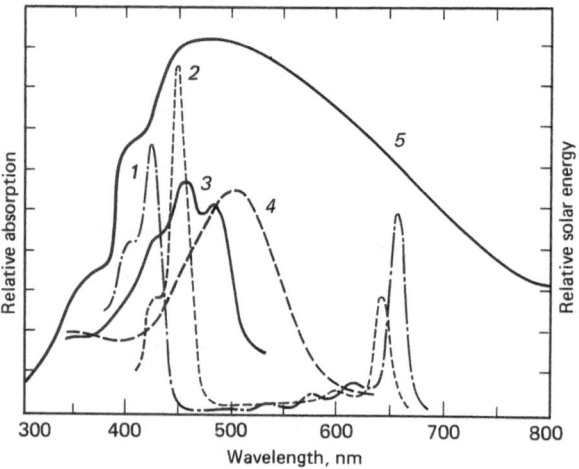

Fig. 2.15. β-Carotene and retinal

Fig. 2.16. Absorption bands of chlorophyll a (1) chlorophyll b (2), β-carotene (3), and rhodopsin (4): The spectrum of sunlight is shown as (5)

absorption serves as a trigger for turning on the protein mechano-chemical system due to conformational changes brought about by the absorption of photons (see [2.17]).

Red halobacteria contain the red protein bacteriorhodopsin, an opsin-retinal complex similar to that present in the eye retina. These bacteria live in very saline lakes. They possess a positive phototaxis to long-wavelength visible light and a negative phototaxis to ultraviolet light. The absorption of light by bacteriorhodopsin leads to a number of events, in particular, to the desalting of

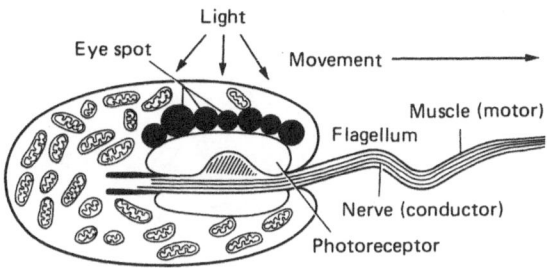

Fig. 2.17. Euglena

the interior of a bacterial cell. Bacteriorhodopsin is not only a photosensor but also a photogenerator which produces a sort of photocurrent.

Euglena infusoria has a positive phototaxis, i.e. it is attracted by light. In this case, in contrast to halobacteria, the photosensitive substance is concentrated in a special organ, namely the eye spot. The spot in *Euglena* is orange-red and consists of a few granules with a size of 0.1 to 0.3 microns. It is connected to a flagellum, by means of which *Euglena* moves. Here, just as in the case of *Phycomyces*, an interaction similar to the neuromuscular interaction in multicellular animals occurs. The corresponding scheme is shown in Fig. 2.17. Two pigments take part in light absorption, one of which is also a carotinoid.

The development of the eye in multicellular animals progressed in different ways. Four stages can be distinguished: a primitive photoreceptor system (the eye spot), the simple eye capable of image formation (the pinhole), the compound eye and the refracting eye. As has often been pointed out, there are only a few possibilities for constructing the organ of vision. Haldane writes [2.43]; "... as far as we can see, there are only four possible types of eye if we define an eye as an organ in which light from one direction stimulates one nerve fibre. There is the insect type of eye, a bundle of tubes pointing in different directions, and three types analogous to three well-known instruments, the pinhole camera, the ordinary camera with a lens, and the reflecting telescope. A straightforward series of small steps leads through the pinhole type to that with a lens and it is quite easy to understand how this should have been evolved several times." Vision provides great advantages; the eye is an adaptive character.

Flatworms, in particular planaria, have simple eyes (pinholes). In an earthworm the photoreceptor cells are located on the surface of the body, each of which has a lens and is surrounded by a neurofibrillar network. The lens is a specialized region of the cuticle (the thickened part of the skin). The structure consisting of a lens and a bundle of photoreceptor cells is called the simple eye (ocellus). It is shown in Fig. 2.18. Many insects also have simple eyes.

We see that light-sensitive regions arise even in unicellular organisms; junctions of the neuromuscular type exist in infusoria. The emergence of a lens was originally associated with a random thickening of surface tissues – the corresponding mutants gained advantages.

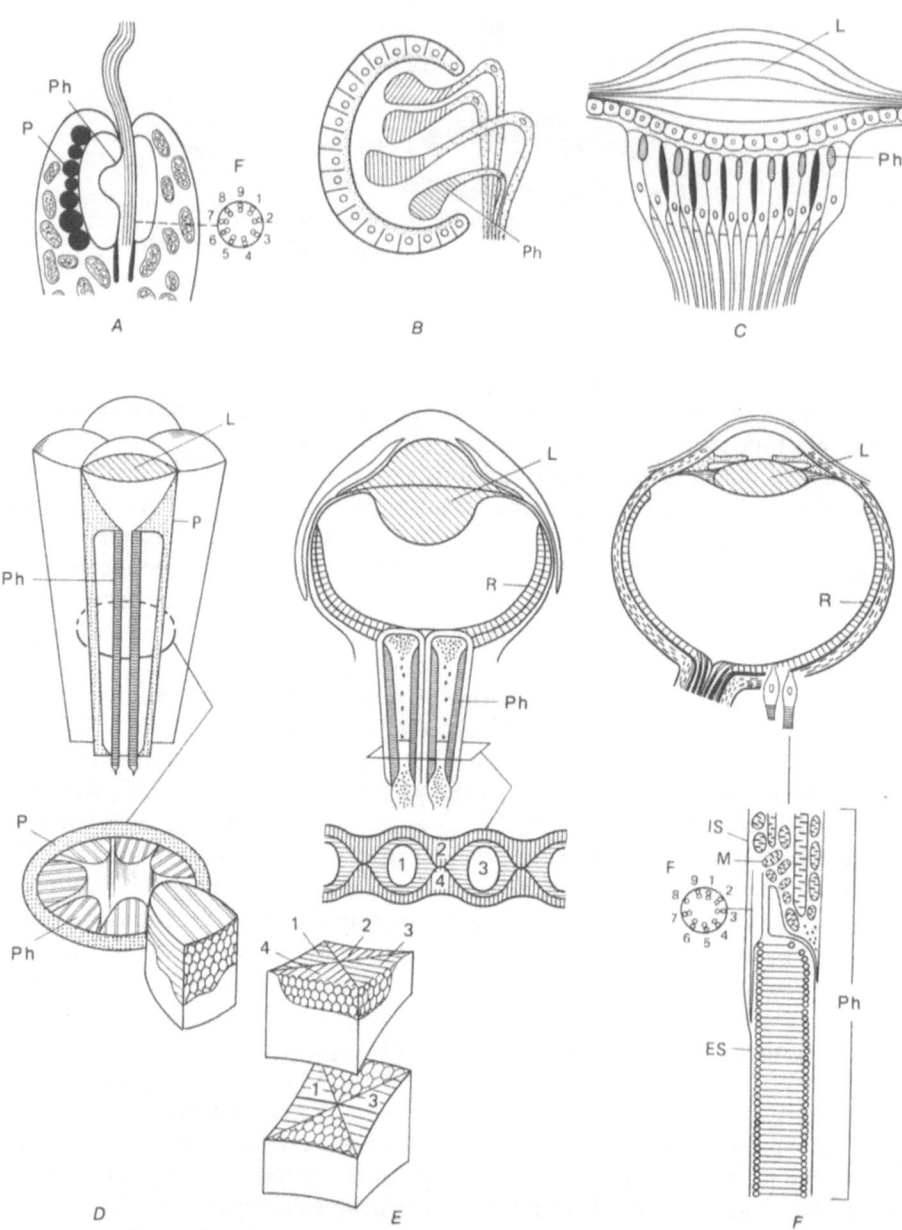

Fig. 2.18. Various types of the eye and their photoreceptors. (*A*) The eye spot, the flagellum in infusoria. (*B*) The simple eye (ocellus) of flatworms. (*C*) The simple eye of an insect. (*D*) The compound eye of an insect. (*E*) The eye of a cephalopod mollusc (1–4 = rhabdomeres). (*F*) The vertebrate eye: *L* = lens; F = flagellum; M = mitochondria; P = pigment granules; R = retina; Ph = photoreceptor. ES and IS are the external and internal segments

The second type of optical system is a bundle of tubules. The size of the image coincides with the object, irrespective of the distance. Good perception of objects is possible if the size of an object is equal to or less than that of the optical system. The distance cannot be estimated. However, if the holes of the tubules are located on the surface of the sphere or its segment, the tubules being directed towards the centre, the system works much better. This principle underlies the structure of the facet eye of a number of arthropods. The compound eye, which forms an image, is known to have existed even in trilobites which lived in the Cambrian period. The structural unit of the compound eye is called an ommatidium. The number of these structural units may exceed 2000 (in dragonflies). Each ommatidium in the compound eye contains a cornea, a crystalline cone and a receptor element called the retinula, which is composed of a central translucent cylinder (rhabdome), around which are located photoreceptor cells (see Fig. 2.18). In *Drosophila* the compound eye contains more than 700 ommatidia.

On other branches of the evolutionary tree, an eye similar to the human eye emerged. Such are the eyes of all vertebrates. In celaphopod molluscs and the octopus, the eye is constructed in the same manner as in vertebrates, the only difference being that the focusing of the image in the octopus occurs just as in the camera, i.e. by way of the movement of the crystalline lens, whereas in vertebrates it is done by means of a change in lens curvature. Figure 2.18 shows the cases mentioned.

The similarity between the eye of molluscs and the vertebrate eye does not itself imply that they emerged from a common ancestor. The general scheme of biological evolution has been given by Valentine [2.44]. From this scheme it can be seen how remote the branches leading to vertebrates and molluscs are. The similarity in the eye structure is a striking example of convergence, of the independent similar development of organisms in similar conditions (in the sea) and of the use of one of the several possible paths of creation of the organ of vision.

In all cases, the photosensitive substances are various rhodopsins, protein compounds of retinal, which are similar to the bacteriorhodopsins of halophilic bacteria. It has been shown [2.45] that the rhodopsin of vertebrates, like bacteriorhodopsin, is a photogenerator. Under the action of light rhodopsin produces a difference in proton concentration inside and outside the cell. A potential difference arises, which is responsible for the generation of a nervous impulse.

In all living things the photosensitive substances are of the same type. The evolution of the organ of vision proceeded in a few different ways specified by the structure of the original type of development, natural selection and by certain molecular factors.

Two key questions arise in connection with the history of the eye. The first is the nature of convergence and the second concerns gradualism and punctualism. The convergence of the characters of unrelated organisms occurs as a result of similar evolutionary constraints and environmental conditions. There are numerous examples of convergence: it will suffice to mention the morphological similarity between many species of marsupials and certain species of placental animals [2.46]. Selection always occurs from a limited number of possibilities.

We have cited the words of Haldane concerning the eye. Long before him Darwin wrote [2.47]: "When we reflect on these facts, here given much too briefly, with respect to the wide, diversified, and graduated range of structure in the eyes of the lower animals; and when we bear in mind how small the number of all living forms must be in comparison with those which have become extinct, the difficulty ceases to be very great in believing that natural selection may have converted the simple apparatus of an optic nerve, coated with pigment and invested by transparent membrane, into an optical instrument as perfect as is possessed by any member of the Articulate Class."

Darwin and Haldane believed that the history of the eye is gradual; they proceeded in the first place from the reasoning that even a very small degree of vision, e.g. the ability to differentiate light from darkness, is already an advantage as compared to the total absence of vision. At the same time one percent of a wing does not yet give any advantages to its possessor. The history of the eye seems to be completely adaptive. Indeed, in nocturnal vertebrates, for example, certain lemurs, the eyes are especially large. On the other hand, vision is completely lost in cave dwellers, which live in complete darkness, e.g. the amphibian *Proteus anguinus*. In the larvae of this animal the eyes are translucent through the skin; in adult protei the eyes are completely hidden by the skin. At the same time, protei are capable of light perception by the entire skin surface. We do not know how and when the atrophy of the eye of the *Proteus* ancestor occurred, whether the process was gradual or punctuational. Palaeontology does not provide direct information about the evolutionary history of the eye, whose structure is practically not preserved in fossils. However, there are no grounds for believing that the evolution of the eye was not punctuated.

3. Directionality of Evolution

3.1 Constraints on Natural Selection

We read in Webster's Dictionary: "Evolution, from Latin *evolutio*, an unrolling or opening, from *evolutus*, pp. of *evolvere*, to unroll; the development of a species, organism or organ from its original or rudimentary state to its present or completed state; phylogeny or ontogeny. To evolve is to develop gradually; to reach a highly developed state by a process of growth and change."

Linguistically, the notion of evolution presupposes development in a certain specified direction. Biological evolution is opposite to a random walk in the space of morphophysiological characteristics. What specifies the direction of evolution?

The theory of evolution proceeds from a disordered, random variability, which is ultimately governed by mutations and recombinations of genomes. No directionality is specified here. Ordered, directional evolution is generated by natural selection, by the interaction of organisms and environmental conditions. The synergetic basis of evolution consists of the directional development of an organized biosphere from the original chaos.

In *The Origin of Species* and in the subsequent construction of Darwinism, the attention is mainly focused on natural selection as the dominant mechanism of evolution. Darwin rejected the notions of the "vital force" and "inner drive" toward adaptive evolution, the nonconstructive notions that had been used by Lamarck (cf. [3.1]). In Darwin's theory the dynamic development of the biosphere emerges from stochastic behaviour. At each stage of evolution natural selection operates.

The palaeontological records, which are becoming more and more complete, give a number of striking and clear-cut examples of evolution directed by natural selection. The directedness of evolution manifests itself in progressive adaptation to the conditions of existence chosen by a particular phyletic branch, to biogeocenosis or an ecological niche. The evolution of horses has been studied very thoroughly (see [3.2] and also [3.3]). The progenitor of the modern horse *Hyracotherium* was a small four-toed animal. The evolution, which lasted about 60 million years, consisted of an increase in body and brain sizes, a change of the structure and mechanics of the extremities (transition to hooves), a change in the teeth associated with the transition from leaf to grass food, etc. Figure 3.1, taken from Simpson [3.2], shows the evolution of the horse's skull.

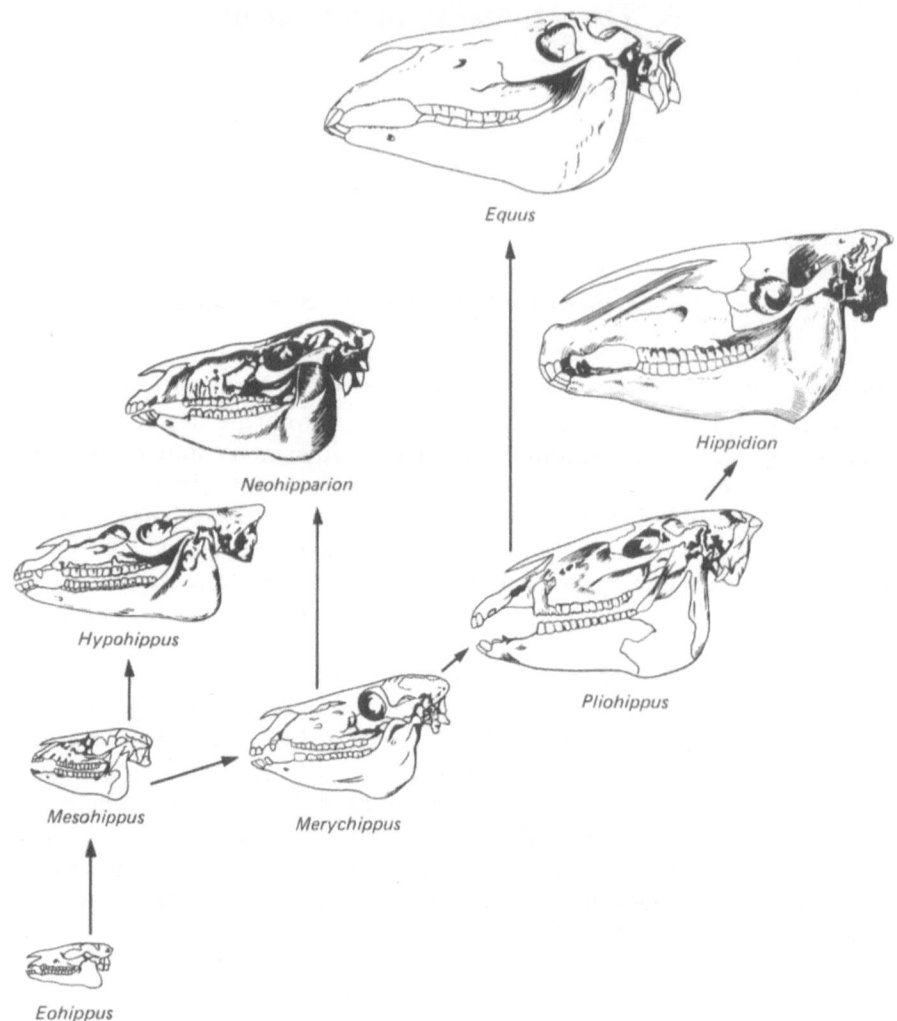

Equus

Neohipparion

Hippidion

Hypohippus

Pliohippus

Mesohippus

Merychippus

Eohippus

Fig. 3.1. Evolution of the horse's skull. All skulls are drawn to the same scale

An evolutionary increase in body size is encountered in many groups of animals. It gives a number of advantages to both predator and prey, being correlated with an increase in brain size; it also improves the conditions of preservation of heat in the organism since the body surface is proportional to the square and the body volume to the cube of linear dimensions.

All these events traced back for many vertebrates and invertebrates are determined by orthoselection. As has already been pointed out, the directedness of evolution is dictated by the interactions of organisms with the environment. Grant writes [3.3], "If environmental selection continues to operate in a given direction for a long time, we can call it orthoselection, and it will produce an evolutionary trend."

One might think that there are no problems here. However, the true situation is more complicated. Natural selection can do many things but not everything. Selection acts under the conditions of stringent constraints imposed by the established structure of evolving organisms and by the possible ways of its change. Darwin understood this very well. In the beginning of his great book he wrote words that are often forgotten [3.4]:

"... we clearly see that the nature of the conditions is of subordinate importance in comparison with the nature of the organism in each particular form of variation, perhaps of not more importance than the nature of the spark, by which a mass of combustible matter is ignited, has in determining the nature of the flames."

Further in the text of *The Origin of Species* this most important idea is not developed further; the stress is placed on the possibilities of natural selection rather than on its constraints. Darwin was mainly concerned with the interaction between organisms and the environment, and with the conditions of existence. In this sense, *The Origin of Species* is devoted to the "spark" rather than to the "flame". Accordingly, even in the best expositions of the theory of evolution (see, for example, [3.3,5]) constraints on natural selection are practically ignored. By the way, understanding what selection cannot do is no less important than understanding its realizable possibilities.

While studying living organisms, we usually ask "what is it for?", meaning that any character is adaptive. The problems associated with adaptation will be discussed in Chap. 4. As we shall see, not every character has adaptive significance.

The "What is it for" question is of limited content. Biology cannot be likened to technology. We rightly ask what this or that part of a machine is intended for. But biological "technology" created by Nature (this should not be confused with biotechnology developed by man) is different from man-made technology.

In the natural sciences, be it biology or physics, the "why?" question rather than the "what for?" one is of scientific significance. The task of science is to reveal the causes of the phenomena and regularities observed, rather than to reduce them to the notion of purpose. Without ascribing teleological meaning to this notion, we can interpret it in physical terms. The purpose pursued by the system arises in an unstable state and the achievement of the purpose is the transition to a more stable state. In the long run, the finalistic conceptions are of causal character. For example, the finalistic principle of least action in mechanics follows from the Lagrange causal equations. It is exactly for this reason that we should ask "why?" rather than "what for?".

Most land vertebrates have four limbs. [Exceptions are apodal amphibians (*Apoda*) and many reptiles: representatives of the infraorder *Anguinomorpha*, the

Fig. 3.2. *Latimeria chalumnae*

suborder *Amphisbaenia* and, of course, snakes (*Ophidia* or *Serpentes*). Some of these retain the rudimentary extremities.]

What are the four limbs for? A person reasoning in the usual way will say that this adaptive character, which is common to vertebrates, provides the best adaptation to environmental conditions, be it the four legs of the horse or hepard, the legs and wings of birds, or the flippers of seals.

But perhaps six limbs, like those of insects or of mythical vertebrates, such as Assyrian winged oxen, centaurs and angels, would have provided greater advantages. Why then four and not six or eight limbs?

Land vertebrates, whose first representatives were amphibians (*Amphibia*), evolved from certain fish. The ancestors of amphibians were fishes belonging to the subclass *Crossopterygii*. The only surviving species from this group, which flourished about 400 million years ago, is the species *Latimeria chalumnae*. Crossopterygians had four lobe fins, from which the four extremities of land vertebrates developed (Fig. 3.2).

This example shows that the developed structure of the crossopterygians imposed certain constraints on further evolution. Natural selection does not discard an already existing structure, but makes use of its various modifications; the limb that was the leg of the dinosaur is transformed into the bird's wing and to the arm of monkey or man.

The number of such examples is virtually unlimited. They point to the prime importance of an already established structure at any stage of evolution. This is valid for all levels of development, beginning with the molecular level. The "rules of movement," (i.e. the material of which an organism is made up, primarily proteins and nucleic acids) were fixed even at the prebiotic stage of evolution. Further development reduces to modifications of existing chemical structures. We may take it to be established that the main types of vertebrate cells have not changed for 500 million years. Nor have their properties changed: the ability to divide mitotically, programmed death, physical chemistry of intercellular contacts, chemotaxis, mechano-chemical properties. The evolution of morphogenesis, i.e. structure formation in an individual development, involves changes in the spatial and temporal utilization of the principal cellular mechanisms, rather than changes of the mechanisms themselves. It may be stated that the evolution of the cells of Metazoa has not undergone significant change, but the distribution and enhancement of functions in individual cells specify a new spatial structure and new behaviour in time.

Thus, the constraints that "vectorize" evolution are imposed by the innate properties of cells themselves. Macroevolution – the appearance of taxa, beginning with a genus – may be treated as departures from these constraints.

We are dealing here with the functional principle of organization of living nature. At all levels, new combinations of a limited number of previously produced physical and chemical building blocks (e.g. amino-acid units of proteins and nucleotide units of nucleic acids, cellular and intracellular membranes, types of cells and cellular mechanisms) occur. Identical systems are used for different purposes at all levels: the contractile protein actin acts both in the muscle of a weight-lifter and in the tail of his spermatozoon; the lung of an amphibian developed from the swimming bladder of fish. This general regularity determines, on the whole, the rate of evolution – the "assembly" of different systems occurs from identical parts.

In other words, living nature generally obeys the same set of laws, whose existence is indeed the prerequisite for developing a scientific theory of evolution. Suppose we had had two hypothetical possibilities: the absence of selection constraints and the inheritance of acquired traits. In either case, even small changes in environmental conditions would have led to a chaotic diversity of traits, to the absence of surviving species and higher taxa. Not only evolutionary theory but also Linnean taxonomy would have been destroyed. This is a good topic for science fiction!

The invariable structural and dynamic properties of organisms may be treated as constraints that channel natural selection, but these properties may also be regarded as independent internal factors that determine the directedness of evolution alongside the external factor, natural selection. Are there any differences between these two approaches? At first sight it might seem that we are dealing here with semantics only. However, this is not the case.

The limbs of vertebrates originated from the four corresponding fins of crossopterygians. But what is the origin of these fins? They also arose by way of evolution, with corresponding constraints at each stage. The driving force of evolution is, of course, natural selection, which began even at the prebiotic, chemical stage. Thus, we should speak not of independent factors but of the constraints on natural selection determined by the structure and possible ways of change of the established systems.

In mathematical language this must mean the following. Evolution may be represented, in principle, by a complex network of Markov chains, i.e. chains of interdependent random events. The probability of a subsequent event depends on what was randomly realized in one or several of the preceding links of the chain. A simple Markov chain, in which only the immediate predecessor is relevant, is determined by a stochastic probability matrix. An element of this matrix p_{ij} is the probability that event j is realized in a given link given that event i occurred in the preceding link. The evolutionary constraints imply that a number of off-diagonal elements of the matrix are equal to zero. Thus, if at the preceding stage of evolution a land vertebrate had four extremities, the probability of appearance of six extremities at the next stage is equal to zero. However, evolution also means changes in the elements of the stochastic matrices themselves; these are especially significant in the case of higher taxa. These general ideas have not yet been implemented – the construction of a rigorous mathematical theory of evolution based on the study of the apparatus of Markov chains has not been realized and is beset with serious difficulties.

The directionality of evolution determined by constraints on natural selection and by the interplay between ontogeny and phylogeny (see below) implies that evolution must progress much faster than it would have done in the absence of constraints. The absence of directionality would have implied selection in the wide diversity of random traits at each stage, the difference of the off-diagonal elements of stochastic matrices from zero. The directionality of evolution, which is also evidenced by other factors to be discussed later, provides, in principle, an answer to the question of the time period required for the evolutionary development of the present-day biosphere. We are still far from creating a quantitative theory that would allow us to estimate this time period but qualitatively we understand the essence of the problem (see Sect. 9.4).

The course of evolution is similar to the running of a self-winding watch. Out of an infinite number of random, disordered movements the mechanism makes use of those which wind the spring. In evolution changes occur which correspond to its direction, i.e. changes that are determined by the previous stages of the structural changes of organisms. Just as the hands of the clock do not run backwards, evolution is irreversible, i.e. directed. As will be shown in Chap. 9, directionality accelerates evolution considerably.

The inseparable link between ontogeny and phylogeny, i.e. between individual development and biological evolution, seems to be obvious.

Evolution is experienced by organisms, each of which is a historical, developing system; a multicellular organism travels a programmed genetic path from zygote to maturity to death. Hence, ontogeny as a whole is subjected to evolution and changes in its program are of decisive evolutionary significance.

The ontogenies of new organisms are made up of the ontogenies of their progenitors, and thus the ontogenies of the ancestors channel evolution. New organisms retain many specific features of their precursors. These features are the expressions of genomes and developmental mechanisms preserved in evolution. The retention of the various features of ontogeny is the simplest way for the formation of the directedness of evolution by a historical precedent. Evolutionary changes are combined with ontogeny, from which they arose. The genome and the mechanisms that support it and change its structure govern evolution; the regularities imposed by them on changes of the genome are independent of any direct pressure from the environment. Thus, evolution is channelled even inside an organism.

These propositions expressed in modern language are fully consistent with the words of Darwin cited previously.

The point is that the genome is a complex integrated system rather than a set of disordered genes. Evolution is directed by the organization of the genome, by the nature of the genome regulatory mechanisms and, of course, by mutational possibilities.

From what has been said, it follows that the study of the relationship between ontogeny and phylogeny is absolutely necessary. The presence of this relationship was quite obvious to Darwin (see "Development and Embryology" in Chap. XIV of *The Origin of Species*) and to his forerunners, notably von Baer.

An extensive article devoted to constraints in evolution determined by development ("Developmental constraints and evolution") has been written by a group of eminent biologists, who are far from sharing the same ideas [3.6]. It is worth dwelling on this article.

A constraint determined by development biases production of phenotypic variants or limits phenotypic variability caused by the structure, character, composition, or dynamics of the developmental system. There are many sources of such constraints. These are the properties of the materials of which organisms are composed, the requirements governing the storage and retrieval of information used during development, particular features of the evolutionarily determined pathways of development exemplified by a group of organisms.

T.H. Huxley wrote in a letter to Romanes (quoted by Spurway, [3.7]): "It is quite conceivable that every species tends to produce varieties of a limited number and kind, and that the effect of natural selection is to favour the development of some of these, while it opposes the development of others, along their predetermined lines of modification." As a matter of fact, such constraints are also implied by the law of homologous variation proposed by N.I. Vavilov (1922), according to which the similarity of developmental pathways in related species causes the appearance of similar variants [3.8]. Speciation is responsible not only for the morpho-physiological characteristics of a species but also for its evolutionary future.

The authors of [3.6] consider the question of the causes of stasis: whether it is determined by developmental constraints or by stabilizing selection. In other words, they ask whether we can distinguish the results of constraints from those of natural selection. This problem is associated with one of the most important problems in biology, the problem of the relationship between genotype and phenotype, realized in the dynamics of the developmental process.

The authors of [3.6] make a distinction between universal and local constraints. The former are determined by the general laws of physics and by the general physico-chemical nature of life phenomena, in particular by the fact that in all living things the legislative role is played by nucleic acids and the executive role by proteins. For example, the chirality of these biopolymers is a universal constraint. The fact that a living organism is a dissipative system may also be treated in the long run as a universal constraint. Conversely, local constraints are inherent in a given taxon or in a given evolutionary branch, such as the four limbs of land vertebrates.

In fact, the reasonings concerning the universal constraints are of no special meaning. Evolution is to a certain extent similar to a chess game, which of course is played according to specific rules: a board of 64 squares, chessmen and individual rules of movement are given. It is clear that these are not constraints but rather the initial predetermined conditions of the game, and the rules cannot be changed during the game; a chess game is not transformed into a croquet match from Lewis Carrol's "Alice in the Wonderland". Local constraints, on the other hand, have meaning and require consideration. For example, the opening move of a chess game and the specific features of ancestral species in biology are local constraints.

Evolutionary constraints are specified by the mechanisms of individual development, ontogeny, which will be considered at a later time. Undoubtedly, the mechanisms of development of a particular taxon render certain phenotypes more accessible than others. This is what constitutes the essence of Vavilov's law of homologous variation.

Many neutral populations are morphologically quite homogeneous. On the other hand, populations containing mutants that disturb the important characteristics of morphology are variable. Waddington [3.9] explains this difference by the fact that the characteristic structure of wild-type was subjected to stabilizing selection in a large number of generations, while the mutant structure was not. Certain developmental constraints are by themselves the results of selection and are not determined automatically by the preceding structure.

In this respect we must immediately say the following. All evolutionary processes are associated in one way or another with Darwinian natural selection. The appearance of constraints is dictated by selection, the possibilities of which then become limited by constraints.

The authors of [3.6] consider the possibility of differentiating the characters determined by developmental constraints from those which arise due to selection. It is hardly of any special significance. Recall the words of Darwin given earlier. The diversity of variants inherent in a particular population is specified by its ontogenetic structure. The selection of variants is determined by adaptation processes, which are of dual character. First, the environment selects organisms that find their way into pre-formed ecological niches. Second, organisms choose an environment, creating or finding new ecological niches. Both organisms and environments have many levels, beginning with the molecular level. Adaptation to the intracellular medium also occurs, and gives rise to characters that are far from being adaptive with respect to the external environment.

The question of the characters specified by selection constraints and by selection itself is directly connected with the problems of adaptation (Chap. 4). Natural selection is directed towards an increase in adaptation of both an adult organism and its embryo to external conditions. Selection constraints are dictated by the structure of an organism and its ontogenetic development, i.e. the dynamic self-consistency of structures and processes. This self-consistency may be treated as adaptation to the conditions of an internal environment. Let us return to the four limbs of land vertebrates. A constraint is imposed by the original structure of crossopterygians. Natural selection transformed their lobe fins to legs, flippers, wings, and arms; these are the results of adaptation to various ecological niches. However, in each of the emerging lines new constraints are produced which determine the subsequent events.

The directionality of evolution rules out direct transformations of one species into another, e.g. oats to wild oats, wheat to rye, or a robin to a cuckoo. As we know, the occurrence of such transformations was claimed by Lysenko. These fantasies have nothing to do with science.

3.2 Ontogeny

Ontogeny is the development of an individual. For organisms that reproduce sexually, ontogeny begins with the emergence of a zygote and ends up with the inevitable death of the adult organism.

Raff and Kaufman wrote at the beginning of their monograph [3.10]: "The essential position is that there is a genetic program that governs ontogeny, and that the momentous decisions in development are made by a relatively small number of genes that function as switches between alternate states or pathways. The significance of this view, if correct, is that evolutionary changes in morphology occur mechanistically, as a result of modifications of these genetic switch systems. If our prediction that there are a relatively small number of such gene switches is correct, then the potential exists for geologically rapid and dramatic evolutionary changes. Such macroevolutionary events are apparently associated with the origins of new groups of organisms."

The genetic program of development does not mean that the genotype of an organism is its plan or blueprint. The situation here is more complicated. Structural genes program the synthesis of proteins, the diversity of which is very large. For example, the shell of silkworm eggs (chorion) contains 180 different proteins, the apparatus of bacterial flagella 120 proteins, the protective shell of phage T4 80 proteins and the ribosome of an animal 50 proteins [3.11]. But the genome also carries information about the subsequent processes of embryo self-organization realizable at the levels of macromolecules, cells, tissues, organs, and of an organism as a whole. These processes are determined by proteins, and not only the protein structure is essential to them but also its amount, site and the time of its appearance. At the cell level three types of cell activity arise: the formation of the specific form, cell division and cell localization. Intercellular contacts, which are of chemical and mechanical nature, modulate these processes [3.12]. Precise regulation of development at all levels in time is of fundamental importance [3.13].

The notion of the genome as a blueprint for a future organism is easily refuted. Having taken apart a man-made machine, say an automobile, we can recover the drawings according to which the machine was constructed. To reconstruct the genome on the basis of what we know about the structure of the organism is impossible. The opposite, i.e. the prediction of the structure of the organism on the basis of the knowledge of the genome, is also impossible. We cannot do this not only because the organism is much more complex than a solid-body machine made by man, but also because self-organization occurs during ontogeny. In this sense the genome may be compared not with a drawing but rather with a recipe specifying the time and place for introduction of the necessary ingredients.

Developmental processes are thus under genetic control, and evolution should be treated as the result of changes in structural and regulatory genes. The diversity of developmental phenomena in multicellular organisms (Metazoa) is extremely large. The Metazoa differ from the Protozoa and prokaryotes in embryonic development. The situation with prokaryotes is much simpler – their hereditary characters are basically chemical and reduce predominantly to the synthesis of enzymes, which is regulated by the gene operon system.

The embryo of a multicellular organism is a complex macroscopic system. The ontogeny processes are strongly interdependent and integrated.

K.E. von Baer, the founder of embryology (or developmental mechanics, as it is called today) formulated as far back as 1828 his own laws of ontogeny (cited from [3.10]):

1. The more general characters of a large group of animals appear earlier in their embryos than the more special characters.
2. From the most general forms the less general are developed, and so on, until finally the most special arises.
3. Every embryo of a given animal form, instead of passing through the other forms, becomes separated from them.
4. Fundamentally, therefore, the embryo of a higher form never resembles any other form, but only its embryo.

Ontogeny consists of cell differentiation, morphogenesis and growth. The tissue form at one stage of embryogenesis determines morphogenesis at the next stage. In the complex macroscopic structure of the embryo, the spatial position of molecules, cells and tissues is a factor which determines development to a considerable extent. Wolpert has developed a theory of positional information [3.12,14,15]. The basic proposition of the theory is that during the evolution of multicellular organisms, it is not so much cells that change as their spatial organization. In the Wolpert theory, the mechanism of spatial organization of cell differentiation is based on the premise that the cells are assigned positional values in a certain co-ordinate system. These values are interpreted in accordance with their structure and history. The cells of the same histological class may have different positional values and may therefore be non-equivalent. Positional information is eventually determined by the chemical and mechanical relationships between cells.

The key question of the theory of development and evolution is the nature of cell differentiation and morphogenesis and also the nature of changes of these processes during evolution.

Phenomenologically these processes are described on the basis of the concept of the morphogenetic field introduced by Gurvich [3.16]. It is the field of a group of cells that exhibit concerted behaviour in shape-forming processes. In this sense the whole embryo at the early stages of development forms a morphogenetic field; each of its cells is characterized by a certain position and serves as a source of chemical and mechanical signals to which the other cells respond. According to Wolpert, this signalling may have threshold values; a "positional" signal activates a gene and if the corresponding protein stimulates its own product (autocatalysis), then with a threshold concentration of the chemical signalling substance morphogen the system will discontinuously pass to a new state.

Positional morphogenetic fields do not extend very far – only to a distance corresponding to less than 100 cells (approximately 1 mm). The time of field setting is of the order of hours. Wolpert maintains that the signals in all vertebrates are the same but that during evolution their interpretation has changed (see also [3.10,17]).

The regulatory mechanisms organized by the genome exert influence on its evolution. The genome and mechanisms of its maintenance and rearrangement directly determine evolution. The regularities of genomic changes are to a considerable extent independent of the direct pressure of the environment, hence the canalization of evolution is specified inside the organism [3.18]. The corresponding notions have been developed in the molecular drive concept described in Chap. 7. We are speaking here of the "flame" rather than of the "spark".

Certain changes in the course of development are adaptive, others may be the inevitable results of physical constraints and architecture of development. No innovation is fixed if it is not functionally integrated into an existing system. A certain path of development characteristics of a given taxon restricts the region of subsequent adaptations. In this sense an organism is a mosaic of relatively new adaptations incorporated into the framework of old constraints [3.19].

In certain cases the small changes of even a single gene may have far-reaching evolutionary consequences. Raff and Kaufman cite, as an example, the dwarfism in *Homo sapiens* associated with the balance of the hypophyseal growth hormone. Dwarfism can develop rapidly and widely in isolated populations with the selection pressure being strong. Presumably this has led to the appearance of populations of pygmies in South Africa, this being the result of the fixation of alleles responsible for the production of the inefficient growth hormone [3.10].

An embryo, just as the organism which develops from it, is an integral entity. It is a very complex chemical machine, which differs from a man-made machine primarily by the nature of signalling. In non-living machines the signals, their sources, transducers and receptors are mechanical, electrical or magnetic, whereas in a "living machine" the signals are molecules and ions and their sources, transducers and receptors are molecular devices. For example, an enzyme may be looked upon as a transducer of the input signal (the substrate) to the output signal (the product). But this is not all. Of prime importance for development and life activity at all levels is mechanics based on chemistry in living nature, i.e. mechanochemistry. We are speaking here of the direct conversion of chemical energy to mechanical energy, a conversion that by-passes the thermal stage (see [3.20,21]). At the molecular level, mechano-chemical processes occur as a result of electron-conformation interactions (ECI; see [3.20,21] and Chap. 5). At the cell level mechanochemistry manifests itself in changes of the shape of cells, their division, and mechanical interactions. Finally, all organisms and their parts move as a whole.

The integration of cells that arise by way of the successive divisions of the zygote is determined by chemical and mechanical interactions between cells. Even the first cell division leads to a decrease in symmetry as compared to the maternal zygote. The non-equivalency of cells, which leads to their differentiation, arises after several cell divisions simply because the cells located within a cellular sphere have contact only with one another and not with the external medium.

The regulation of differentiation and morphogenesis realized by chemical and mechnical interactions between cells is changed during ontogeny. In the sequential stages of embryonic development the cells become more and more specialized. This means that an ever-increasing number of genes present in the genome are regulated and, in particular, are repressed. Of special importance here are the classical experiments carried out by Spemann on amphibians. He transplanted portions of an embryo at the various stages of its development. If at the blastula stage a portion corresponding to the presumptive abdominal epidermis (i.e. a portion from which this epidermis must develop) is swapped with a portion present in the region of the presumptive brain, then the embryo will develop in a normal way. This means that the corresponding cells are totipotent at the brastula stage and are not specialized, and their subsequent fate depends on the medium into which they find their way. A similar experiment carried out at the gastrula stage gives quite a different result. The presumptive epidermis transplanted to the site of the presumptive brain develops only as epidermis; the presumptive brain develops into the brain even when transplanted into the region of the epidermis. If the transplantation of the presumptive eye is carried out at the stage of the later blastula or the early gastrula, then the presumptive eye forms this or that organ, depending on the site to which it is transplanted. Conversely, at the neurula stage the presumptive eye develops into an eye at any site (Fig. 3.3). The potentialities of cells and tissues are narrowed and specialized; totipotency is replaced by unipotency [3.22,23]. Raff and Kaufman [3.10] also describe more elaborate experiments that disclose the dynamics of morphogenesis.

Mutations that exert an influence on development play a central role in evolution. It may be considered to be an established fact that the changes of a small number of genes can bring about significant changes in morphology.

The most important role in ontogeny is played by time factors, i.e. the time relations in the syntheses of various proteins and the formation of various cells and tissues. Even small changes in the "timetable" lead to considerable morphological alterations. At the molecular level the appearance of various

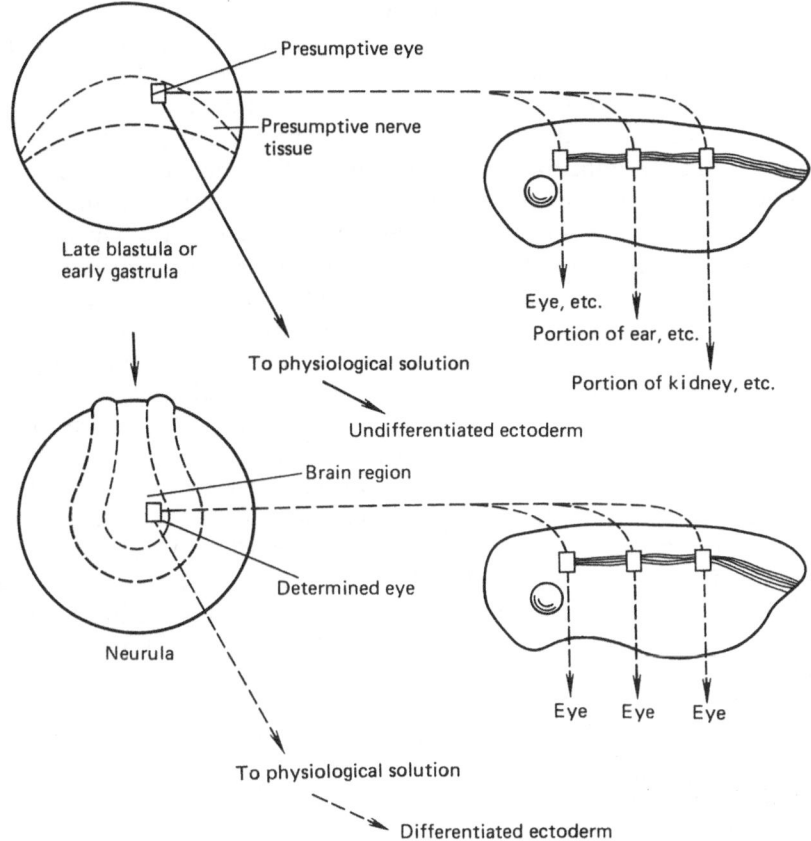

Fig. 3.3. The fate of the presumptive and determined eyes

proteins at various stages of formation of phage T2 particles was first observed by Khesin [3.24]. Changes in ontogenetic events with time (heterochrony) are of great importance to evolution. The time factors will be discussed in the next section.

Both ontogeny and phylogeny occur at each stage in such a manner that structures and processes that have already developed are not rejected but are used in subsequent development. Jacob, who drew attention to this most important position, defined it as the creation of a new system by way of tinkering with the old one (in French it is called *bricolage*) [3.25]. Such tinkering, however, does not imply a smooth, gradual progress of ontogeny and phylogeny. In both processes we encounter phenomena similar to non-equilibrium phase transitions. It might seem that the idea of tinkering contradicts this. However, phase transitions, both equilibrium and nonequilibrium, are realized in homogeneous media. A fur collar can be reshaped into a fur cap, after all.

A detailed study has recently been carried out which investigates specific mutations that change the process of ontogeny without interrupting it. These are so-called homoeotic mutations (see Chap. 8 of [3.10]) which manifest themselves in *Arthropoda*, in particular in the famous *Drosophila melanogaster*. Fruit flies (and also more complex animals) are built up of compartments, in which key genes determine the development of a number of cell clones [3.26]. The mutations of these homoeotic genes bring about sharp macroscopic changes in the developmental programme. The mutation *Antennapedia*, for example, transforms the determination of the antenna disc to that of a leg disc. The regulatory genes control the operation of a set of structural genes. The control is effected by means of regulators, namely polypeptides containing 60 amino-acid residues. The presence of such regulators has been proved not only for *Drosophila* but also for vertebrate animals, including human beings (see, for example, [3.27]).

Thus, evolution is directed by constraints on natural selection which are determined by ontogeny. Let us cite the words of Gould [3.28], which correlate with the words of Darwin quoted earlier: "Organisms are not billiard balls struck in deterministic fashion by the cue of natural selection, and rolling to optimal position on life's table. They influence their own destiny by interesting, complex, and comprehensible ways." In a later work [3.29] Gould writes: "The constraints of inherited form and developmental pathways may so channel any change that though selection induces motion down permitted paths, the channel itself represents the primary determinant of evolutionary direction".

The directedness of evolution determined by an earlier established structure may be illustrated by a caricature by Daumier (Fig. 3.4). In a series of pictures drawn by the great artist the king of France Louis Philippe is transformed to a

Fig. 3.4. Evolutionary transformation of Louis Phillippe of France into a pear

pear. This transformation is gradual but it is characteristic that the king is transformed into a pear, not into an apple.

The modern theory of pre-biotic molecular evolution constructed by Eigen clearly shows that this evolution is directed. The Eigen theory is described in detail in Chap. 8.

3.3 The Time Factors of Embryogenesis

The basic proposition of evolutionary embryology was formulated in *The Origin of Species*: "... the various parts in the same individual embryo, which ultimately become very unlike and serve for diverse purposes, being at an early period of growth alike." The transition of embryonic rudiments from totipotency to unipotency, which is evidence of an increase in indispensability in individual development, is in effect an expression of this basic regularity.

The forerunner of Darwin in the analysis of facts of evolutionary embryology was von Baer [3.30]. Darwin agrees with von Baer, who stated that "the embryos of mammalia, of birds, lizards and snakes, probably also of chelonia are in their earliest states exceedingly like one another, both as a whole and in the mode of development of their parts; so much so, in fact, that we can often distinguish the embryos only by their size. In my possession are two little embryos in spirit, whose names I have omitted to attach, and at present I am quite unable to say to what class they belong. They may be lizards or small birds, or very young mammalia, so complete is the similarity in the mode of formation of the head and trunk in these animals. The extremities, however, are still absent in these embryos. But even if they had existed in the earliest stage of their development we should learn nothing, for the feet of lizards and mammals, the wings and feet of birds, no less than the hands and feet of man, all arise from the same fundamental form."

None of this has been refuted by modern science. However, von Baer was not an evolutionist; his theory of increasing differentiation calls only upon a conservative principle of heredity to preserve stubbornly the early stages of ontogeny in all members of a group. Evolutionary changes at the later stages are ignored.

The anti-evolutionist Agassiz stated that the ontogenetic stages repeat the adult forms of animals. In this respect he was a direct forerunner of the evolutionist Haeckel, who enunciated his famous biogenetic law (theory of recapitulation) shortly after the work of Darwin was published in 1866 ([3.31]; see also [3.32]): "Ontogeny is the short and rapid recapitulation of phylogeny ... During its own rapid development ... an individual repeats the most important changes in form evolved by its ancestors during their long and slow palaeontological development."

Ontogeny recapitulates phylogeny. This brief and expressive proposition has been known to us since our school years. Not long ago the Haeckel law was still taught in school, but many recent books on evolutionary theory state that the Haeckel law is erroneous and that no recapitulation in his sense exists. A clear-cut and convincing disapproval of the biogenetic law is contained in the monograph by Gould [3.33]. What follows in this section is based on this book.

Haeckel's theory requires a change in the timing of developmental events as the mechanism of recapitulation. The biogenetic law asserts that the change was all in one direction, i.e. a universal acceleration of development, pushing ancestral adult forms into the juvenile stages of descendants. This is not always the case, however; apart from recapitulation, i.e. the retention of structures of adult ancestors in the embryos of descendants, there is also pedomorphosis, the appearance of ancestral juvenile traits in adult descendants. Gould distinguishes two types of pedomorphosis: neoteny (literally "holding youth"), i.e. cases involving the retardation of somatic development, and progenesis, i.e. cases of pedomorphosis produced by accelerated maturation. In the former case, there occurs acceleration, and in the latter, retardation of the somatic development of parts and organs. The most widely known case of neoteny is that of the axolotl. The axolotl is the larva of the amphibian *Ambystoma tigrinum*, which lives in water and breathes with gills, like all tadpoles. However, the axolotl can reproduce sexually without turning into the Ambystoma. The axolotl means "playing in water" in the Aztecan language. The metamorphosis of the axolotl to the Ambystoma takes place in warmer waters and after feeding on thyroid extract.

Gould cites the poem by Garstrang devoted to axolotl [3.33]:

The Axolotl and the Ammocoete

Amblystoma's* a giant newt who rears in swampy waters,
As other newts are wont to do, a lot of fishy daughters:
These Axolotls, having gills, pursue a life aquatic,
But, when they should transform to newts, are naughty and erratic.

They change upon compulsion, if the water grows too foul,
For then they have to use their lungs, and go ashore to prowl:
But when a lake's attractive, nicely aired, and full of food,
They cling to youth perpertual, and rear a tadpole brood.

And newts Perennibranchiate† have gone from bad to worse:
They think aquatic life is bliss, terrestrial a curse.
They do not even contemplate a change to suit the weather,
But live as tadpoles, breed as tadpoles, tadpoles altogether!

(Garstrang, 1951, p. 62, [3.34])

---

*The generic name of the axolotl, properly spelled Ambystoma.

† Several genera of salamanders are permanently larval in morphology; no metamorphosis has ever been observed. Because they retain external gills throughout life, they are called perennibranchiate [3.33].

As has been noted by many investigators (in particular, Severtsov), phylogeny is a sequence of ontogenies. No Haeckel recapitulation exists since the embryonic features inherent in all vertebrates, such as gill slits in the early embryos of reptiles, birds and mammals, do not represent the previous state of the adult organism (fish?); they represent only the invariable resemblance of the early development.

The embryos of vertebrates preserve the ancient characters since they are protected either in the egg or in the mother's womb from the direct action of natural selection. At the early stages of development there are no factors contributing to specialization; natural selection exerts only an indirect influence. During ontogeny, specialization is increased.

Another striking example is provided by the larvae of crustaceans. There is a great diversity of adult representatives of this class, which includes more than 20,000 species. In most cases, crustaceans have similar larvae (nauplii) which swim freely in water. The crustaceans *Cirripedia*, including the acorn barnacle (*Balanus*), which are permanently fixed in the adult stage, have nauplii. Darwin wrote: "... even the famous Cuvier did not understand that the acorn barnacle is a crustacean: one look at its larva will suffice to prove this." Nauplius-forms are also exhibited by the shrimp *Penaidae* and the parasitic sacculus.

While rejecting recapitulation as the universal and only mechanism of the relationship between ontogeny and phylogeny, one has to introduce a more general concept of heterochrony: evolution due to changes in the relative time of appearance and rate of development for characters already present in ancestors. Heterochronic changes are regulatory effects, in contrast to the appearance of new characters at any stage of development due to a change in structural genes.

There are thus two basic parallels between the stages of ontogeny and phylogeny: recapitulation and pedomorphosis. These two cases are shown schematically in Fig. 3.5, which was borrowed from the book by Gould [3.33]. In the direct parallel of recapitulation, the ontogeny of the most advanced descendants repeats the adult stages of the phyletic series of ancestors. In the inverse parallel of pedomorphosis, the ontogeny of the most remote ancestor goes through the same stages as the phylogeny of adult stages read in the reverse order. The progressively more juvenile stages of ancestors become the adult stages of successive descendants.

Various cases of heterochrony are listed in Table 3.1 [3.33].

The direction and scale of heterochrony may be illustrated by the ratio of the size and shape of an organism. The shape of an organism may be represented by the relative sizes of individual organs. The empirical relationship between the size of an organ y and the size of an organism x is given by the following expression:

$$y = bx^a \; . \tag{3.1}$$

We make use of this formula in order to describe heterochrony. We have

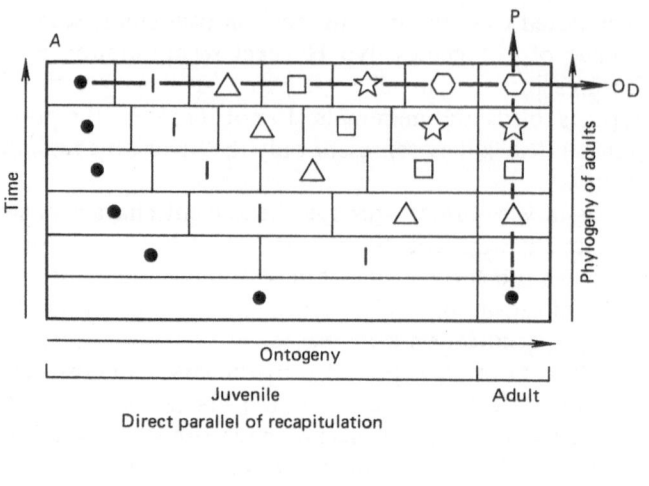

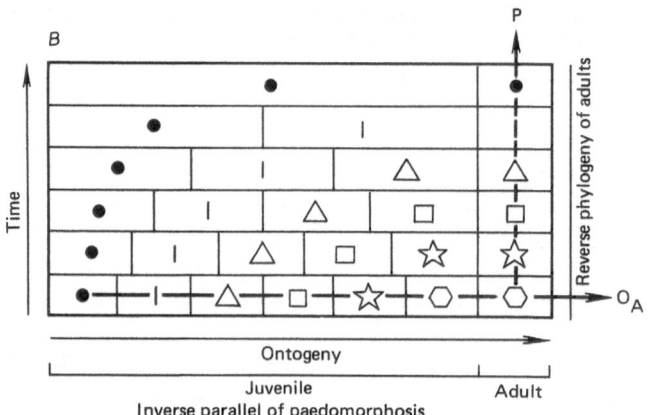

Fig. 3.5. Two types of parallels between ontogeny and phylogeny according to Gould: P = phylogeny; O_A = ancestral ontogeny; O_D = descendant ontogeny; A = direct parallel of recapitulation; B = inverse parallel of pedomorphosis

$$\log y = \log b + a \log x \ . \tag{3.2}$$

This dependence is given in Fig. 3.6. The solid straight line represents the situation for ancestral ontogeny. The shape of an adult organism is denoted by the point C. Let us draw through this point a line with slope $a = 1$. This is a line of isometry because the shape represented by the ratio y/x is constant at all its points. The sizes and shapes are dissociated if the standard shape C appears with a different size in descendants. If the shape C appears with smaller sizes in descendants, there is an acceleration of shape relative to size (the line segment AC in Fig. 3.6); if C is shifted to larger sizes, we have retardation (the segment CP).

Table 3.1. The categories of heterochrony

Timing		Type of heterochrony	Morphological result
Somatic features	Reproductive organs		
Accelerated	—	Acceleration	Recapitualation (by acceleration)
—	Accelerated	Pedogenesis (=progenesis)	Pedomorphosis (by truncation)
Retarded	—	Neoteny	Pedomorphosis (by retardation)
—	Retarded	Hypermorphosis	Recapitulation (by prolongation)

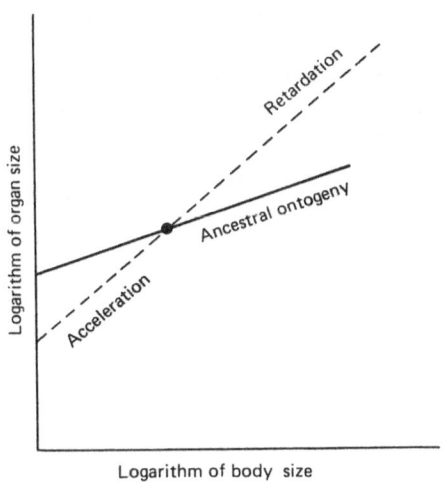

Fig. 3.6. Graphical representation of heterochrony – the Gould clock

Let us consider two straight lines (3.2) which differ only by the value of b:

$$y = b_1 x^a \quad \text{and} \quad y = b_2 x^a .$$

In Fig. 3.6 these straight lines are parallel. For any point on the line $\log y = \log b_1 + a \log x$ there is only one point on the line $\log y = \log b_2 + a \log x$, to which the same shape corresponds, i.e. the same ratio y/x:

$$\frac{y_1}{x_1} = \frac{y_2}{x_2} . \tag{3.3}$$

At these points

$$y_1 = b_1 x_1^a \quad \text{and} \quad y_2 = b_2 x_2^a \ . \tag{3.4}$$

From (3.3) and (3.4) we find

$$\frac{b_1}{b_2} = \left(\frac{x_1}{x_2}\right)^{1-a} . \tag{3.5}$$

The ratio x_1/x_2 expresses a relative difference in size with the same shape. This is a measure of heterochrony, which is equal to

$$s = \frac{x_1}{x_2} = \left(\frac{b_1}{b_2}\right)^{1/(1-a)} . \tag{3.6}$$

The measure s shows the relative difference in size with the same shape of ancestors and descendants – it is a measure of geometric similarity.

In what follows we shall consider the heterochrony model, known as the Gould clock model.

Heterochrony is of direct evolutionary and ecological importance. Both recapitulation and pedomorphosis are the consequences of two processes with different evolutionary meaning. The first process is hypermorphosis, the extension of ancestral ontogeny. The second is acceleration in the development of somatic characters.

Recapitulation occurs because advantageous characters can often augment their selective value by appearing earlier in ontogeny.

Pedomorphosis (neoteny) has direct evolutionary significance, leaving a number of genes "jobless" and, hence, available for subsequent modifications.

The direct ecological and, hence, evolutionary importance of heterochrony

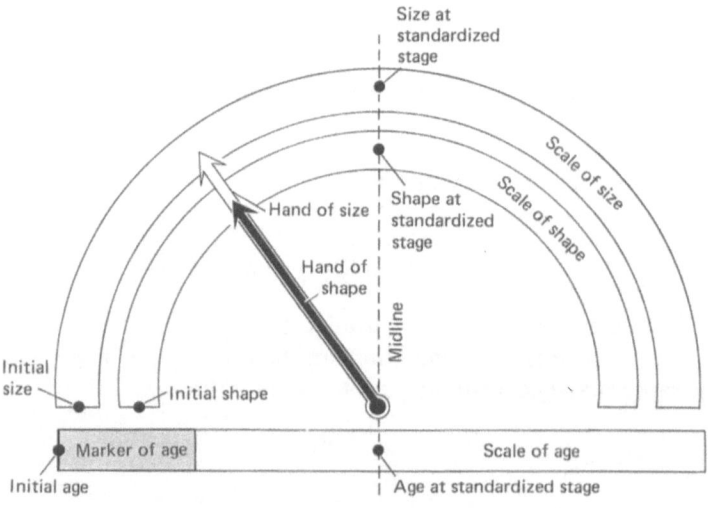

Fig. 3.7. The Gould clock model of heterochrony

is distinctly manifested in two types of natural selection: K-selection and r-selection.

K-selection prevails under the conditions of an increasing population of one genotype at the expense of another. K-strategy is characterized by slow (relatively!) reproduction, late maturation, long lifetime and great parental care of offspring. The characteristics of r-strategy are the opposite. As a matter of fact, there is a continuous gamut of mixed strategies.

Since the timing of maturation is of prime importance for the two strategies, heterochrony has a direct relation to ecological adaptation. The primary processes of heterochrony are acceleration and retardation. The former is associated with the r-regime and the latter with the K-regime. Gould has put forward a hypothesis, according to which progenesis is associated with r-strategy and neoteny with K-strategy. The timing indices of ontogeny are of primary value for phylogeny and ecology.

The model of heterochrony consists of the following [3.33]. Suppose we have a "clock" in the form of a semicircle with two hands and three scales. One of the scales shows the numerical characteristics of size and the second the characteristics of shape. There is also a linear scale which presents time, i.e. age. The Gould clock is shown in Fig. 3.7. Various types of heterochrony will be represented by differences in the ticking of the clock.

In order to establish the three scales, Gould uses the example of the fossil mollusc *Gryphaea angulata*. Table 3.2 gives the size ratios (lengths of the flat valve) and the shape ratios (coiling of the shell expressed by the ratio of valve lengths).

The size and shape vary symbatically. The calibration of the clock according to Table 3.2 is shown in Fig. 3.8.

Figure 3.9 shows the domains of heterochrony – accelerated and retarded on the age scale, paedomorphic and recapitulatory on the shape scale, and smaller or larger on the size scale.

Gould describes the course of the heterochrony clock for a number of cases:

Table 3.2. Relationship of size (length of flat valve) and shape (coiling as expressed in ratio of valve lengths) in ancestral *Gryphaea angulata* (from Gould [3.33])

Size	Shape
2.03 (average size at beginning of coiling)	0.41
5	0.81
10	1.38
15	1.89
20	2.35
30 (average adult size of ancestor)	3.21
38 (maximum size of ancestor)	3.84
40	4.00
46 (maximum size of descendant)	4.45

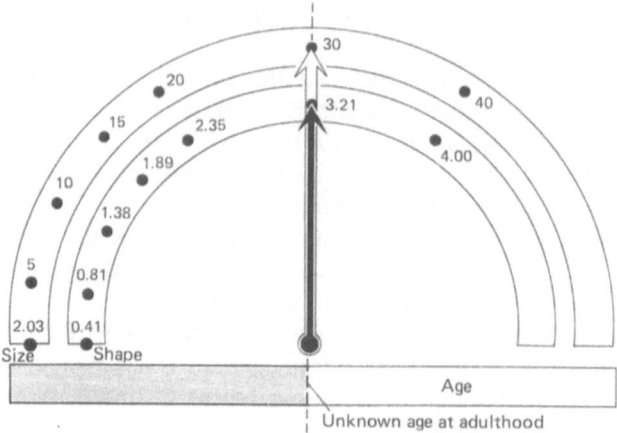

Fig. 3.8. Calibration of the Gould clock model for *Gryphaea angulata*

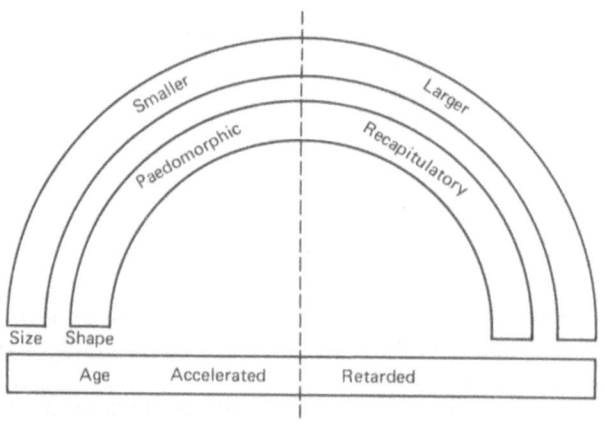

Fig. 3.9. The domains of heterochrony on the Gould clock: accelerated and retarded on the age scale, pedomorphic and recapitulatory on the shape scale, and smaller or larger on the size scale

progenesis due to truncation of ontogeny with early sexual maturation; hetero-morphosis due to a delay in maturation and an increase in size; acceleration due to the speeding-up of somatic development; proportioned dwarfing caused by slow growth with a constant rate of development; proportioned giantism due to rapid growth. We shall confine ourselves to neoteny, in which case the sizes are normal and the shape is retarded (Fig. 3.10). Neoteny is inherent in the species *Homo sapiens*. Maturation is retarded, the size increases and the shape remains in the realm of juvenile ancestors (Fig. 3.11). The monkey baby bears a much closer resemblance to man than the adult monkey does (Fig. 3.12 [3.35]).

The various types of heterochrony characterize the specific directedness of evolution determined by the relationship between ontogeny and phylogeny.

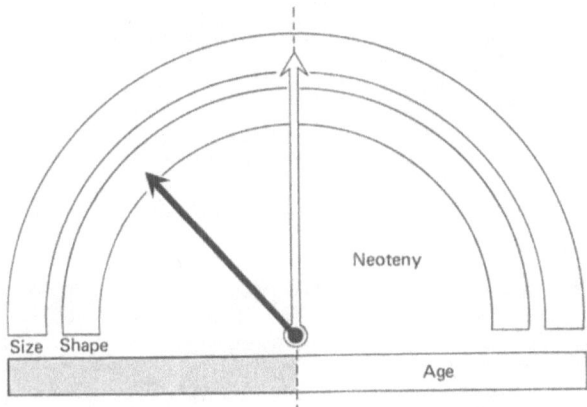

Fig. 3.10. The Gould clock in the case of neoteny

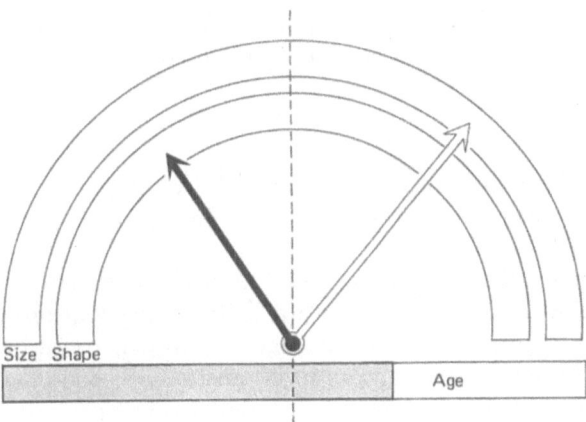

Fig. 3.11. Representation of human neoteny. Maturation is retarded, size increases, and shape remains in the realm of juvenile ancestors

It is obvious that heterochrony can be described in physico-mathematical language. The Gould clock model deserves a corresponding algorithm which has not yet been realized.

3.4 Theoretical Approaches to Developmental Dynamics

An organism at any stage of its development is a dissipative, synergetic system, i.e. an open system which is far from equilibrium.

The development of an organism from the original zygote may be treated,

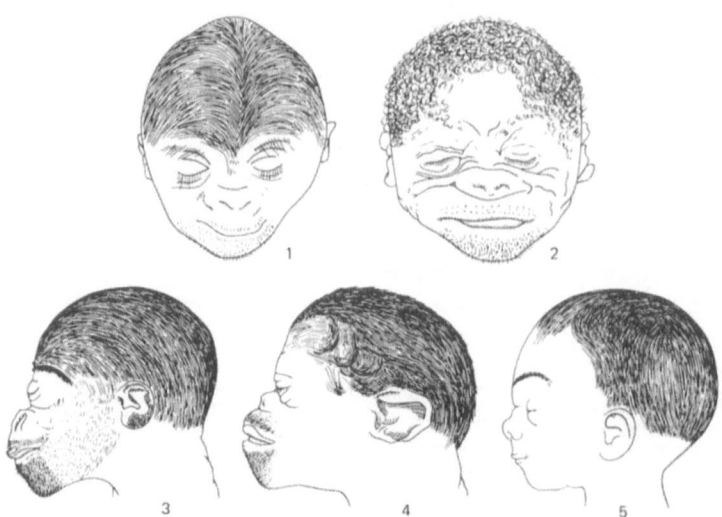

Fig. 3.12. The embryonic head of a gorilla, full face (*1*); the embryonic head of a chimpanzee, full face (*2*); the embryonic head of a gorilla in profile (*3*); the embryonic head of a chimpanzee in profile (*4*); the embryonic head of a man (*5*) [3.35]

first of all, as the appearance of spatial order in the original homogeneous environment. A corresponding theoretical model was first worked out by Turing [3.36,20,21].

In the volume of a certain chemical reactor reactions occur and the reactants diffuse. A stationary state with a uniform distribution of all reactants throughout the entire volume of the reactor can exist. However, under certain conditions such a state appears to be unstable. The system manifests cooperative behaviour: in response to even the slightest perturbations it passes from the initial state to a new state. Turing considered a one-dimensional two-component system in which autocatalysis and diffusion take place, the diffusion being characterized by various coefficients for two different components. The size of such reactive–diffusion dissipative systems is determined by macroscopic variables, namely diffusion coefficients and kinetic times:

$$l \sim \sqrt{D\tau} \ .$$

If τ is the clock, which is characteristic of cell dynamics, and the diffusion coefficient corresponds to molecules of organic substances with a molecular mass of 10^2–10^3, which move in a medium with a viscosity close to the viscosity of the cell contents, then l is of the order of 1 mm [3.37]. This corresponds to cell assemblies, including about 100 cells. It is exactly such assemblies that form morphogenetic fields.

In the Turing model a non-uniform distribution of reactants occurs; in other words, spatial self-organization of a dissipative system takes place. Thus,

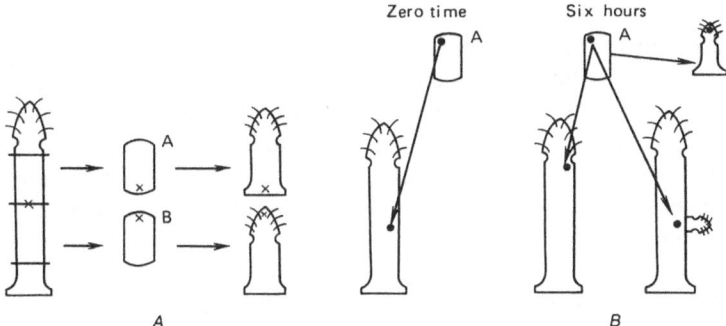

Fig. 3.13. Schematic representation of regeneration and transplantation experiments on the hydra

positional information about the morphogenetic field is generated spontaneously and is specified by the distribution of the concentrations of reactants,
which are called morphogens. Thus, morphogenesis can be reduced entirely to
chemical processes, as shown in the works of Gierer and Meinhardt [3.38, 39]
(see also the review [3.40] and the monograph [3.41] by Belintsev). The chemistry reduces to local autocatalysis and further inhibition.

These models have been successfully applied to treat the self-organization
of a relatively simple hydra organism (*Hydra vulgaris*). The self-organization of
the hydra has been studied experimentally in transplantation and regeneration
tests, the scheme of which is shown in Fig. 3.13 [3.42]. Fragments (*A* and *B*)
were cut out of the hydra organism and placed in a nutrient medium. From
these fragments grew whole animals of reduced size. The "head-sole" axis is
reproduced, which is evidence of a certain polarity, i.e. asymmetry of the tissue.
It appears that the same cells can develop into the sole or the head. The position
of the organs is determined by the morphogen concentration gradient. Figure
3.13b shows transplantation tests. If cells from a trunk section are transplanted
immediately after removal into another hydra trunk, they will be integrated
into the organism to produce a normal individual. If, however, the trunk section
is left alone for 6 hours between removal and transplantation into the new
trunk, one end will start to develop as a head, meaning that after transplantation the new organism will have an extra head growing from the side of the
trunk [3.43]. Morphogens were identified: an activator, which is a peptide with
a molecular mass of 1300, and an inhibitor, which is a substance of non-protein
nature with a molecular mass of less than 500 [3.44].

The hydra is an excellent experimental model for the study of ontogeny,
since it is composed of a relatively small number of differentiated tissues organized in one dimension.

Multicellular animals contain, generally speaking, a large number of various types of cells. In principle, an interesting exception is provided by organisms
of the Mesozoa type: small (not larger than 1 cm, usually smaller than 1 mm)
flatworms, which are parasitic inside some marine worms, molluscs and echino-

derms. The number of cell types in the Mesozoa is relatively small (some tens) but these organisms are unfortunately difficult to study.

No less success has been achieved in the application of the concepts of chemical dissipative systems to the examination of the early ontogenetic stages of *Drosophila*. The model in this case is two-dimensional. It shows that even at the stage of formation of the blastoderm (the formation of a specific layer of cells on the egg surface) a separation of the embryo into compartments, i.e. discrete, nonoverlapping zones occurs.

Self-organization consists of the spontaneous breaking of the symmetry of the macroscopic state under symmetric initial and boundary conditions. Local control of the global behaviour of the system exists. The mechanisms of its separation can be reduced to a smooth deformation of the macroscopically homogeneous states and to jumps of finite amplitude from a homogeneous state.

Detailed investigations of chemical models of the type described have been carried out, but we cannot dwell on them here.

Thus, models of the chemical marking of the morphogenetic field have been constructed. They are insufficient, however. While imitating well the formation of morphogenetic structures, these models appear to be inaccessible for direct experimental testing. The possibility of self-organization, which was first demonstrated by Turing, is determined by the synergetic nature of the system under consideration, i.e. by the possibilities of macroscopic enhancement of small fluctuations under instability conditions. The problem is one of synergetics, of the physico-mathematical model of a dissipative system rather than of chemistry [3.40, 41].

In a study of morphogenesis, chemistry must be combined with mechanics. An organism is formed by mechanochemical processes, that is divisions and movements of cells. Of substantial significance are the mechanical (plastic and elastic) properties of cellular systems. A cell, which was treated in the past as being composed of a nucleus and a viscous fluid (the protoplasm) appears in the light of modern investigations to contain a sufficiently rigid mechanical system, namely the cytoskeleton built up of fibrillar proteins [3.45]. The forces that form the embryo are generated by the cytoskeleton and reduce to viscoelastic stresses.

The simplest biological object that can be used for the study of morphogenetic phenomena is the cellular slime mould *Dictyostelium discoideum* (Dd). In this case morphogenesis is observed directly. If there is sufficient food, Dd is an aggregate of single-celled amoebas which multiply by division. *Dictyostelium* feeds on bacteria. In times of food deficiency the amoebas unite into an organized collective form. The individual cells become centres of aggregation, combining about 10^5 cells each. Chemotaxis occurs, i.e. a directed movement of cells induced by a gradient of the concentration of a certain substance in the external environment. This substance, called the attractant, is cyclic adenosine monophosphate (cAMP) and is secreted by cells.

There is a certain similarity between the situation described above and the behaviour of people in a shop. If there is sufficient food in the shop, people enter

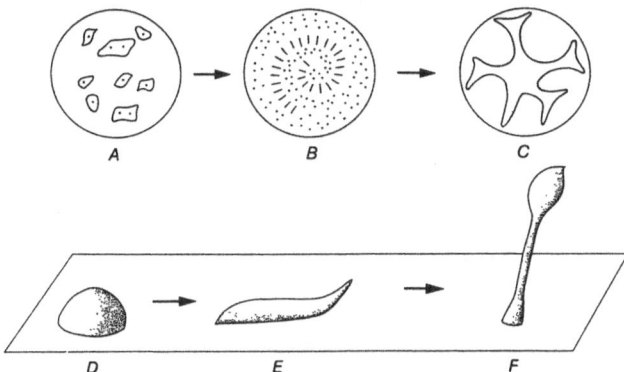

Fig. 3.14. Phases of the life cycle of the slime mould *Dictyostelium discoideum*: (*A*) individual amoebas; (*B*) initial stage of spiral aggregation; (*C*) later stage of aggregation; (*D*) early aggregate; (*E*) migrating slime mould; (*F*) fruiting body (at the top is the sac with spores)

and exit, independently of one another. If, conversely, there is a shortage of food, an organized, aggressive structure (a queue) is formed. The role of the attractant is played by verbal communication: the questions "What have you got?" and "Who's next?" and the answers to them.

As a result of the union of amoebas, a migrating slime mould arises, which has a sleeve-like shape and is capable of moving as a whole towards a light source (phototaxis). Such a state of *Dictyostelium discoideum* lasts for a few days. The Dd then forms a fruit body containing spores, by means of which the slime mould multiplies. There are two kinds of cells: generative (spores) and vegetative (stalk). Both morphogenesis and differentiation take place [3.46]. These phases of the life cycle of Dd are shown in Fig. 3.14. Table 3.3 characterizes the phenomena that take place during the aggregation period of development of Dd [3.41].

A detailed physical theory of the behaviour of Dd has been worked out by Belintsev [3.41]. This theory is concerned with all the events presented in Table 3.2. At the stage of dispersed amoebas a continuous secretion of substance c, i.e. cAMP, occurs, as do chemotactic and diffusive movements of cells. The non-linear kinetic equations take account of the two processes. The homogeneous state appears to be unstable and a regular separation takes place; the aggregates of cells are separated by intervals of the order of the diffusion path of c. The level of secretion of c plays the role of a critical parameter.

At the next stage a large-scale aggregation occurs, which follows from the basic equations. The theory describes the release of the "nucleus" of the corresponding territory, which is the region in which the cells realize a pulsed secretion of c. Two regimes are observed: synchronous oscillations of c and periodic vortex autowaves.

Macroscopic control of differentiation and divergence of cells then occurs. The initial macroscopic homogeneity is disturbed spontaneously upon aggrega-

Table 3.3. Aggregation period of development of *Dictyostelium discoideum*

Time in hours from beginning of starvation	Observed form of co-operative behaviour	Form of individual activity of cells	Biochemical prerequisites
0	Homogeneous cellular field of non-interacting amoebas	Non-directed movement of amoebas	
4	Small-scale aggregation	Continuous secretion of cAMP, chemotaxis per cAMP gradient	Growith of activity of adenylate cyclase (AC)
8–10	Large-scale wave aggregation, waves of attractant and chemotactic activity in a homogeneous field	Ability of cells to perform induced secretion of cAMP	Pulsed regime of AC activity
10	Centripetal stream aggregation, radial and spiral fluxes of cells	Stable intercellular contacts; appreciable fraction of cells autonomously secrete cAMP	Oscillatory regime of AC activity

tion of stalk cells due to their mutual attraction. Compact aggregates of cells are formed, and spatial separation takes place as a result of a directed chemotactic transfer of stalk cells to the top of the aggregate. Of substantial significance is the asymmetry of the boundary conditions for c, i.e. at the surface and along the contact boundary between the aggregate and the substrate. The axis of the slime mould is set up. Control of these processes is the result of the diffusion of two morphogens. Self-organization of the spatial form is determined by two non-chemical factors: surface tension and hydrostatic pressure.

Belintsev formulated the unified view of these historical events which involve a regular change of the dynamic modes of signalling. Belintsev has also developed a theory on the same conceptual basis for self-organization during the morphogenesis of the epithelium. The scheme of these events is shown in Fig. 3.15. In the originally passive cellular layer "marking" occurs: regions are singled out, in which the cells differ morphologically from the surrounding cells by having an extended form. Active movement of these cells leads to a deformation of the layer as a whole. Belintsev points out that the formation of domains of such polarized cells, which are separated from the surrounding tangentially extended cells, may be regarded as a universal beginning of almost all morphogeneses of multicellular animals, be they sand dollars or amphibians.

The self-organization of a developing organism occurs for the following reasons.

1.   The cells of embryonic epithelia possess mechanochemical activity during the period of morphogenetic activity. Due to the chemical sources of meta-

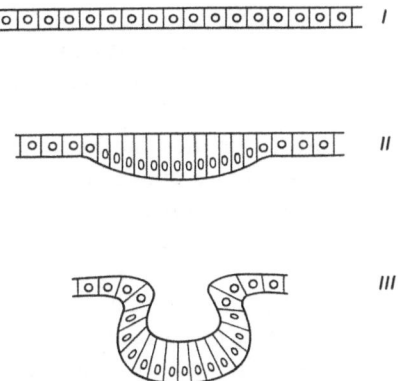

Fig. 3.15. Schematic representation of successive events in epithelial morphogenesis: (*I*) initial state; (*II*) formation of a domain of colony-forming cells; (*III*) local bulging of the layer

bolic energy they are capable of spontaneous deformation, while possessing two stable forms: an isotropic and a polarized form.

2. Active deformations of the cells give rise to elastic stresses in the embryo. The rate of relaxation of these stresses is substantially higher than the rate of the most active deformations.

3. Elastic stresses modulate the process of active deformation of the cells, so that the observed long-range order of cellular polarization disappears upon relaxation of these stresses.

4. Contact polarization takes place. This is the transfer of mechanical activity between the contacting cells.

The behaviour of a cellular layer described above is represented by nonlinear equations, in which a scalar dynamical parameter appears, which is equal to zero in the isotropic state and which is positive for cells polarized normal to plastids. The equations also take account of the effect of contact polarization of cells and of the effect of tangential elastic stresses. A deformation field of the cellular layer is determined, which corresponds to the facts observed. The Belintsev theory leads to rather general conclusions. It explains the group regimes of the mechanochemical activity of cells in layers bound by elastic linkages to the passive external material. Such an interaction is essential during the morphogenesis of integumentary structures and also during the development of limbs. The presence of incipient instabilities accounts for the formation of patterns typical of integumentary morphogenesis: with long-range order (feathers, squama) and with short-range order (hair). One might think that the punctuated appearance of the plumage of birds was determined by small changes in the controlling parameters in the instability region.

Mechanochemistry, i.e. a direct conversion of the metabolic chemical energy usually stored in adenosine triphosphate (ATP) into mechanical work, is of fundamental importance in developmental processes. It should be noted in this connection that the effect of propagation of the front of contact polarization of cells in embryonic material on the efficiency of protein synthesis has been proved experimentally [3.47]. The degree of extension of cells exerts an influence on protein synthesis [3.48].

Roughly speaking, chemistry creates mechanics. But, of course, reverse phenomena also take place; mechanics creates chemistry. Mention should be made of the classical experiments conducted by Loeb, in which an ovicell pricked by a needle begins to divide. The biological, in particular enzymic, activity of proteins is determined by the interaction of electronic (i.e. chemical) and conformational degrees of freedom (see Chap. 5 and [3.20,21]). As was first pointed out by Volkenstein [3.49], the forced mechanical deformation of an enzyme brings about a change in its activity. This has been proved by experiment [3.50].

A detailed description of the physico-mathematical theory of ontogeny, i.e. differentiation and morphogenesis, goes beyond the scope of this book. The theory has been expounded in the monograph of Belintsev [3.41]. The theory in question is a model, phenomenological theory, but it also takes account of chemistry and mechanics, and the specific behaviour of a dissipative system, which leads to its self-organization as a result of instabilities. This topic is within the area of synergetics.

The key problem of evolutionary biology is the relationship between genotype and phenotype, between heredity and natural selection. The experimental and theoretical solution of this problem is still a matter for the future. However, as has been shown above, the physical approaches to ontogeny are promising. It may be said that the key concepts which must underlie our understanding of the phenotype-genotype relationship are in the realm of autocatalysis, mechanochemistry and synergetics.

The historical nature of genotype and phenotype is determined by constraints on natural selection and, hence, by the directedness of evolution.

3.5 Nomogenesis

The directedness of evolution was the main idea of the theory proposed by L.S. Berg in 1922 and known as nomogenesis [3.35]. This term implies evolution based on regularities rather than on contingencies; nomogenesis is contrasted to Darwinian evolution based on chance, which was called "selectogenesis" or "tychogenesis" by Berg and Lyubishchev [3.51].

Today the Berg theory is mainly of historical interest. It had been developed before the enunciation of the synthetic theory of evolution. The modern understanding of the pivotal problems discussed by Berg is significantly different from the way they were seen at the turn of the century. A number of difficulties encountered by the theory of evolution in its early stages no longer exist. Of course, new difficulties have appeared instead; the science is developing. Nomogenesis, alongside the erroneous anti-Darwinism, contains a number of important ideas and positions, which found their place in science at a later time. It is necessary to remove "the husk from the kernel", since many of the scientists guided by the Berg ideas today are not capable of doing this.

Let us begin with the "husk". Berg maintains that variability exists only within narrow limits and therefore directive natural selection is impossible; there is nothing to choose from. He wrote:

"In order for natural selection to exert and influence in the presence of a certain direction in evolution, (1) a certain direction must last for an infinitesimal period of time and (2) there must be an infinitely large number of evolutionary directions. Neither of these exists".

This is erroneous reasoning. Both classical Darwinism and the synthetic theory of evolution recognize the constraints on natural selection that have been discussed above, but this does not lead to either of the positions indicated. Selection can act within a limited range of changes and with sufficiently prolonged directed evolution.

Berg refuted the divergence of species, stressing the importance of convergence. From this follows polyphyletism in speciation, in contrast to Darwinian monophyletism.

Instead of Darwinism, which is based on the random variability and natural selection of the fittest organisms, Berg wrote about the "rationalness" inherent in organisms, about their innate ability to adapt to environmental conditions.

The above positions were at one time regarded in Russia as evidence not only of anti-Darwinism but also of Lamarckism, idealism and agnosticism. *Nomogenesis* was attacked with severe and tactless criticism. In 1939 in one of the leading newspapers there appeared an article under the title "No place for pseudoscientists in the Academy of Sciences" signed by A.N. Bakh, B.A. Keller, Kh.S. Koshtoyants and V.I. Milovanov. The article was published in connection with the nomination of L.S. Berg as an academician. Not only Berg but also N.K. Koltsov was attacked. The nomination of Berg was delayed until 1946 when he was elected not for his biological works but in the field of geography. It is appropriate here to cite the words of A.A. Lyubishchev [3.51]:

"In those cases where this or that form of teleology becomes part of the reigning ideology, so that the loyalty of a thinker or scientist is judged by his adherence to the dominant teleology, it becomes a direct tool of true obscurantism."

The Darwinism of the critics of Berg was really of teleologic character, i.e. they did not burden themselves with scientific argumentation and were guided by the dogma.

After 53 years a second edition of Berg's work *Nomogenesis* appeared. The introductory article to this edition, written by K.M. Zavadsky and A.B. Georgievsky, contained a detailed critique of nomogenesis. Berg was again criticized for idealism, agnosticism, anti-Darwinism and Lamarckism, this time with detailed argumentation. But it was only the "husk" that was noted, while the positive content of Berg's works was left unnoticed; it was not clear why his book was published again.

In an extensive co-operative work *Development of Evolutionary Theory in the USSR* (Leningrad: "Nauka", 1983) the repeatedly cited works of Berg devoted to evolution were not analyzed; they were only labeled anti-Darwinian alongside the works of Lysenko.

However, the actual situation was different. Berg may have made some mistakes, but even his valid views were distorted by critics.

Idealism in biology is always associated with vitalism, with its separation from physics and chemistry. Berg was by no means an idealist. He proceeded from the finite physicochemical essence of biological phenomena. He wrote:

"Everything that occurs in organisms obeys exclusively the laws of physics and chemistry. There has never been observed and, I think, will never be observed in any living creature any contradiction to the laws that govern the course of processes in the inorganic kingdom."

"Research in the area of natural sciences can be done successfully only with the aid of forces known to physics."

"No wonder is possible in the world: nature works exclusively by means of the laws of physics and chemistry."

As has been pointed out, these positions are in full agreement with the present-day status of natural sciences and with reasonable natural philosophy.

The words cited above were not only declarative. Berg really predicted the future of science when he wrote that "it is possible to construct a natural system of the animal and plant kingdoms ... in the sense of 'chemical' similarity .... It will be possible to group organisms into series and systems, like chemical compounds...."

"Differences in species (or, in general, differences between two forms) lie, first of all, in the stereochemical grouping of molecules of proteins or, in general, of substances that make up an organism .... Heredity consists ... in the transmission of a known grouping of molecules ...."

"What is inherited is the ability to assimilate and exchange substances equally."

"... Since the chemical composition of proteins in closely related forms is similar, it is not surprising that the descendants realize forms similar to ancestral forms."

"To the evolution of forms there must correspond the transmutation of the chemical composition of proteins."

Berg was the first to predict the future of the molecular theory of evolution. He was right; the evolutionary trees are built today on the basis of the structure of proteins and nucleic acids. It is clear that Berg has nothing to do with idealism. Molecular biology could likewise have been accused of idealism. People who have nothing to do with science could have accused Berg of reductionism or mechanism, but not of idealism or vitalism.

The accusations of agnosticism were based on the following words: "To trace the origin of rationalness in living things means to find out the essence of life. And the essence of life is as intractable as the essence of matter, energy, feeling, consciousness, volition."

As a matter of fact, this "essence" is similar to Kant's "thing in itself". However, these are mere words. As we have seen, Berg maintained that the future progress of science will allow one to disclose the physico-chemical, molecular nature of heredity and evolution. Will this mean the penetration into the "essence of life?" The question is philosophical rather than scientific. From the standpoint of contemporary biology the answer is positive.

The accusations of Lamarckism are associated with the principle of the "original rationality of all living things", which implies the presence in organisms of "heredity, variability and self-preservation". As regards Lamarckism, Berg wrote: "Lamarckism, except when it lapses into tautology, gives nothing to account for the evolutionary process. It only states the fact of presence of rationality in organisms."

The rationality concept as put forward by Berg is only a definition: "Only those things should be called living, which, as a rule, respond rationally to irritations and systematically convert heat into work."

The basic content of *Nomogenesis* is not associated with this notion. Concluding his book, Berg lists 10 points of difference between Darwin's theory and nomogenesis. The word "rationality" does not figure in these positions. Thus, the accusation of Lamarcksm is also groundless.

Speaking of the conversion of heat into work, Berg implies the specific thermodynamic properties of open living systems (see Chap. 1).

There remains anti-Darwinism. The errors made by Berg are significant, as we have seen, and his nomogenesis would not have deserved to be mentioned here if its content reduced only to the negation of directive natural selection.

The main proposition of nomogenesis is the directedness of evolution specified by the innate properties of organisms, i.e. by established structures and by the limited possibilities of their change. Berg was the first to point out the constraints on natural selection (see Sect. 3.1). The conclusion made by Berg as to the absence of selection was erroneous, but the notion of its limitedness is valid and important. Let us now consider the "kernel".

"Darwin maintained that the variation of characters occurs in all directions . . . . We think that the change of characters is limited by certain boundaries."

"Not just any forms arise, from which natural selection would have to choose the fittest ones; on the contrary, only those organs emerge which must be formed due to the constitution of an organism and external conditions."

"Selection has to work with an established tendency rather than with chaotic, random characters that arise without any regularity."

This is quite correct and Darwin understood this. We have already cited the words from *The Origin of Species* to this effect. Berg cites two passages from Darwin's work on artificial selection [3.52]:

"Any significant variation is governed by a law and is determined to a far higher degree by the nature or constitution of the being, than by the nature of the changed conditions."

"Although any variation is directly or indirectly caused by some change in environmental conditions, we must never forget that the nature of the organization exposed to an influence is the factor which is much more important for the result."

As a matter of fact, there is no contradiction between Berg and Darwin, only the stress is placed differently. In *The Origin of Species* attention is focused on the effect of the environment and on natural selection.

Many years later Gould wrote [3.53]: "... organisms are not pieces of putty, infinitely mouldable by infinitesimal degrees in any direction, but are, rather, complex and resilient structures endowed with innumerable constraints and opportunities based upon inheritance and architecture (both molecular and morphological)." Natural selection can be the directing force for changes, but the organism takes a substantial part in this process, restricting the directions of possible variations.

In *Nomogenesis* Berg paid due attention to heterochrony. A "pre-adaptation" is possible, i.e. the appearance of characters that originally had no adaptive significance but acquired it in new ecological conditions. Today it is quite clear to us that every organism and every species possesses numerous non-adaptive characters. This subject will be discussed in more detail in Chap. 4.

Berg considered convergence, treating it as the similarity between evolutionarily distant organisms which emerged as a result of independent development in the same direction. The most important manifestations of convergence are thought to be the homologous series in variation discovered by N.I. Vavilov [3.8]. Many of the phenomena of similarity between organisms, which are commonly treated as mimicry, are in fact accounted for by convergence. Independent development is determined by nomogenesis, i.e. by the laws of directed evolution. The convergent similarity of distant organisms is evidence of their polyphyletic origin.

What has been said above is consistent with the modern theory. At the same time Berg made an error by contrasting convergence with divergence. Both phenomena take place. As we shall see, while refuting divergence, Berg contradicted one of the most important of his own propositions.

According to Berg, speciation occurs by way of "mass transmutation", which changes the norm maintained by stabilizing selection. The mechanism of this phenomenon, which is saltational in nature, remains mysterious, and the notion of "mass transmutation" is probably the weakest spot in *Nomogenesis*. Being unaware of population genetics, Berg spoke of mutations but he believed that they are rare and play no role in speciation. At the same time, Berg formulated a rather important proposition in this context.

"The birth and death of individuals, species and ideas occur by a catastrophic process. The appearance of all these categories is preceded by a prolonged, concealed period of development, which proceeds according to certain laws and is then followed by a leap, a saltus, which leads to their appearance, propagation on the surface of the earth and occupation of a place under the sun. The process of conversion of a gas to a liquid is a discontinuous change."

In other words, speciation is similar to a phase transition in the physical sense of this notion. The modern theory of evolution has arrived at these conceptions only recently. This has already been mentioned and will be discussed in Chaps. 8 and 9. Speciation by way of divergence implies a decrease in symmetry, which is similar to a second-order phase transition. This means that Berg, by rejecting divergence, contradicts his own position.

Phase transitions in speciation and species extinction are inevitable consequences of the remoteness of the biosphere from equilibrium. As we have already said, the modern approach to the key problems of theoretical biology is based on synergetics.

We see that Berg was much superior to his critics. Having made a number of errors of anti-Darwinian nature, he still appeared to be the forerunner of new ideas. As a fundamental work, *Nomogenesis*, which is devoted to the factors of directed evolution, contains very important propositions.

Firstly, Berg formulated a fundamental relationship between biology, physics and chemistry; he pointed out the possibility of tracing evolution by the protein structure.

Secondly, he was the first to put forward detailed arguments in favour of stabilizing selection.

Thirdly, he analyzed the constraints imposed on evolution by the internal factors associated with ontogeny, the structure of an organism and with possible ways of its change.

Fourthly, he gave a clear-cut formulation of various heterochrony phenomena.

Fifthly, he carried out a detailed analysis of convergence phenomena.

And, finally, he was the first to draw an analogy between speciation and a phase transition.

These ideas were later partly developed by A.A. Lyubishchev, who did not refute canonical Darwinism [3.51]. The articles written by Lyubishchev are devoted to the methodology and terminology of evolutionary theory. The attitude towards Darwin is not consistent. While considering "natural selection to be important, even creative and necessary, but not the driving force of evolution", Lyubishchev accuses Darwin of extreme inconsistency, which ends up with "complete lack of principles", and speaks of Darwinism as a scientific opium.

The works of Berg have found a response in world science; his book has been translated into English, but Mayr in his fundamental history of biological thought mentions Berg's nomogenesis alongside the aristogenesis of Osborn and the Omega principle of Teilhard de Chardin [3.54]. Lyubishchev's article devoted to evolution has not been recognized by the scientific world.

Nomogenesis has been completely overlapped in the subsequent develop-

ment of the theory of evolution. We have paid attention to nomogenesis, however, not only because it is one of the brightest pages of the history of biology.
The point is that even today in this country, the works of Berg are either completely refuted or accepted without reservations together with erroneous anti-
Darwinism. Neither of these standpoints is correct. Nomogenesis is often referred to for non-serious criticism of the synthetic theory of evolution. Criticism
of this kind has become fashionable in this country. This is understandable from
the psychological standpoint. The forbidden fruit is sweet. The books and articles written by Berg and Lyubishchev were regarded as heretical and were forbidden for a long period of time. They are now available and, of course, they
have gained recognition. The crude dogmatic criticism of Berg's works has
caused the opposite effect, proving that they are correct. At the same time, true
Darwinism has been distorted by Lysenko and his colloborators, who put forward the so-called creative Darwinism, which had nothing to do with science.
Even today many scientists confuse Darwinism with its caricature.

The role of the theory of nomogenesis in the understanding of directed
evolution has been significant.

4. Adaptation

4.1 Fitness

The full title of the great work of Darwin known as *The Origin of Species* is more detailed and spectacular: *The Origin of Species by Means of Natural Selection or the Preservation of Favoured Races in the Struggle for Life*. This title expresses Darwin's key idea: the natural selection of organisms that are best adapted to environmental conditions. The 'favoured races' are exactly those which are the fittest. Chapter 4 of the book is called 'Natural Selection or the Survival of the Fittest'.

The definition of the term 'adaptation' that appears in present-day textbooks of evolutionary theory [4.1] reads:

"Adaptations must be defined as the origination and development of particular morphophysiological properties, whose importance to an organism is unambiguously associated with the various general or particular environmental conditions."

A similar definition is given by Schmalhausen [4.2]:

"Adaptation is the morphophysiological manifestation of the relationships between an organism and the environment recognized through their changes."

Darwin's idea was met with objections and is still disputed. What is the criterion of fitness? The commonplace answer is survival. But then the statement of the survival of the fittest becomes a mere tautology: the survival of those that survive (survival of the survivor). It is exactly in these terms that Darwinism was treated by the philosopher Karl Popper, whose works have gained wide popularity among naturalists [4.3].

In fact, there is no tautology here and the situation is much more complicated. The unit of selection at the species level is a population, whose survival is not identical to the survival of an individual since the existence of a population and, hence, of a species depends on the reproduction of organisms as well. The criteria of fitness are connected with survival, but they may be formulated independently. Let us cite some examples. Marine mammals (seals and whales) possess a streamlined body, which allows them to move quickly in water and to hunt successfully. Nocturnal animals (e.g. some lemurs), being active in semi-darkness, possess very big eyes. Flies living on the Falkland Islands, where strong winds often blow, are devoid of wings and are therefore not carried away to the sea. It is more difficult to understand why the female winter moth (*Hybernia defoliaria*) has no wings. The phenomena of mimicry and convergence are directly associated with adaptation. We have already mentioned the industrial melanism of the black moth *Biston betularia*. Xerophytes (drought-resistant plants), which grow in dry (xeric) habitats, put out only small leaves or are totally leafless. This is one of the features that reduce the evaporation of moisture. At the same time, xerophytes are characterized by large and thickened root systems, which can extract much of the limited soil moisture.

Adaptedness is relative. At any moment a new form which has better adaptive potential may arise. Lewontin compares two hypothetical situations [4.4]. In one population the fecundity was increased. At the same time, the immature offspring became more vulnerable to epidemic diseases and to attacks by predators. In another population the fecundity did not change; the offspring made better use of food resources, but they perished in larger numbers in the adult stage. It is impossible to say which of the populations was better adapted. There is no tautology here.

As has already been said, *The Origin of Species* is devoted primarily to the interaction between living organisms and the environment. As early as during Darwin's lifetime the philosopher Hartmann pointed out that Darwin's book should have been called "The Origin of Adaptations" [4.2]. One cannot agree to this, but Darwin really wrote chiefly about the 'spark' rather than about the 'flame'.

The organization of living things is always adaptive simply because they live in a real environment. During the process of evolution, adaptation is never complete, but is perpetually taking place. One of the main tasks of evolutionary theory involves undertaking a kind of engineering analysis of the relationships between an organism and the environment [4.4]. According to Darwinism, speciation can be reduced to adaptive radiation. This is the case with a number of bird species, for example *Hedyotis* (Rubiaceae) in the Hawaiian Islands [4.5] and the famous Darwin finches (*Geospizinae*), the latter being confined to the Galapagos Islands [4.6]. One of the finch species, the woodpecker finch (*Camarhynchus pallidus*), appears to be a bird which makes use of a tool. It bores a hole into tree bark with its beak, and then extracts insects from the hole by inserting a twig or cactus needle which it holds in its mouth. This is an excellent example of adaptation. It should be emphasized that the chiselling-out of a chink in tree bark by a woodpecker or by a woodpecker finch requires a special scalp. It has bones which serve as a shock absorber when the beak hits the tree, and thus protect the brain from concussion.

Since we are now talking about a bird using a tool, mention should also be made of a second case, the behaviour of the Egyptian vulture (*Neophron percnopterus*). This bird uses stones in order to break the shell of ostrich eggs. The vulture picks up a stone in its beak, and drops it on the egg from above. Extremely interesting experiments were conducted by Spanish zoologists. The same species of vulture is encountered in Spain, but it is, of course, a different population. There are naturally no ostrich eggs in Spain. When porcelain eggs were made which were filled with the contents of hen eggs, the Spanish vultures were quick in acquiring the behavioural patterns practiced by their African counterparts.

Of course, an organism adapts as a whole, while its individual characters often have no adaptive significance. Neutral and harmful characteristics, since they may be associated with useful characteristics, can be fixed during evolution. The characteristics that are beneficial in one respect may be found to be harmful in another [4.2]. For example, the evolution of mammals proceeds in the direction of increasing size, say in the case of horses (Cope's Rule). Obviously, this improves the conditions of heat preservation in the body, as heat production is proportional to body volume and heat transfer is proportional to body surface area. The growth of the brain and the capabilities of both sides in prey-predator encounters are associated with an increase in body size. At the same time, an increase in body size requires an increase in the amount of food consumed; if food resources are limited, Cope's Rule does not work [4.5].

"Natural selection is the key factor which maintains the required equilibrium between organism and environment and which also determines the origination and development of adaptations" [4.1].

How should we treat the environment? Here we ought to speak of ecological niches or of biogeocenoses. The number of niches is practically unlimited and we cannot consider the niches to be preformed, pre-existent for emerging organisms. The populations themselves change the nature of niches and create new niches. There are no empty niches; such a notion is simply devoid of meaning. In his classical work Schmalhausen writes: "An organism is never passively exposed to the influence of environmental factors. It counteracts these factors actively, following its own laws determined by a historically grounded response base ... The autonomy of life processes is more and more strongly manifested in higher organisms" [4.7]. In Chap. 3 of his work Schmalhausen writes about the constraints on the evolutionary process, which are determined by the already developed structure of the organism, its ontogeny. In his other work [4.2], Schmalhausen states that "the facts of irreversibility of evolution show quite clearly that the evolution of a particular kind of organism cannot proceed in all directions but only in some. The historically developed structure of an organism limits the possible pathways of the evolutionary process to a certain range".

An organism is an integrated system, which is why the co-adaptation of its organs is realized. At the same time, because of heterochrony, the different organs of an organism are found to be adapted to different degrees, and the rudimentary organs, which earlier had adaptive significance, have lost this during subsequent development. During adaptation both "progressive" processes leading to the complication of structure and function (aromorphoses according to Severtsov [4.8]), and regressive processes, so-called "catamorphoses" are possible. The transition to the parasitic way of life and loss of vision by cave organisms (Protei) are examples of catamorphoses.

Adaptational changes are eventually governed by mutations which produce material for evolution by natural selection. As will be shown in Chaps. 6 and 7, mutations cannot be reduced to point replacements of nucleotides in DNA and of amino-acid residues in proteins. But in any case they are mutations, i.e. heritable changes in genome which arise randomly.

In certain cases, adaptation (i.e. fitness) is not due to mutation. A classical example is the "habituation" of pneumococci to antibiotics, say to penicillin. The cessation of the action of an antibiotic on infecting bacteria has been described by physicians as the adaptation of pneumococci to a new factor (penicillin) and the inheritance of this adaptation. This explanation is Lamarckian: the inheritance of an acquired trait. The impossibility of such inheritance for sexually reproducing multicellular organisms is evident. As a matter of fact, there is no mechanism of adequate transmission of acquired somatic traits to gametes. But a pneumococcus is a single prokaryotic cell and one would expect this difficulty to be removed since there is no difference between gamete and soma.

Another possibility is determined by mutations. If a particular population of bacteria contains mutants resistant to an antibiotic, then the latter will act as a selective factor. Unstable cells will perish and only a stable mutant will be left and will multiply.

Since inherited adaptation occurs gradually and in a certain direction, being amplified from generation to generation, and mutation is quite random, statistical analysis of populations must lead in these two cases to different results. Relevant experiments were conducted by Luria and Delbrück [4.9]. They have shown unambiguously that the selection of mutants resistant to an antibiotic occurs, rather than the inherited adaptation to it. The acquired traits are not inherited in prokaryotes either. It is not the adaptation as such which is inherited, but rather a genome.

Later, however, investigations were carried out, the results of which seem to contradict those described above. Cairns and his co-workers [4.10] have reproduced the results obtained in 1952 by Ryan [4.11], indicating the directed appearance of so-called *Lac*-revertants in *E. coli*, which are capable of fermenting lactose in response to its addition to the medium. Similar phenomena have also been observed in other prokaryotic systems [4.12,13]. These results, which have not been universally accepted, were discussed in detail in the pages of the journal *Nature* [4.14–16]. Opinions were expressed, according to which the inheritance of acquired characters is possible in prokaryotes.

The "central dogma" of molecular biology, the impossibility of information transfer from proteins to nucleic acids, is thereby refuted.

The actual situation is more complicated. If the experiments carried out are correct, they by no means imply the information transfer from protein to genome built up of DNA. First, the change of the situation caused by addition of lactose may bring about an increase in mutability and a chance appearance of mutants that were not observed earlier. The second explanation is more general and informative. The introduction of lactose alters the conditions of existence of a population of *E. coli*, and changes the fitness landscape (see Chaps. 8 and 9). A new fitness peak appears which can be reached only if *E. coli* is capable of producing a protein to ferment lactose. Mutations arise randomly, but the presence of the various mutations facilitates the appearance of subsequent ones. For example, the emergence of a double mutant is incomparably more probable for a single mutant than for an unmutated genome. This gives rise to the directedness of molecular evolution, which has been proved for RNA macromolecules and viruses in the works of Eigen and his school (see Chap. 8). The adaptation peak is achieved sooner or later.

What has been said above does not refute the central dogma of molecular biology, and indicates that contrasting the mutation process to the adaptation process is in fact meaningless. Two multidimensional spaces coexist and interact: the space of possible mutations and the space of fitness. Mutations are originally random, but following each other along the "mountainous ridges" of this landscape, they reach adaptive peaks in a directed manner.

The situation with eukaryotes is simultaneously more complicated and simpler. Both spaces are rather inhomogeneous. At the same time the adaptation to external factors clearly points to the selection of stable mutants. The adaptation of insects to insecticides and pesticides and the adaptation of rats to the anticoagulant warfarin are both like this. The latter poison, which kills rodents, exterminated almost all rats in Great Britain in the 1950s, but two decades later the country was again populated by rats that had become immune to warfarin.

Adaptation is the principal idea of Darwinism. In the *Origin* Darwin uses the word 'beautiful' more than 20 times. Darwin applies this word to shapes and traits that indicate the adaptability of an organism to environmental conditions. This is a kind of utilitarian aesthetics. Although aesthetic contemplation is unselfish (Kant), we admire adaptational consistencies in about the same way as we admire the streamlined shape of a modern automobile. Genuine aesthetics is not confined to works of art and literature; aesthetic appraisal may also be made with respect to nature, science, technology, sports and, finally, human (or animal) behaviour [4.17].

A number of phenomena of living nature make a direct aesthetic impression on us. We admire flowers, birds and butterflies, without pondering on the functional significance of the sparkling feathers of a drake or of the iridescent blueness of the wings of the butterfly *Morpho*. Incidentally, these are also adaptive features that play an important role in sexual selection. The bright colour of flowers implies adaptation, as it attracts pollen-transporting insects. Here is a case of co-adaptation and co-evolution; many insects and flowering plants undergo evolution jointly, realizing efficient pollination.

One of the most important environmental factors is temperature. Aleksandrov has revealed the molecular basis of temperature adaptation for a number of plants and animals, in particular frogs. The Northern grass frog *Rana temporaria*, the geographic range of which extends from 43° to 70° Northern latitude, requires a temperature range from 13° to 26°C. A closely related species, the lake frog *Rana ridibunda*, inhabits an area from 40° to 60° Northern latitude and prefers temperatures from 18° to 28°C. The denaturation temperatures of functional proteins of these species accordingly differ, i.e. they are higher for the lake frog than for the grass frog. These temperatures by far exceed the temperatures at which frogs sustain their life. But the lower denaturation temperature means the lower rigidity of the protein molecule, its greater conformational lability (see Chap. 5). The heat-loving nature of frogs stems from the conformational properties of their proteins ([4.18,19], see also [4.20]).

In a number of cases one should speak of pre-adaptation, of the presence of the original traits, which, while undergoing modification, can provide adaptation to new niches. For example, the four lobe fins of a crossopterygian may be regarded as being a pre-adaptation to dry land, on which amphibia, reptiles, birds and mammals with four limbs appeared. This is devoid of special meaning. Of course, crossopterygians did not know that their progeny would come out of the sea to dry land. The four lobefins determined the subsequent evolution of vertebrates by the very fact of their existence, which imposed decisive constraints on selection (Chap. 3).

As has been said above, there is no array of empty ecological niches to which the various organisms appear to have been pre-adapted. An adaptation arises as a result of the interaction of organisms with the environment; in some cases the environment chooses organisms and in others it is vice versa. Punctuated evolution implies that many characters of new species and higher taxa do not have adaptive significance at all. The relevant questions are discussed in Sect. 4.4.

The difference between human society and biogeocenosis, which is determined by the development of the brain of *Homo sapiens*, consists of the adaptation of man to environmental conditions achieved by technical means, and of the specific co-adaptation of the members of one population, one country or a group of countries. The words adaptability and adaptor as applied to man have a negative meaning. Under conditions of suppression of man by a state, "adaption" means conformism. However, in a society novel ideas and new humans appear (in the way that mutations appear in a species). These eventually change a "niche", thus demonstrating "social ecology". Any actions of man are adapted in varying degrees to conditions of the social environment. A work of fiction or art is created in certain circumstances, which are reflected in one way or another in his creative activity. Very promising is the consideration of social adaptation on the basis of information theory (see below and Chap. 10).

4.2 Biochemical Adaptation

In the preceding section we dealt with adaptation at the morphophysiological level and proteins were mentioned only in discussing the heat-loving nature of frogs. As a matter of fact, a cell or an organism is a complex open system, a chemical machine, which exchanges matter and energy with its surroundings. At the molecular level the basis of life is biochemistry. We may say that biochemistry is adapted to life, to ontogeny and phylogeny. In the long run, biochemistry is responsible for the global properties of the external world; the oxygen of the atmosphere was produced by photosynthesis. Each stage in the elaborate chain of biochemical reactions catalyzed by enzymes participates in one way or another in adaptation processes.

In essence, the nature of biopolymers, i.e. proteins and nucleic acids, and also of other biologically functional molecules and ions might also be looked upon as adaptation to life. However, this is not entirely accurate. According to this interpretation we might also say that the rules established for moving chess pieces are adapted to the course of a chess game. In all probability the opposite is true.

Biochemical adaptation has been described and analyzed in detail in two monographs written by Hochachka and Somero [4.21,22]. The treatment in this section is based on these books.

Hochachka and Somero write about the strategies of biochemical adaptation. Three kinds of strategy are given [4.21]:

1.   The types of macromolecules (primarily enzymes) present in the system may be changed.
2.   The amounts or concentrations of macromolecules may be adjusted.
3.   The functions of the macromolecules of the cell may be regulated in an adaptive manner.

In the second book [4.22] the three strategies look different:

1.   The adaptation of the macromolecular components of cells or fluids of an organism.
2.   The adaptation of the microenvironment in which macromolecules function.
3.   The adaptation at the functional level when a change in the efficiency of macromolecular systems, especially enzymes, is not associated with a change in the number and type of macromolecules contained in the cell.

Adaptation is achieved due to the fixation of the structure (and hence of the dynamic properties) of biopolymers and other biologically functional substances; this happens as a result of their action in the required amount, at the required time and in the required parts of the organism. Naturally, biochemical functioning depends on interactions of all substances participating in the process.

Table 4.1 lists the main stages of chemical and biological evolution [4.21].

Living organisms exist under diverse conditions. Biochemical organization must provide the energy needed, the adaptation to the molecular and ionic composition of the environment, to temperature and to pressure. Biochemical adaptation consists primarily of the adjustment to abiotic factors. At the same time it provides the required functionality with respect to food chains, to the pheonomena of parasitism and symbiosis. Let us consider some examples.

Different types of bacteria are adapted to different salt concentrations in an external aqueous environment. There are extreme halophiles, which grow

Table 4.1. The major stages of chemical and biological evolution

Time (millions of years ago)	Stages of evolution
1000	Multicellular organisms. Eukaryotic cells. First aerobic bacteria – oxidative phosphorylation
2000	Green plant photosynthesis. Bacterial photosynthesis. Phosphorylation: light energy into ATP
3000	First cells (prokaryotic). Pentose phosphate cycle. Fermentation: a chemical source of energy. Release of CO_2
4600	Formation of the Solar system

under conditions close to saturation with mineral salts. In extreme halophiles the intracellular concentrations of sodium and potassium salts are close to those in the environment, e.g. in the Elton and Baskunchak lakes or in the Dead Sea. The proteins of such bacteria as *Halobacterium halobium* are noted for a high content of the amino-acid residues Glu and Asp and a relatively small content of the hydrophobic residues Leu, Ile, and Val. The specific feature of these proteins is their extreme instability; they become stable only in the presence of a high content of K^+ ions, that is up to a few moles per liter. This refers to both cytoplasmic proteins and ribosomal proteins. Halophiles cannot exist in a medium with a small salt concentration [4.22].

There are freshwater and sea fishes. In freshwater fishes the ion concentration in the blood is maintained at a higher level than in the surrounding water. The organism is protected against excess osmotic suction of water. The physiological regulation of this state is achieved by uptake of Na^+ ions through the gill surface and by release of excess hypotonic urine, in which the salt concentration is reduced. Ion transport effected by means of Na^+K^+-ATPase (ATPase activated by Na^+ and K^+, see Chap. 10 in [4.20]) is polarized in such a way that the resulting transpot of Na^+ is directed inwards; the transport system has a high affinity for Na^+. For ionic pumps to operate, energy is required. Adenosine triphosphate serves to store chemical energy. Upon dephosphorylation of ATP, which occurs in the presence of the ATPase, free energy is released in an amount of the order of 10 kcal/mole. It is this energy that is utilized in ionic pumps.

In fishes living in salt water, the structure and performance of the Na^+K^+-ATPase are different. The gill apparatus pumps off Na^+ actively, keeping the Na^+ concentration in the blood constant. As a result, the ion composition of the blood of freshwater fishes is very close to that of sea fishes. Fresh water contains 0.2 mM Na^+ and 0.1 mM K^+ and freshwater-fish blood contains 130 mM Na^+ and 3 mM K^+. Sea water contains 500 mM Na^+ and 10 mM K^+ and the blood of sea fishes contains 150 mM Na^+ and 5 mM K^+. Figure 4.1 shows schematically the role of various organs in the maintenance of the salt and water balance in freshwater and marine fishes. The ATPase present in the gills of the latter organizes an active transport of Na^+ and K^+ ions outwards. A change in the external K^+ concentration alters sharply the rate of active release

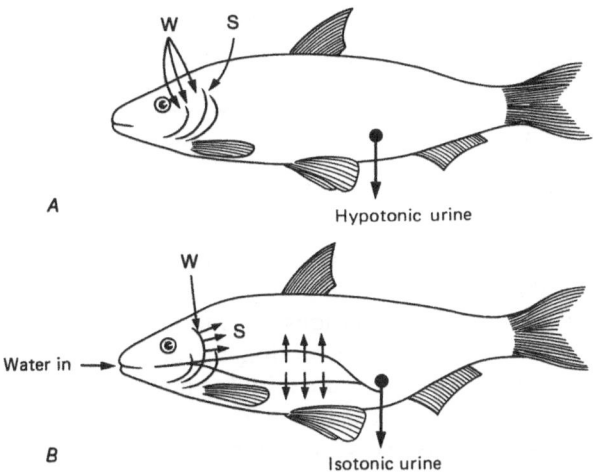

Fig. 4.1. The role of various organs in the maintenance of the salt (S) and water (W) balance in freshwater (*A*) and marine (*B*) bony fishes [4.22]

of Na^+. An exchange of equal amounts of Na^+ for K^+ occurs. The ATPase in the gills of freshwater fishes does not require K^+ ions.

Many species of fish (e.g. salmon and eels) travel from one habitat to another. The young salmon move out from the streams where they were born into sea, where they live and reach sexual maturity. In order to spawn they return to their native streams. I will never forget the sight of the mass movement of the sockeye (*Oncorhynchus nerka*) moving up the Ozernovsk River on Kamchatka from the Okhotsk Sea. Big fish spring out of the water, moving in dense schools. Upon migration from the river to the sea and back again the ATPase cation pump undergoes a change. It is possible that the kinetic properties of the pump are altered as a result of a change in the phospholipid component [4.21].

The muscle work of animals requires energy sources. The energy is always supplied by adenosine triphosphate (ATP). It is produced in the course of oxidative phosphorylation due to the oxidation of organic compounds during breathing (see Chap. 13 of [4.20]). Upon oxidation of one mole of glucose, a carbohydrate, 36 moles of ATP is formed, this being equivalent to the release of metabolic energy of the order of 4 kcal/g. The same amount of energy is produced by the oxidation of proteins. A much greater amount of energy, of the order of 9 kcal/g, is provided by the oxxidation of fats and lipids.

The main products of oxidation reactions are carbon dioxide and water. As has been pointed out by Hochachka and Somero [4.21], for desert dwellers the formation of metabolic water is no less important than the storage of energy. The amount of water liberated upon oxidation of lipids is twice as much as that released upon oxidation of carbohydrates (see Table 4.2).

During flight birds utilize the energy of lipid oxidation. It has been pointed out [4.21] that the maximum muscle activity is displayed by the hummingbird,

Table 4.2. Formation of oxidation water, or metabolic water, when different foodstuffs are metabolized

Energy source	gm H_2O formed/gm food	liters O_2 consumed/gm food
Carbohydrate	0.556	0.828
Lipid	1.071	2.014

a tiny creature with a wing frequency of 100 wing beats per second. It is to be noted in passing that such frequent wing beats are an auto-oscillatory process (see [4.20], Sect. 12.5). Auto-oscillations arise in non-linear systems due to the forces that depend on the state of motion of the system itself; the range of auto-oscillations is independent of the initial conditions. This means that each wing beat is not a response to the transmission of the nerve impulse. Humming-birds perform nonstop flights over distances of the order of 1000 km. Their flight muscles contain a large number of mitochondria (in which ATP is produced by oxidative phosphorylation) and a considerable reserve of fat droplets.

In the flight-muscles of two-winged insects, the source of carbon and energy is not a lipid but the polymeric carbohydrate glycogen, which is oxidized by atmospheric oxygen. In locusts, fats serve as a "fuel" during long-distance flights.

The metabolism of nitrogen compounds leads to the formation of ammonium, a compound which blocks a number of important biochemical processes and which is in fact a poison. In the course of evolutionary adaptation, ways of conversion and removal of ammonium from the organism have developed. In many land invertebrates, and also in birds, ammonium is removed in the form of uric acid.

When living creatures began to venture out from the sea onto land, the problem of removal of ammonium became especially important. In aquatic organisms ammonium is easily released into the surroundings during breathing. In land vertebrates ammonium is converted to urea, $(NH_2)_2CO$, which is removed in urine. In land molluscs the main excretion is uric acid. Urea and uric acid play an important role in the maintenance of the water balance in these organisms and in the formation of their shells.

A number of marine organisms have biochemical systems which serve to detoxify hydrogen sulphide (H_2S), an external poison, which is released during the anaerobic metabolism of bacteria. In these systems H_2S is oxidized to free sulphur or to $S_2O_3^{2-}$, SO_3^{2-} and SO_4^{2-} ions. Such processes play an essential part in some deep-water biocenoses. In deep-water hydrothermal sources, in so-called "black smokers", the original water contains a large amount of dissolved hydrogen sulphide. This water becomes mixed with ordinary sea water. Sulphur-oxidizing bacteria function in the system, converting HS^- ions to SO_4^{2-}. This oxidation leads to the accumulation of energy in the form of ATP. Colonies of sulphur-oxidizing bacteria are contained in the organisms of certain pogonophora; such symbiosis provides the means of existence for these and other invertebrates (Chap. 15 of [4.22]). On the other hand, hydrogen-sulphide bacteria severely limit life in the Black Sea, beginning at a depth of 150–200 m.

We have already mentioned the adaptation of two related species of frogs to temperature, which eventually reduces to the conformational flexibility of proteins. Naturally, the progress of all molecular processes depends on temperature.

In cold-blooded, poikilothermic animals there are biochemical mechanisms which provide a muscle temperature higher than the environmental temperature (tunas and a number of other bony fishes). Metabolic heat is generated. This kind of thermoregulation is displayed by many insects. For example, in moths of the family *Sphingidae* the thoracic temperature is maintained at about 41°C with the ambient temperature varying from 15 to 35°C. The source of heat is again ATP, the heat being liberated upon splitting of ATP into ADP and inorganic phosphate, which occurs under the action of appropriate enzymes. Processes of this kind are also characteristic of other insects, in particular bumblebees and honeybees. Honeybees, which are social insects, regulate the beehive temperature in a remarkable way. If it is cold, they gather in dense swarms, maintaining the internal temperature between 18° to 32°C with the outside temperature ranging from −17° to +11°C. If it becomes too hot in the hive, the nonforaging bees cease to gather pollen and nectar from the workers, which start bringing water droplets instead. The workers in the hive are located at a certain distance from one another and flap their wings, providing the rapid evaporation of water and a decrease in temperature. What a brilliant example of social physiology!

The thermogenesis in the muscles of vertebrates also occurs due to the dephosphorylation of ATP and is activated, in particular, by thyroid hormones. An important source of heat for mammals is brown fatty tissue in which the oxidation of fatty acids may be uncoupled with phosphorylation. The energy released is not stored in ATP and is liberated as heat.

The freezing of blood and other fluids in an organism leads to the decay of cells and tissues and thus to death. For animals which live in cold climates it is essential that the body fluids, i.e. aqueous solutions, should not be crystallized or converted into either an overcooled or a glassy state. Up to 20% of the blood of Arctic insects consists of glycerol, a substance which easily vitrifies but practically never crystallizes. The blood of polar fishes contains dissolved glycoproteins and lipoproteins, which serve as antifreeze.

In a number of animals, active metabolism ceases during summer periods when water basins are dry, and thus the animal "hibernates". Many animals living in temperate or polar areas hibernate in winter. In these cases, glycerol plays an important role, its concentration increasing when the organism is dehydrated. The adaptation of animals to the state of inactivity occurs as a result of a substantial reconstruction of the biochemical cycle (Chap. 7 of [4.20]).

Very important biochemical events take place during ontogeny. It would seem that we may speak of biochemical adaptation to the different stages of ontogeny. But, as a matter of fact, this is just the ontogeny itself with its own biochemistry.

An insect larva differs from an imago to a much greater extent than the imagoes of various species from one another. A larva is a quite different organism. It is clear that complete metamorphosis of insects occurs as a result of substantial changes in biochemistry, which have been thoroughly studied in a number of cases.

The development of mammals occurs through a number of stages (Chap. 8 of [4.21]):

1. The zygote receives nutrients directly from the surrounding tissue; the blood supply through the placenta is not yet sufficient.
2. The placenta and the blood circulation system are well developed; the foetus feeding system has been established and the foetus receives all its nutrients from the blood; the greater part of metabolism waste is removed in the same way.
3. In the early postnatal period, the rearing period, the newborn is fed with the mother's milk.
4. The animal enters the weaning period when it is still fed with the mother's milk but consumes other food to an ever-increasing degree.
5. The animal reaches the point of totally independent feeding.

Each stage of development has its own biochemistry. This has been described in detail by Hochachka and Somero [4.22]. Biochemical adaptation, of course, occurs both in a developing organism and in its mother's body. The composition of the mother's milk, for example, is accordingly changed. During embryonic development the main source of energy is carbohydrates. After birth the baby organism is not fed for a certain short period of time. When being fed with mother's milk, the organism receives far more fats than carbohydrates. The digestion of substances supplied to the developing organism occurs via biochemical systems, which undergo a corresponding change.

We will confine ourselves to a few examples. Biochemical adaptation on the whole convincingly supports the general idea of Darwinism concerning adaption in living nature. It may be said that there are two categories of corresponding mechanisms: genetically programmed mechanisms and mechanisms that arise due to environmental conditions. The second category reduces eventually to the first; it is in fact adaptability rather than fitness that is genetically programmed. Therefore, this division into two categories given by Hochachka and Somero is rather provisional.

Biochemical adaptation reduces, in general, to a limited number of varieties which make use of a limited number of functional motives. At the same time the number of enzymes operating in the biochemistry of cells is extremely large.

As noted in Sect. 4.4, the majority of morphophysiological traits have no direct adaptive significance. At the biochemical level, however, the needs of organisms have been met by adaption.

4.3 Adaptability

Adaptability means the ability of organisms, populations and species to adjust to environmental conditions, both biotic and abiotic. Fitness is the result of adaptability; conversely, adaptability itself may be treated as fitness. A logical and model physico-mathematical analysis of these notions has been made by Conrad in his monograph [4.23]. Here we will not carry out such an analysis, but will confine ourselves to the basic concepts. Conrad defines adaptability as "the ability to cope with unexpected disturbances of the environment."

The ontogenetic and phylogenetic origin of adaptations at all levels, beginning with the genetic level, determines the possibility of evolution. Without adaptability the biosphere could not have developed.

It has been repeatedly noted in this book that an organism is a chemical machine. In this context it is necessary to consider its basic features.

Each part of a man-made machine is designed to perform a certain function. Pre-destination is different from adaptability since the former is static and does not change with time, while the latter is dynamic in character.

A chemical machine (a cell, an organism, a population) can change its behaviour in accordance with environmental conditions. An artificial machine is not capable of doing this in most cases. A living machine is an open, far-from-equilibrium system, i.e. a dissipative system. An artificial machine is also open; the internal combustion engine exchanges matter and energy with the surroundings. However, it is not too far from equilibrium; it is not a disspative system. In a living machine there are clearances and tolerances, factors that determine its ability to change its behaviour. At none of the levels of organization is there any one-to-one correspondence between the elements of the system. Here we are dealing with degeneracy, beginning with the degeneracy of the genetic code. In other words, there are numerous ways of dissipation and of export of entropy to the surroundings. For a given influx and efflux of energy there are a number of different states of a dissipative system.

Bionics is a branch of science and technology dealing with the application of biological principles to the study and design of engineering systems. Perhaps the first work in this field was the attempt made by Leonardo de Vinci to design

Table 4.3. Modes of adaptability

Compartment	Modifiability component
Biota	Migration, extinction, speciation; routability of matter and energy in food webs
Population	Culturability; social and topographical plasticity
Organism	Morphological plasticity (number of organs); morphological flexibility (including behavioural plasticity)
Organ	Morphological plasticity (size of organ); histological plasticity
Cell	Plasticity of cellular morphology; differentiability
Cellular phenome	Physiological plasticity of cells and body fluids
Genome	Gene pool diversity

wings with which a man might fly. There have been some achievements in bionics, but an artificial dissipative system possessing adaptability has not yet been constructed and the ways to the solution of this kind of problem are still obscure.

We may speak only of three components of adaptability expressed by behavioural uncertainty, tolerance for decorrelation, and tolerance for error [4.23]. Table 4.3 lists the modes of adaptability.

A dissipative system cannot predict a change in the state of the environment, but it can adapt to it.

Conrad rightly notes that aging and death are necessary features of biological adaptability of both an organism and a species. For example, if a soma is developing from sexual cells, the parent organisms must die because they are not capable of self-reproduction or correction of defects. Immortality would have meant the lack of evolutionary adaptability, living space would have been overcrowded by old organisms, and birth would have been sharply reduced. Accordingly, genetic variability would have been reduced to zero [4.23]. One involuntarily recalls the "struldbrugs", the immortal inhabitants of the Laputa in Swift's *Gulliver's Travels*, who lost all human features and were devoid of the ability to think and feel. Immortality would have meant the impossibility of generating new information, i.e. the impossibility of creative activity of Nature and man.

The aging of an organism and a species means, in the first place, the loss of adaptability. Death is the inevitable and necessary consequence.

A living dissipative system appears to be capable of maintaining its stable state, i.e. homeostasis (or homeorhesis according to Waddington), under changing conditions. This is what adaptability and fitness mean. An ecological niche for an organism or a population is a set of biotic and abiotic features of the environment, with which the organism or population interacts in some way.

We have already considered the differences between two forms of species, namely generalists and specialists. The adaptability of the former is higher than that of the latter. The ecological niches of specialists are narrow and those of generalists are expansive.

In a number of cases the generalists include species with a relatively large brain mass. The limiting position here is occupied by *Homo sapiens*, possessing the highest adaptability. Conrad discusses the relationship between the higher nervous system and the elementary responses of finding food and procreation (p. 276 of [4.23]). Two statements may be cited:

"1. Thought and other modes of adaptability exist in order to maintain or improve adaptations, such as features which allow organisms to eat and reproduce.

2. Adaptations, such as features which allow organisms to eat and reproduce, exist in order to maintain thought and other modes of adaptability."

The expression "exist in order to" has a teleological connotation, including the notion of the purpose. The expression should be changed to "make it possible". Which is more important – thinking, or feeding and reproduction? Conrad righteously compares this situation with the old chicken-and-egg riddle: "Which came first, the chicken or the egg?" The role played by the higher nervous activity in adaptability and fitness is as important as that played by the simpler responses. Therefore, either of the two statements is valid when rephrased.

The loss of the ability to recover organs by higher vertebrates is in essence the payment for the integrity of the organism controlled by the nervous system. Such integrity is the highest level of development achieved by nature.

The concept of adaptability is directly associated with the efficiency concept. Table 4.4 gives the main factors that affect the efficiency (p. 248 of [4.23]).

The division into two groups of factors A and B is provisional to a certain extent since a population itself is an essential component of a niche, i.e. biogeocenosis, and in this sense there is a feedback between organisms and cenosis.

The adaptability concept is dynamic. Evolutionary theory is not the theory of adaptability, but rather the theory of mechanisms of historical development of dissipative systems being adjusted to the conditions of their existence.

Table 4.4. Principal factors affecting efficiency

A. Organism-dependent factors
  1. Adaptability
  2. Internal specialization (number of component types and disproportionality of their occurrence)
  3. Behavioural diversity
  4. Biomass turnover per unit volume*
B. Niche-dependent factors
  1. Macroscopic parameters of the niche (e.g. temperature, salinity)
  2. Thermodynamic richness of the niche (e.g. free energy, nutritive quality)
  3. Structural (topographic) diversity of the niche
  4. Biotic diversity of the niche
  5. Diversity of niche states
  6. Range and timing of environmental disturbances

* Must be considered during periods of population growth and decline.

Adaptability is a necessary feature of the social behaviour of human beings. *Homo sapiens* is a social animal and must adjust to abiotic and social conditions of existence. The optimal state of a society corresponds to the maximum efficiency and the maximum diversity of mental activity, both speculative and emotional. In a totalitarian society the adaptability to changing situations is replaced by subordination to the stable conditions of a static system. It is clear that a totalitarian society based on such a replacement cannot evolve, and is thus doomed to aging and death.

4.4 Non-adaptiveness

While describing the heritable characters of an organism determined by mutations rather than by modifications, biologists usually ask "what for?". What is this or that trait intended for? It is tacitly assumed that any trait is adaptive, being the result of natural selection for fitness.

The formulation of the question in such a form is in itself a statement that requires proof. The "what-for" question is not a scientific question. In the natural sciences, i.e. physics, chemistry, biology etc., one ought to ask "Why?"

The majority of traits of any organism or any species has no adaptive significance. These traits do not result from natural selection. Let us consider two examples.

The red colour of blood. The red colour of vertebrate blood is a well-defined trait. In itself it has no adaptive significance since the blood is within the body and does not interact with light. The question "What is the red blood intended for?" is meaningless. Why the blood is red is known. The blood colour is determined by the haemoglobin of erythrocytes, the most important protein, by means of which the organism is supplied with oxygen. The colour of haemoglobin is due to the haem group – the porphyrin ring with an iron ion in the centre. Thus, the red colour of blood is a subsidiary trait, which plays no part in physiology. Conversely, the green colour of grass or leaves has direct adaptive significance since it is determined by the presence of chlorophyll, a substance which is also built up of a porphyrin ring but with a magnesium ion. The green colour of chlorophyll means that it absorbs light in the red region of the visible spectrum of the sun, with a wavelength close to 700 nm. It is light absorption that is the cause of photosynthesis, which determines the life of plants and, indirectly, all life on earth.

The external genitalia of the female spotted hyena. The external reproductive organs of the female spotted hyena *Crocuta crocuta* hardly differ from those of the male hyena. What for? This feature has no adaptative advantage. On the contrary, it is somewhat inconvenient for copulation. It appears to be possible to answer the "why" question. For this species the female is appreciably taller than the male, which probably has adaptive significance in birth and breeding.

The larger height of the female animal is associated with unusually high levels of the male sex hormone androsterone in the female organism. The structure of the external genitalia of the female appears to be a subsidiary trait.

The brain of Homo sapiens sapiens. The brain of human beings of the Palaeolithic era did not differ from the brain of modern humans. Both Cro-Magnon and modern man, irrespective of race, belong to the same species and the subspecies *Homo sapiens sapiens,* in contrast to the Neanderthal subspecies *Homo sapiens neanderthalensis.* Thus nature endowed the man of the Palaeolithic era with a brain capable of solving not only current problems but also abstract problems, and of creating science, art and religion. Did Cro-Magnon man need a brain capable of abstract thinking? It is evident that he cannot have done. The amazing capabilities of the human brain originally had no adaptative significance. They emerged as a subsidiary trait as a result of the adoption of upright posture and the development of speech.

These and many other examples show that absolute selectionism, which attributes adaptive significance to all traits of animal organisms, is unacceptable. This was clearly understood by Darwin.

In 1880, in a letter to the journal *Nature,* Darwin wrote (cited from [4.24]): "Unfortunately, as I have found out, Sir Thompson does not understand the principle of natural selection... Had he understood it, he would not have written the following phrase in *Introduction to Challenger's Travels:* 'The nature of the abyssal fauna does not provide the slightest support for the theory that ascribes the evolution of species to extreme variations governed by natural selection alone.' This is stereotyped critique, which is often practised by theologians and metaphysicians when they write on scientific subjects, but this is something novel to a naturalist... Can Sir Thompson name someone who has ever stated that the evolution of species depends only on natural selection? As for myself, I think that no one has presented so many observations of the results of the use and disuse of parts as I have done in my *Variability of Animals and Plants in Domestication*; and these observations were made for this special purpose. I also added an important collection of facts, demonstrating the direct impact of environmental conditions on organisms." Darwin adhered to the pluralistic position. We have already cited his words concerning the spark and the flame. From these words it follows that the natural selection of the best-adapted organisms is not the only evolutionary factor.

The article by Gould and Lewontin [4.24] is called "The spandrels of San Marco and the Panglossian paradigm: A critique of the adaptationist program". The spandrels of San Marco Cathedral in Venice were designed by narrowing triangular spaces formed by the intersection of circular arcs at right angles. These structures are absolutely independent of the surrounding architecture. The authors emphasize that it is meaningless to propose the "adaptive history" for each separate trait. As regards the Pangloss paradigm from *Candide* by Voltaire, it consists of the following: "Things cannot be different from what they are... Everything is made for the best purpose. Our noses were made to carry spectacles, so we have spectacles. Legs were clearly intended for breeches, and we wear them."

Gould and Lewontin refute and ridicule absolute selectionism, i.e. adaptationism. As these authors state, the *Tyrannosaurus* male could have used its small forelegs to tickle the female, but this does not explain why these forelegs were so small (one should visit the Palaeontological Museum of the Academy of Sciences in Moscow where one will be convinced of the size differences between the forelegs and hindlegs of giant pangolins; what a strong impression it makes). In this case too it is impossible to answer the "what-for" question.

According to Gould and Lewontin, the adaptationist programme consists of the following principles

1. If some adaptation argument appears to be invalid, another one should be resorted to.
2. If some adaptive argument is invalid, it should be presumed that another one must exist.
3. The lack of a good adaptive argument is accounted for by an insufficient understanding of an organism's conditions of life.
4. One should proceed from direct usefulness and rule out other characteristics.

The non-adaptiveness of many characters was noted much earlier by Chetverikov, who as far back as 1922 wrote: "Population systematics knows of thousands of examples where species differ not by adaptive but rather by indifferent (in the biological sense) characters; to try to look for adaptive significance in all of them is both a non-productive and thankless work, and more often than not one does not know which is more surprising, the unlimited wittiness of the authors or their belief in the unlimited naivety of the reader" (see [4.25]).

The wittiness mentioned by Chetverikov implied, in fact, the search for anthropomorphic explanations of the various characters which stem from adaptationism. No doubt there are characters whose adaptative significance has not been understood, though they are the result of natural selection. But the presence of a multiplicity of characters which have no relation to adaptation is also beyond doubt.

The genetic differentiation and fixation of characters can occur irrespective of natural selection, as will be pointed out in Chaps. 6 and 7. Punctuated speciation means non-adaptationism; after 2 species have diverged, which takes a relatively short period of time, they simply have no chance to adapt to existing conditions and look for or create new ecological niches for themselves. The synthetic theory of evolution, while proceeding from gradual allelic substitutions, is of course based on the adaptationist programme. However, at the present stage of development of evolutionary theory, new conceptions have acquired fundamental importance.

Punctuated changes at all levels imply non-adaptive changes. Evolution is governed not only by natural selection but also by historical, developmental and structural constraints (Chap. 3). Punctuated speciation is the result of chromosomal rearrangements (molecular drive, Chap. 7) rather than of point substitutions in DNA (molecular drift, Chap. 6). The speciation itself, reproductive isolation, has no adaptive character. How did the wings emerge in fossil reptiles? What advantage could two percent of a wing have given to the dinosaur? The non-adaptiveness of a number of characters is associated with violations of gradualism, i.e. punctuated evolution [4.26].

The problem of adaptation and non-adaptationism has been examined in detail by Gould and Vrba [4.27]. The authors note that the acceptance of adaptation rests on two different criteria: on historical origin, when we speak of features arising as a result of natural selection, and on the current usefulness of characters that enhance fitness, irrespective of their origin. Incidentally, the latter characters might originally have been unadapted but accessible and only at later stages might they have acquired adaptative significance. Gould and Vrba cite Darwin's words pertaining to such a character: "The sutures in the skulls of young mammals have been advanced as a beautiful adaptation for aiding parturition, and no doubt they facilitate, or may be indispensable for, this act; but as sutures occur in the skulls of young birds and reptiles, which have only to escape from a broken egg, we may infer that this structure has arisen from the laws of growth, and has been taken advantage of in the parturition of the higher animals."

Gould and Vrba propose the term "exaptation" for such situations. The following definitions are given.

1. The action of a useful character that has not been produced by natural selection for current utilization may be called an effect.
2. The characters that have been developed for other applications (or have no applications at all), and then co-opted for their current role, are called exaptations.
3. Adaptations have functions and exaptations have effects.
4. A general static phenomenon consisting of fitness is called aptness (from Latin *aptus* meaning fit, suited). Table 4.5 characterizes various cases of fitness.

One of the examples of exaptation is the adaptation of feathers first to thermoregulation, and only later to other purposes such as the catching of insects, improvement of flight conditions, and decorations that play a role in

Table 4.5. A taxonomy of fitness

Process	Character	Usage
Natural selection shapes the character for a current use: adaptation	Adaptation	Function
A character, previously shaped by natural selection for a particular function (an adaptation), is co-opted for a new use: co-optation	Exaptation	Effect
A character whose origin cannot be ascribed to the direct action of natural selection (a non-adaptation) is co-opted for a current use: co-optation	Exaptation	Effect

sexual selection. A large set of non-adaptations is a reservoir and a source of evolutionary flexibility. Gould and Vrba maintain that this is a central phenomenon in evolution.

Another term that characterizes the same phenomena, adoptation, has been proposed by Dover on the basis of the molecular drive concept considered in Chap. 7.

Vrba noted [4.28] that the adaptations of organisms, which vary in different species, are selected for direct applications, but they can also determined by chance the various rates and tendencies of speciation and species extinction.

From what has been said above it is obvious that the notion of pre-adaptation is in fact devoid of meaning. The more instructive notion of exaptation overlaps it completely.

The social existence of man is radically different from his biological existence. Nevertheless, certain ideas of theoretical biology can contribute to a better understanding of human social life.

Creativity in the fields of science and art always has exaptive character, using the term proposed by Gould and Vrba. It is clear that the development of science is necessary for achieving progress in life conditions, engineering, agriculture, medicine etc. However, the advancement of sciences is not adaptive on the whole; true science produces results not for the purpose of their practical application but following the inner logic of science itself. Faraday discovered the laws of the electromagnetic field without thinking of a dynamo-machine. Maxwell built up his electromagnetic theory of light without thinking of radio. When Einstein discovered the equivalency of mass and energy, he was not trying to make an atomic bomb. Creative scientific work is not adaptive.

Part II

Molecular Basis of Evolution

5. Proteins

5.1 The Protein Structure

At the molecular level of life processes we encounter informational biopolymers, i.e. macromolecules of proteins and nucleic acids. The role of nucleic acids is "legislative": they are responsible for the biosynthesis of proteins. The role of proteins is "executive". They can do everything; they are only incapable of self-synthesis. In other words, all functions of a cell, organism or species are determined at the molecular level by proteins. The structure of proteins and their functional dynamics are thus subject to natural selection.

Resorting to the molecular foundations of evolution, it is necessary to become acquainted with the physics of proteins, with their structure and properties. In this chapter we shall consider the main features of globular proteins, which are responsible for enzymic reactions.

In what follows it is assumed that the reader is acquainted with basic information about biopolymers. For those who are not, I recommend my book [5.1].

As is known, only the amino-acid sequence in the protein, its primary structure, is encoded genetically. Natural selection works with respect to the biological properties of proteins, which depend directly on the spatial structure of the molecule, i.e. the tertiary and quaternary structures. Molecular biology, and in particular molecular genetics, would be meaningless if there were no correlation or correspondence between the primary and the spatial structure. Of course, a correlation does exist, but, as we shall see, it is not unambiguous, contrary to the original concepts.

The first, primary feature of proteins (and also of nucleic acids) is the macromolecular structure. The facts that can be explained by this structure are of general importance.

The macromolecules of synthetic and biological polymers exhibit conformational flexibility, i.e. they are capable of assuming various conformations as a result of the rotations of atomic groups about single σ-bonds. This does not apply to polymers, whose chains consist of conjugated σ- and π-bonds, say polyacetylene:

$$\cdots-CH=CH-CH=CH-CH=CH-\cdots$$

Macromolecules of this kind are deviod of conformational flexibility and absorb
light in the visible region of the spectrum. Polyacetylene is black in colour.

The macromolecules of a flexible polymer in solution naturally fold up into
a random coil. Maximum entropy corresponds to the coil-like state since it can
be realized in a large number of ways, while the most stretched state is the least
probable; it is realized in only one way. A coil is a loose, fluctuating entity which
is permeable to a solvent. Its size and electrical and optical properties (which
depend on the chemical structure) can be calculated on the basis of the conformational theory of macromolecules, which was developed by our team in
Leningrad in the 1950s [5.1–4]. After the publication of these works I came to
the very wrong conclusion that there were no important theoretical problems to
be solved in the physics of macromolecules and that further work must be done
in the field of either technology or biology. I chose biology and I do not regret
it now. But a little later, the outstanding theoretical physicists, notably I.M.
Lifshitz, discovered a new field of macromolecular physics [5.5,6].

The distribution of end-to-end distances in the coil is Gaussian. The coil
size $\sqrt{h^2}$ is expressed by the quantity $Z^{1/2}b$, where Z is the number of freely-
jointed segments having a length b. Of course, the macromolecular chain is not
freely-jointed; certain restrictions are imposed on it, which are determined by
fixed valence angles and by a discrete set of possible conformations of each
chain link. However, the relative positions of two rather remotely-spaced monomeric units are not correlated in any way. Hence, the macromolecule may be
represented by a set of freely-jointed segments, whose length increases with
increasing rigidity of the chain. Evidently, the volume of the coil is of the order
of $h^3 = Z^{3/2}b^3$. In this volume there are Z segments. Their concentration is of
the order of $c \sim Z/Z^{3/2}b^3 = Z^{-1/2}b^{-3}$. The value of c decreases with increasing
Z. Hence, the coils is really a rather loose system with low density and strong
fluctuations. The smallness of c implies that the greater part of the coil volume
is occupied by solvent molecules. It can be shown that the correlation radius of
segment concentration fluctuations in the coil has the same order of magnitude
as the coil size [5.6].

Ordinary macromolecules with no conjugated bonds are devoid of specific
electronic properties. They are insulators rather than semiconductors. Polyacetylene, on the other hand, for example, has semiconducting properties.

Under the action of an external field of force, or as a result of sufficiently
strong interactions between non-adjacent chain links, the macromolecular coil
can "collapse" into a globule. The theory of the coil-globule transition has been
developed by I.M. Lifshitz, A.Yu. Grosberg and A.R. Khokhlov [5.5,6]. The
difference between these two states is determined by the character of fluctuations (I.M. Lifshitz, 1968). In the globule there are less fluctuations than in the
coil and their correlation radius is much smaller than the size of the macromolecule. The globule is a condensed state. There is every ground for regarding the
states of the globule and coil as phase transitions of the macromolecule, and for
treating the globule-coil transition as a second-order phase transition. At the
same time, there is a certain similarity with a solid-gas transition. The coil may

be likened to a real gas and the globule to a solid. The fundamental distinction between the polymeric system and a set of independent particles is determined by linear memory, i.e. the fact that monomer units are linked into a single chain.

After these preliminary remarks concerning macromolecules, let us turn to proteins. The most widespread biological function of proteins is enzymatic. Enzymes serve as specific catalysts for all biochemical reactions. In most cases we have to deal with globular proteins, which are in an aqueous environment in a cell.

Like macromolecules of synthetic polymers, protein chains exhibit conformational flexibility, that is they do not have conjugated bonds. It is for this reason that they fold up into globules. A protein is a heteropolymer built up of 20 common amino-acid residues; it is a "text" written in a 20-letter alphabet. Apart from linear memory, which is inherent in any homopolymer chain, the protein chain, which has a primary structure, exhibits "informational memory".

A mental transposition of any two links in a homopolymer does not in any way alter its structure or properties. On the contrary, such a transposition in a protein can bring about significant changes.

The globular structure is the result of conformational flexibility; conformational motions determine the dynamics of the protein, and hence its biological properties, especially enzymatic catalysis (see Sect. 5.2). Landau said that the task of theoretical physics consists of establishing new relationships between objects and phenomena that might seem to have nothing in common. A striking example of such a new relationship is the similarity between the high-elasticity of natural rubber and enzymatic catalysis; the two phenomena are governed by the conformational mobility of macromolecules.

A second fact which emerges from the macromolecular nature of a protein is its dielectric properties. Proteins (and also nucleic acids) are "*honest*" insulators. Molecular biology was developed simultaneously with the elaboration of the modern trends in solid-state physics. Naturally, a number of physicists tried to apply the corresponding ideas to biology. This led to new conceptions of the semiconductor, ferromagnetic and even superconducting properties of biopolymers and of special quantum effects in biology. The authors of these concepts, including eminent physicists, have nothing to do with biology and have overestimated their possibilities. Had proteins been semiconductors, they would have been devoid of conformational flexibility and would not have been able to fold up into globules. At the same time, they should have absorbed light in the visible region of the spectrum. The absorption bands of proteins lie at around 200 nm and those of nucleic acids at about 260 nm.

The electrical conductivity of proteins observed in a number of cases is accounted for by the presence of ionic impurities, which are not easy to remove.

Because of the "textual" nature of the protein chain, the presence of various interactions between amino-acid residues, and also reactions between these residues and the solvent, each protein globule has a specific structure.

The specific structure of a globule is formed by the non-bonded interactions between non-adjacent links and the interactions of the macromolecule with the surroundings, namely water. Proteins that function in biological membranes are in a lipid rather than an aqueous environment.

These interactions are as follows (cf. [5.1]):

1. Chemical S—S bonds between cysteine residue (Cys or C).
2. Electrostatic interactions between oppositely-charged amino-acid residues (the anions Asp (D), Glu (E) and Tyr (Y) carry negative charges and the cations His (H), Lys (K) and Arg (R) carry positive charges).
3. Electrostatic interactions with metal ions present as co-factors in a number of proteins.
4. Coordination bonds to metal ions: Zn^{2+} and transition-metal ions.
5. Van der Waals interactions (orientational, induction, and dispersion forces) of amino-acid residues.
6. Hydrogen bonds between peptide groups (as shown below),

and between the peptide groups and the side groups of amino-acid residues; also hydrogen bonds between chain links and water. We must also point out the presence of hydrogen bonds between N—H and O—H groups and the π-electrons of aromatic rings, which appear in phenylalanine Phe (F), tryptophan Try (W) and histidine His (H) residues [5.7].

7. An especially important role is played by hydrophobic interactions between a protein and water. The immersion of a non-polar group (say a hydrocarbon residue) in water brings about a decrease in entropy and a corresponding increase in free energy. Therefore, a globule is built up in such a way that the non-polar (hydrophobic) residues appear to be localized predominantly in the interior of the globule, whereas the polar (hydrophilic) groups are located on its surface, in contact with water. Since the residues are united into a single chain, they cannot be separated completely; part of the hydrophobic residues are left on the surface of the globule and part of the hydrophilic residues in its bulk.

These interactions, except the S—S bonds and co-ordination bonds to metal ions, are characterized by free energies of the order of a few kcal/mole.

These are weak, non-chemical interactions. The same order of magnitude is shown by the energies of the barriers to internal rotation, which separate the conformers. Changes in the globule conformations are associated with the same weak interactions.

The presence of interpeptide hydrogen bonds determines the secondary structures of protein chains in the globule. These are known (see [5.1]) to be alpha-helices and parallel and antiparallel β-ribbons. The study of the topological structures of protein globules has shown that they obey the following rules [5.8–10]:

1. Portions of the secondary structure that are adjacent in the sequence are often in contact with one another in three dimensions too.
2. Portions of β-X-β units (where β represents parallel ribbons in the same β-layer, though not necessarily adjacent; X is an alpha-helix or a strand in another layer or an extended chain portion) are right-handed.
3. Portions of secondary structures do not intersect and do not form branch points in the chain.

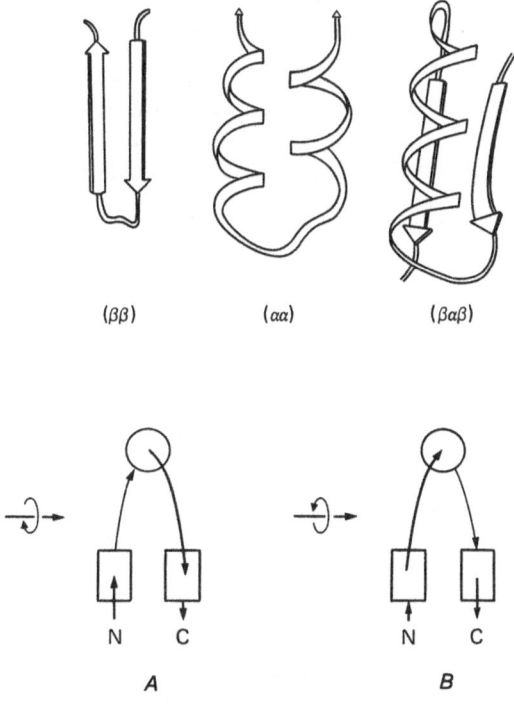

$(\beta\beta)$ $(\alpha\alpha)$ $(\beta\alpha\beta)$

N   C   N   C

A   *B*

Fig. 5.1. Topological features of proteins: $\alpha\alpha$, $\beta\beta$, and $\beta\alpha\beta$ = supersecondary structures often encountered in proteins; a and b represent $\beta\alpha\beta$-structures; circles represent cross-sections of α-helices; rectangles represent cross-sections of β-ribbons; (*A*) left-handed structure; (*B*) right-handed structure (after Chothia [5.9])

Figure 5.1 shows the characteristic arrangement of the secondary structures in proteins. These structures are oriented in the globule in such a manner that the hydrophobic residues are located on the internal side of the alpha-helix or β-ribbon; they project towards the interior of the globule. Such regular elements must have a continuous hydrophobic surface.

Protein globules are tightly packed formations. This is proved by a direct proportionality between the volumes of amino-acid residues inside the protein and the volumes of free amino-acids in the corresponding crystals (Fig. 5.2).

Four classes of the structure of globular proteins have been established [5.8,9]:

1. Proteins built up of α-helices. They are characterized by the presence of bundles consisting of four α-helices, which are nearly parallel or slightly turned. The energetics of such bundles has been studied by Scheraga and co-workers [5.11]. Typical α-helical proteins are globins. Figure 5.3 shows the structure of myoglobin.
2. β-Proteins which mostly contain collinear or orthogonal β-layers.
3. α/β-Proteins in which the α- and β-segments nearly alternate along the chain. These structures usually contain βαβ-clusters (see Fig. 5.1).
4. (α + β)-Proteins. These are heterogeneous structures with an irregular arrangement of units of the two types.

These classes have been identified as a result of a detailed investigation of the three-dimensional structures of proteins by means of X-ray diffraction analysis. This technique has been used to study more than 200 proteins. The recent progress in the development of the so-called two-dimensional nuclear magnetic resonance has made it possible to establish the structure of proteins

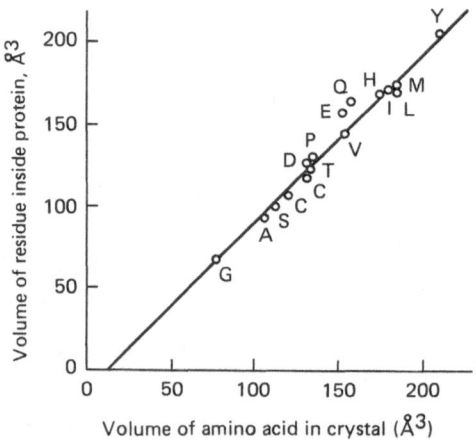

Fig. 5.2. The average volume of amino-acid residues inside protein as a function of the volume of amino-acid residues in crystals [5.9]

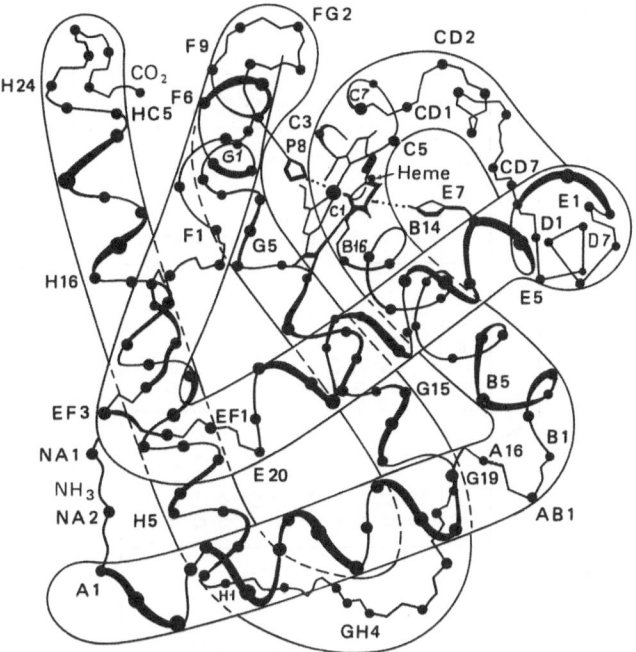

Fig. 5.3. The structure of myoglobin

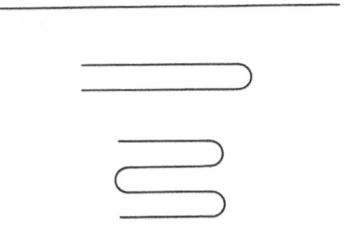

Fig. 5.4. Folding of the hydrophobic chain

(small ones, so far) directly in solution. This does not require the crystallization of proteins, which is often a difficult task.

Golovanov et al. [5.12] have examined theoretically the conformational twisting of a hydrocarbon hydrophobic chain in an aqueous solution. The chain is twisted in such a way that its surface, which is in contact with water, is minimal. Accordingly, a short chain is folded in half and a long chain must have additional bends at distances of a quarter and three quarters of its length, followed by distances of 1/8, 3/8, 5/8, 7/8, etc. (Fig. 5.4). It has turned out that this crude relationship is also valid for proteins. A comparison of data from X-ray diffraction analysis of proteins has shown that the chain folds are really located near the positions indicated. These data are collected in Table 5.1.

Table 5.1. The positions of folds in protein molecules*

Protein	N	Bend point — Simple values of n/N						
		$1/8 = 0.125$	$2/8 = 0.25$	$3/8 = 0.375$	$4/8 = 0.50$	$5/8 = 0.625$	$6/8 = 0.75$	$7/8 = 0.575$
Pancreatic inhibitor of bovine trypsin	58		0.26		0.46			
Ubiquitine	76	0.11	0.25		0.53	0.61	0.76	0.85
Cytochrome c of yeast	108		0.22		0.49	0.57	0.71	0.83
Myoglobin	146	0.13	0.27	0.38	0.53	0.65	0.80	0.87
Haemoglobin (α-chain)	153	0.14	0.26	0.37	0.52	0.63	0.77	0.82
Bovine β-lactoglobulin	162	0.08	0.29	0.39	0.49	0.63	0.75	0.82
Phage T4 lysozyme	164	0.13	0.22	0.34	0.49	0.65	0.71	0.87
Cyanobacterium C-phycocyanin	174	0.10	0.19	0.40	0.59		0.72	
Aspartate proteinase	325	0.12	0.24	0.35	0.50	0.63	0.74	0.89
Apolactate dedhyrogenase	331	0.14	0.26	0.40	0.47	0.63	0.76	
Cytochrome P 450	414	0.12	0.24	0.36	0.51	0.64	0.76	0.86
Average:		0.12	0.25	0.37	0.51	0.63	0.74	0.85

* Values of n/N, where N is the number of amino-acid residues in the molecule; n is the number of the residue at the point of turn.

Of course, one should not expect greater accuracy. The folds indicated in Table 5.1 are located between units of the secondary structure.

Presumably, the observation of such a simple relationship is not a consequence of hydrophobicity alone. Not all amino-acid residues in the protein are hydrophobic. The folding in half corresponds to the maximum compaction of the globule.

In many cases, globules are made up of sufficiently distinct, well-defined domains.

The protein-enzyme molecule consists, roughly speaking, of two parts. The first portion is an active site at which a catalyzed chemical reaction occurs, which converts the substrate into a product. The second portion is the remainder of the globule, the skeleton so to say, into which is built an active site. No chemical events take place in this "passive" portion of the molecule, but its conformational mobility is necessary for a biological function, namely for enzymatic catalysis.

Co-factors, in particular metal ions and prosthetic groups (see [5.1]), are present in the active protion and play a very important part in the functioning of the protein.

5.2 Dynamics and Function

Despite the close packing, the protein globule exhibits conformational mobility, this being of greatest importance in molecular biology.

A macromolecular globule is a solid body. What are its general features?

In Schrödinger's words, an organism is an aperiodic crystal [5.13]. The meaning of these words, which are extended in modern literature to the protein molecule, is that an organism and a protein are solid bodies, which differ from crystalline bodies by the absence of an atomic-molecular periodicity. We shall see at a later time that the concept of the aperiodic crystal is important from the point of view of information theory. This concept, however, also requires a definition in the thermodynamic sense. The word "crystal" implies a thermodynamic equilibrium. The entropy of a perfect crystal at absolute zero is equal to zero; it is the Nernst theorem. What can be said about the aperiodic crystal?

Frauenfelder and his co-workers [5.14–18] treat the state of the protein globule as the glassy state. Ordinary glasses are condensed amorphous bodies, which are in a non-equilibrium state with no minimum free energy. The conformational lability of the globule leads to the formation of a mixture of "substates", which differ relatively little from one another. We are speaking here of different conformations.

Such a "glass" must possess high entropy at 0 K. Landau once said to me: "What can a physicist say of glass? Only that its entropy at 0 K is different from zero." As a matter of fact, much can be said of glass. In particular, the temperature dependence of the heat capacity of glass C_p near absolute zero is not gov-

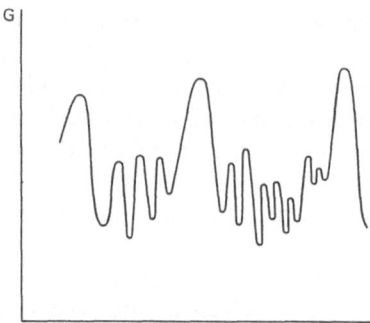

Fig. 5.5. The energy landscape of protein according to Frauenfelder

erned by the T^3 Debye law, which is valid for a crystal. The dependence of the C_p of glass on T contains also linear and quadratic terms. The heat capacity of protein at low temperatures also contains terms proportional to T and T^2. This, however, cannot serve as an argument in favour of the glassy state of glass.

Frauenfelder treats a protein as a biological machine that must possess two different states, at least in order to perform its function. An example would be the states of free myoglobin (Mb) and of myoglobin that has added on a ligand, say CO (MbCO). In turn, each of the states exists in a large number of conformational substates, which participate in the fulfilment of the same function, but possibly at different rates. The corresponding energy landscape is shown in Fig. 5.5.

There are experimental data pointing to the coexistence of different conformational substates in proteins. These data were obtained in the form of non-uniformly broadened bands in the spectra, or specific Mössbauer effects in iron-containing proteins [5.19]. The most striking effect is observed in the IR-spectrum of MbCO. The valence vibration of CO is represented here by three bands at 1933, 1945 and 1966 cm^{-1}, which correspond to three conformations of MbCO differing by the angles between the CO molecule and the normal to the haem plane. The estimation of the angles based on the infrared dichroism of MbCO gave the following values: $\alpha_1 = 15° \pm 3°$, $\alpha_2 = 28° \pm 2°$, and $\alpha_3 = 33° \pm 4°$ [5.15,18].

But can we state that protein is like glass, i.e. that it is not in an equilibrium state? Structurally, glass is analogous to a liquid that has been frozen without being crystallized. Glass can be crystallized but this requires much time since the viscosity is too high and the molecules have no time to be reoriented. Antique glass vessels that can be seen in museums have partly lost transparency; partial crystallization has taken place in them, which has taken thousands of years. The softening temperature of glass T_g depends on the rate of heating. The following condition is valid (provided that there is a single relaxation process with a relaxation time τ):

$$\left(\frac{d\tau}{dT}\right)_{T=T_\mathrm{g}} = \frac{1}{q} \qquad (5.1)$$

where $q = (dT/dt)_{T_\mathrm{g}}$ is the rate of temperature change. If

$$\tau = A \exp(E/kT) \qquad (5.2)$$

where E is the energy barrier which is overcome upon relaxation, then T_g is found from the expression

$$\frac{AE}{kT_\mathrm{g}^2} \exp\left(\frac{E}{kT_\mathrm{g}}\right) = \frac{1}{q} . \qquad (5.3)$$

Vitrification and softening of glass are kinetic rather than thermodynamic processes [5.2,20].

Sharonov has studied the temperature course of the heat capacity $C_\mathrm{p}(T)$ of polymeric glasses. A maximum in the heat capacity is observed at T_g: its position (i.e. T_g) and height strongly depend on the rate of heating [5.21].

Phenomena of this kind have not been studied in the case of proteins. No direct proof has been obtained for the nonequilibrium of protein. It may be asserted that the native structure of proteins is of a definite character; this compact structure cannot be regarded as a frozen liquid. Protein is a highly organized amorphous solid body. The distinction between protein and glass is fundamental. Any glasses, polymeric, silicate, etc., are capable of forming periodic crystals. If this does not occur, it is only because the relaxation times are very high. On the other hand, the protein molecule will never become a periodic crystal; it remains in the amorphous aperiodic state for indefinite periods of time and here we need not speak of the relaxational attainment of the equilibrium. This has already been attained as the globule is compact. The protein globule has no analogues; it is a construction, properly speaking.

If we melt glass and cool it down, its microstructure will undergo a radical change. Nothing like this occurs with protein. Upon slow "annealing" of the thermally denatured protein, its native structure is restored.

Protein denaturation may be likened to the melting of a crystal. It is a solid-liquid transition rather than a solid-gas transition. The denatured protein retains a number of order elements inherent in the native state and is capable of renaturation. It is possible that what has been said of glass is applicable to a certain extent to the denatured state of protein, but not to its native state.

The thermal motion of atoms in a protein globule, which is of fluctuational nature, does not upset the tertiary structure. The presence of three vibrational bands of CO in MbCO indicates that there are three stable conformations of this complex. The relative intensities of these bands undergo change upon heating ("annealing").

The unambiguous definiteness of the three-dimensional structure of the protein globule changes little upon transition from one conformation to another. Nevertheless there is no doubt that different conformations of protein

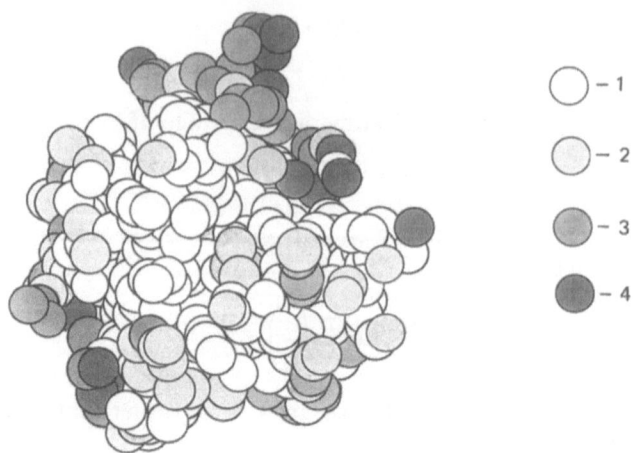

O – 1
O – 2
◑ – 3
● – 4

Fig. 5.6. The structure of ribonuclease C_2 from *Aspergillus clavates* according to X-ray diffraction data with a resolution of 1.55 Å. Differently shaded circles represent atoms with different mobilities, i.e. different mean-square displacements $\langle U \rangle = \langle U^2 \rangle^{1/2}$: (1) $\langle U \rangle < 0.5$ Å; (2) 0.5 Å $< \langle U \rangle < 0.6$ Å; (3) 0.6 Å $< \langle U \rangle < 0.7$ Å; (4) $\langle U \rangle > 0.7$ Å. Values of $\langle U \rangle$ for atoms in α- and β-structures are considerably lower than those for atoms in irregular portions [5.22]

exist (sub-states according to Frauenfelder), which play a very important role in the functioning of the protein. At the same time, from what has been said it follows that the conformational entropy of the protein globule in the native state is very low at absolute zero.

The methods of the modern theory of spin glasses may, in principle, be applied to the study of proteins, but this does not mean at all that the protein is a glass.

The fluctuational mobility of the protein globule is directly observed by means of X-ray dynamic analysis, Mössbauer spectroscopy, nuclear magnetic resonance and other methods. Figure 5.6 shows the structure of ribonuclease, the different mobilities of the atoms being indicated [5.22].

The fluctuational mobility of the macromolecule is primarily conformational in character. The rotations of the atomic groups about single bonds require considerably less energy than the stretching of a valence bond or the deformation of the valence angle. Any chemical reaction involves both a change in the states of electrons and a displacement of atomic nuclei. We may speak of the interactions of the electronic and nuclear degrees of freedom. The displacements of nuclei in macromolecules are predominantly conformational. Accordingly, electron-conformation interactions (ECI) must play a determining role in chemical processes that involve macromolecules [5.1,6,23–25].

Let us consider enzymic catalysis. Simple physical considerations show that the conformational mobility of protein determines its catalytic activity, which reduces to a decrease in the activation energy of a biochemical reaction.

We shall make use of the old, classical model of electrons, the harmonic

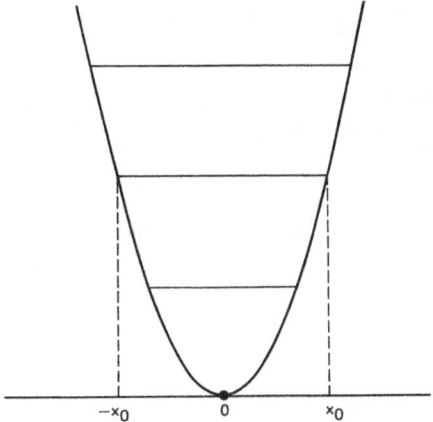

Fig. 5.7. The potential curve and energy levels of electrons – the harmonic oscillator

oscillator model. This means that electrons are located in a potential well of parabolic shape (Fig. 5.7). The well is specified by atomic nuclei and it may be assumed that the abscissa in Fig. 5.7 is a conformational co-ordinate. Without solving the Schrödinger equations, we make use of the de Broglie waves in order to find the quantization conditions [5.26]. The energy of the oscillator is made up of potential and kinetic energies:

$$E = \frac{bx^2}{2} + \frac{mv^2}{2} \; , \tag{5.4}$$

where b is the elastic coefficient and m is the mass of the oscillator. An electron vibrating with an amplitude x_0 given by

$$x = x_0 \cos 2\pi vt \; , \tag{5.5}$$

where the vibrational frequency is equal to

$$v = \frac{1}{2\pi} \sqrt{\frac{b}{m}} \tag{5.6}$$

is moving in the well along a portion of length $L = 2x_0$. The velocity of this motion is not constant, it depends on the coordinate x. At points $x = \pm x_0$ the velocity is equal to zero; it changes sign. Hence, since the energy is constant, it follows that

$$\frac{bx^2}{2} + \frac{mv^2}{2} = \frac{bx_0^2}{2} \; , \tag{5.7}$$

whence

$$v = \left\{\frac{b}{m}(x_0^2 - x^2)\right\}^{1/2} = 2\pi v (x_0^2 - x^2)^{1/2} \ . \tag{5.8}$$

The de Broglie wavelength corresponding to the motion of the oscillator, $\lambda = h/mv$, is not constant either. Let us make a crude assumption, namely that inside the well between the points x_0 and $-x_0$ over the entire length the total electron energy is equal to the kinetic energy. In other words, we treat an electron in the well as a free particle. On the walls of the well the total energy is equal to the potential energy

$$E = \frac{bx_0^2}{2} \ . \tag{5.7a}$$

Hence, the distance between the walls is

$$L = 2x_0 = 2\left(\frac{2E}{b}\right)^{1/2} \ . \tag{5.9}$$

Assuming that inside the well over the entire length L the kinetic energy remains constant, i.e.

$$\frac{mv^2}{2} = \frac{bx_0^2}{2} \ ,$$

we surmise that the velocity v is also constant. For such an electron there are corresponding standing de Broglie waves with wavelengths given by

$$\lambda_n = \frac{2L}{n} = \frac{4}{n}\left(\frac{2E}{b}\right)^{1/2} \ , \quad n = 0, 1, 2 \ldots \ . \tag{5.10}$$

Hence, the possible values of electron velocity are

$$v_n = \frac{h}{m\lambda_n} = \frac{nh}{4m}\sqrt{\frac{b}{2E_n}} \tag{5.11}$$

and the oscillator energy is found from the condition

$$E_n = \frac{mv_n^2}{2} = \frac{m}{2}\left(\frac{nh}{4m}\sqrt{\frac{b}{2E_n}}\right)^2 \ . \tag{5.12}$$

So, we obtain (cf. (5.6))

$$E_n = \frac{nh}{8}\sqrt{\frac{b}{m}} = \frac{\pi}{4}nhv \ . \tag{5.13}$$

As is known, the rigorous solution has the form

$$E_n = \frac{hv}{2} + nhv \ . \tag{5.14}$$

An *a priori* non-rigorous solution coincides with the rigorous one in the main point: the energy levels of the oscillator are equidistant. In the solution (5.13) the spacing between the adjacent levels is given by

$$E_{n+1} - E_n = \frac{\pi}{4} h\nu \ . \qquad (5.15)$$

In the rigorous solution

$$E_{n+1} - E_n = h\nu \ . \qquad (5.16)$$

The difference is only in the numerical factor and also in the absence of the zero-point energy of the oscillator ($E_0 = h\nu/2$) in the non-rigorous solution.

The possibility of deriving a correct quantization law for the oscillator (the energy is proportional to $n h\nu$) by means of such a simple and crude model is determined by the virial theorem, according to which the average kinetic energy is proportional to the average potential energy in the presence of a power potential function.

Standing de Broglie waves, i.e. electrons in the well, exert pressure on its walls. Its force f is easy to calculate. From formulae (5.9) and (5.12) it follows that

$$f = \left| \frac{dE}{dL} \right| = \frac{n^2 h^2}{4mL^3} \ . \qquad (5.17)$$

In the state of equilibrium these forces are balanced. Let us consider the corresponding model of enzymic catalysis [5.1,27]. It is shown in Fig. 5.8. The

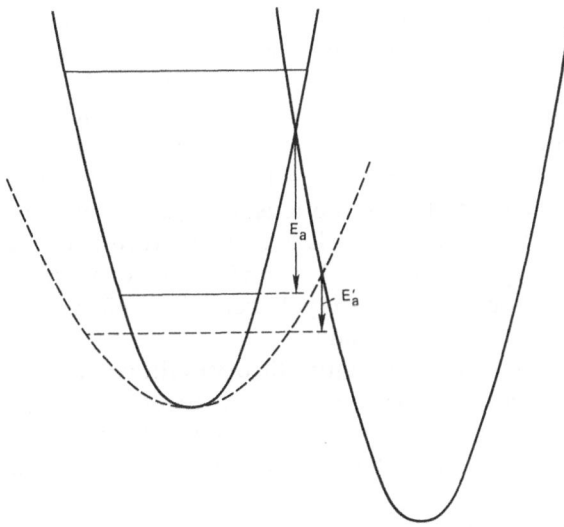

Fig. 5.8. Diagram explaining the meaning of electron-conformation interactions

initial state of an enzyme-substrate complex is depicted by a parabola on the left and the final state, i.e. the enzyme-product complex, by a parabola on the right. The solid curves and lines refer to the system without electron-conformation interactions (ECI), i.e. conformational dynamics. The activation energy of the process in this case is equal to E_a. However, the interaction of the substrate with an active site perturbs the electronic states of the corresponding amino-acid residues. The perturbed wavefunctions are represented by linear combinations of the unperturbed functions of both the ground state and all perturbed states. Hence, there is an admixture of such states, i.e. there is electronic excitation in the system with a certain statistic weight. This gives rise to unbalanced pressure of the electrons, de Broglie waves, on the walls of the well, and the new state of equilibrium corresponds to an extended parabola shown by a dashed curve. The energy levels of electrons decrease in turn (dashed lines). The free energy, i.e. the spacing between the lower level and the point of intersection of the parabolas, falls off to the value E_a'. A kind of feedback is realized, i.e. change in the electronic state causes a change in the conformation, which in turn alters the electronic state.

The model described above is, of course, oversimplified and is largely of didactic value. However, it outlines the principal features of the physics of the process. Moreover, simple calculations show that a small admixture of the excited states of electrons is sufficient for such a decrease in activation energy, which leads to an increase in the reaction rate by many orders of magnitude.

Addition of an electron to the system, say the reduction of cytochrome c, must bring about similar changes.

As has been said above, the protein globule is a highly ordered amorphous solid body. The displacement of electrons (electron density) in a solid is specific. For example, in calculations of the electronic conductivity of an ionic crystal one has to take into account that it is not a free electron which moves in a crystal, but a quasi-particle, which is known as a polaron. This is an electron plus the polarization of the surrounding ions. The mass and charge of the polaron are different from those of a free electron. In a similar way, upon displacement of electron density in a biopolymer or in a supramolecular biological structure (a membrane), not the electrons but rather so-called conformons (i.e. electrons plus conformational motions of the surrounding groups) move [5.1,24,28]. The conformon (it is clear that the word is derived from "conformation") differs from the polaron by rapid energy dissipation, which is determined by the aperiodicity of the biopolymer. The conformon moves not farther than a few links of the protein chain.

The construction of a quantitative theory of the conformon has been started in the works of Shaitan and Rubin [5.29]. Evidently, there are grounds for believing that conformational motion is a limited continuous diffusion in an extended potential well of small depth. Relatively large conformational motions are possible in the protein globule due to the presence or appearance in it of rare, free, empty portions ("holes") and also of shifts similar to dislocations. The times of conformational motions vary within wide limits. They range from times

Table 5.2. Characteristic amplitudes and times of thermal fluctuations of the elements of the protein structure at 300 K

Process	Amplitude [nm]	Time [s]
Valence vibrations	0.001–0.01	10^{-12}
Microconformational motions of side groups	0.03–0.1	10^{-8} to 10^{-9}
Bending motions of α-helices and β-ribbons	0.2–0.7	10^{-7} to 10^{-8}

less than picoseconds to times exceeding seconds. Table 5.2 gives some relevant data. Slower conformational motions involving whole protein domains also occur.

At present a great diversity of the dynamics of chemical reactions in condensed phases has been established [5.30]. Contributions to this dynamics are introduced by friction, electronic non-adiabaticity and a flux of intramolecular energy. A substantial role in a number of cases is played by the tunnel effect (in particular, in cytochrome c). In crystals that have a periodic structure, chemical reactions propagate as autowaves.

However, as far as proteins are concerned, we should resort to the concepts of electron-conformation interactions (ECI). A genuine quantitative theory of electronic-conformational interactions has not yet been developed: for such a theory to be constructed wide use must be made of quantum mechanics and quantum chemistry (see, for example, [5.25]). But the occurrence of electron-conformation interactions and their important role have been established. The results of some experimental investigations have been reported (see also [5.1,6]).

Krakoviak and co-workers [5.31–33] have demonstrated correlations between the chemical reactivity of macromolecules of synthetic polymers and their conformational intramolecular mobility, which is determined by the method of polarization luminescence. Striking examples of ECI have been provided by Sharonov and his colleagues [5.34], who studied the magnetic rotation of the plane of polarization of the light (Faraday effect), and magnetic circular dichroism (MCD) of haem-containing proteins. These phenomena are examined in the region of intrinsic absorption of the haem group entering into the composition of globins, cytochromes, etc. They characterize the electronic state of this group. At the same time, the magneto-optic properties of haem appear to be strongly dependent on the conformational state of a haem-containing protein. The magnetic rotation and MCD correlate with the conformational stability of complexes of haem proteins with ligands, i.e. with the stability of these complexes towards denaturation. A number of authors [5.35,36] have carried out a detailed study of the contributions of electron and conformational energies to the stereochemistry of haem. The conformation of protein complexed with a ligand is different from the conformation of free protein. At helium temperatures near $0°K$ the complexes underwent a photochemical dissociation and the altered conformation remained frozen.

Crystallized adenylate kinase from pig muscle has two different conformations, which undergo interconversion as a result of the protonation and deprotonation of the His-36 group present in the bulk of the active site. Both the conformation of the protein and its packing in the crystal are changed. It has been found that these conformations play an important role in the binding of the substrate in solution and also in an enzymic process [5.37].

Sinev et al. [5.38] have established, by means of small-angle X-ray scattering, a correlation between the enzymic activity and the conformational dynamics of 3-phosphoglycerate kinase.

We have given only a few examples indicating the determining role of electron-conformation interactions in the enzymic function of proteins. The number of such examples is practically unlimited. Electron-conformation interactions play a very important role in other functions of proteins too, in particular in their mechanochemistry.

These properties of globular (and also of fibrillar) proteins depend in the long run on the genetically coded primary structure of polypeptide chains. It is exactly these structures that are changed in the course of biological evolution. Evolutionary changes in proteins take the form of point substitutions of amino-acid residues (their additions and deletions) and of more significant rearrangements of the primary structures. The point mutations of proteins and the more radical rearrangements are determined by the corresponding events in genomes and are fixed by natural selection. The key problem which we encounter in a study of proteins reduces to the relationship between the primary structure of protein and the three-dimensional structure of its molecules, which is responsible for their biological functions. This problem can be solved in two ways. First, by carrying out experimental investigations with a view to elucidating the role of point and larger changes and, second, by constructing the corresponding theoretical models. We begin with such models.

5.3 Modelling of Proteins

The important problem of the evolutionary biology of proteins coincides on the whole with the analogous problem at the level of an organism, species, higher taxon: To what extent is a particular character resulting from natural selection adaptive? Since the selection of proteins proceeds with respect to their biological functions governed by the three-dimensional structure of their molecules, it is clear that this structure is more or less adaptive. However, two questions arise: Is the position of each amino-acid residue in the chain selected? Is the primary structure of protein too under the pressure of natural selection? A detailed answer to the latter question is given in Chap. 6. At this point we will analyze the situation with the aid of theoretical models.

It is known that averaging over the primary structures of proteins does not yield statistically reliable distinctions from a random distribution of amino-acid

residues along the chain [5.39–41]. The percentage content of amino-acid residues approximately correlates with the number of codons by which they are
encoded; Trp and Met residues, each of which is represented by a single codon,
are encountered very seldom. This does not mean, however, that the position of
each residue is not fixed by natural selection.

Ptitsyn [5.42] has studied a theoretical model of a "protein" which is built
up from two types of residues, polar and nonpolar (hydrophilic and hydrophobic). As has already been pointed out, the surfaces of the regular portions of
protein (alpha-helices and beta-forms, which are in contact with each other)
consist mainly of non-polar groups. In order for a protein chain composed only
of two types of amino-acid residues (instead of twenty) to assume a regular
structure, it is necessary that alpha- or beta-portions with at least one continuous hydrophobic surface be formed in it. A computer generation of random
amino-acid sequences of twenty natural residues (with account taken of their
average content in globular proteins) has shown that in such sequences clusters
of nonpolar groups are often formed, which give rise to alpha and beta-portions
with continuous non-polar surfaces [5.43]. This also appears to be valid for
random seuqences of polar and non-polar residues united into a single chain,
whose fractions are assumed to be equal. An alpha-helix of a minimum length
must consist of two non-polar residues, which are separated by three residues,
$(i, i + 3)$, and a β-ribbon must be composed of two nonpolar residues in positions $(i, i + 2)$. It is obvious that the shorter the alpha or beta-portion, the more
probable its appearance in the random sequence. Figure 5.9 [5.42] shows the
dependence of the probability of appearance of an alpha- or beta-portion on
its length obtained in these calculations. For comparison, Fig. 5.9 also shows the

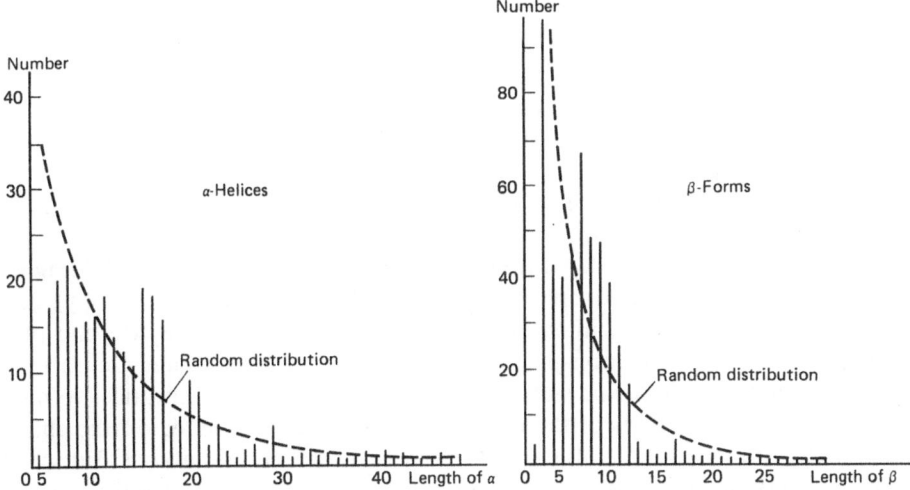

Fig. 5.9. Probability curves for appearance of α- and β-forms in protein models and histograms for
real proteins (after Ptitsyn [5.42])

histograms of the distributions of α- and β-segments in 62 globular proteins. The curves were normalized to the total number of the corresponding portions in these proteins – 274 α-helices and 446 β-ribbons. The crude agreement between the probability distribution for a simple model and the distributions in real proteins is evidence that the concrete nature of the amino-acid residue is not so important as far as the structure of the globule is concerned; the important factor is whether the residue is hydrophobic or hydrophilic. The average lengths of the segments in real proteins (alpha-helices consisting of 13 units and beta-forms made up of 6.8 units) are very close to those in the model under consideration: 12 units for alpha, and 7 units for beta. Similar results have been obtained in calculations of the distributions of the number of α-helices, β-forms and $\beta\alpha$-pairs among the $(\alpha)_n$, $(\beta)_n$ and $(\beta\alpha)_n$ sequences containing n such segments located sequentially along the chain. The average number of α-helices in 77 proteins is 2.5 and the average number of β-forms is 3.8, which practically coincides with the values obtained for the simple model (2.5 and 3.7). However, for $\beta\alpha$-pairs there is a difference: 2.6 in proteins and 1.6 in the model.

Based on these results of a model investigation, Ptitsyn suggests that the protein be treated as "an edited random co-polymer". The "editing", which arises as a result of natural selection, involves only a few amino-acid residues located at and near an active site of the protein-enzyme. This region, which may be called active, is located in the "passive" backbone, which is of predominantly random nature. Evidently, the behaviour of the active and passive portions of protein must be different during evolution. The electron-conformation interactions occur in the active portion. In the passive part, in the backbone, no electronic events take place, but conformational changes occur. For enzymic activity the whole globule is required. The reports in the literature indicating that a considerable portion of protein can be cut off without impairing its enzymic activity have proved erroneous.

From these results it follows that many of the point mutations in the passive part of protein, which consist of the replacement of those residues which do not alter their class (since polar residues are substituted for polar ones and nonpolar residues for nonpolar ones), cannot exert a substantial effect on the structure and functions of protein.

Another model approach to the structure of globular proteins has been undertaken by Dill and co-workers [5.44–48].

The main factors considered above, which are responsible for the folding of protein into a globule (hydrophobic nature and conformational entropy), are non-specific. Hydrophobic interactions are non-specific since they depend primarily on the surfaces of non-polar residues buried in the globule. Conformational entropy is non-specific since it is determined by steric hindrances to motion in the compact states of the macromolecule; it also depends on the chain length and volume of the globule, and is low, as has been said above. But what forces are responsible for the internal architecture of the globule, for the specific secondary structures? Usually, the formation of alpha and beta forms is ascribed to hydrogen bonds. It can be shown, however, that the number and forces of

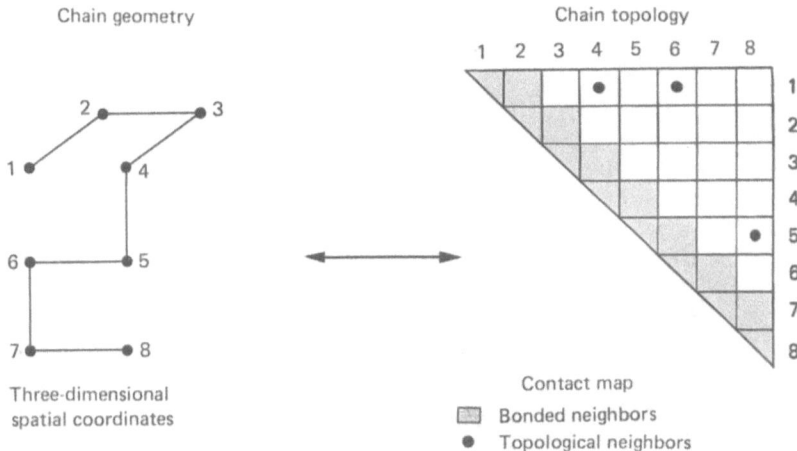

Fig. 5.10. The model of the protein chain and topological contacts (Chan and Dill [5.44])

hydrogen bonds in a globular protein are approximately the same as in an unfolded chain. The answer to the question of the nature of the forces that create ordered periodic α- and β-structures has been obtained with the aid of the so-called lattice model. It has been found that any flexible macromolecule must form an appreciable number of secondary structures if a compact structure is built due to the forces of attraction between the monomers.

The monomers are located as lattice points in the cubic-space lattice. The monomers adjoining one another in the chain in positions $i, i + 1$ are bound. The monomers that are through-space neighbours and non-bonded are "topological" neighbours $(i \neq j - 1, j, j + 1)$. An alpha-helix is formed through a series of contacts $(i, i + 3)$, $(i + 1, i + 4)$, $(i + 2, i + 5) \ldots$, a parallel β-sheet by a series of contacts (i, j), $(i + 1, j + 1)$, $(i + 2, j + 2) \ldots$, and an anti-parallel β-sheet by a series of contacts (i, j), $(i + 1, j - 1)$, $(i + 2, j - 2) \ldots$. The forces that operate between the bound and topological neighbouring monomers are different. Figure 5.10 shows a typical chain conformation in the lattice and the corresponding "map" of contacts.

Of course, the true alpha-helices cannot be reproduced in the cubic lattice. However, helices sufficiently similar to alpha-helices and characterized by periodically recurring cylindrical conformations can be obtained. This is shown in Fig. 5.11, which was borrowed from Dill et al. [5.44]. The modelling of beta-forms as well as chain rotations are realized directly, this being shown in Fig. 5.12.

Chan and Dill define the compactness of the macromolecule as the ratio of the number of intramolecular topological contacts (the number of dots on the contact map) to the possible maximum number of such contacts:

$$\varrho = \frac{n}{n_{\text{max}}} , \quad 0 \leqslant \varrho \leqslant 1 . \tag{5.18}$$

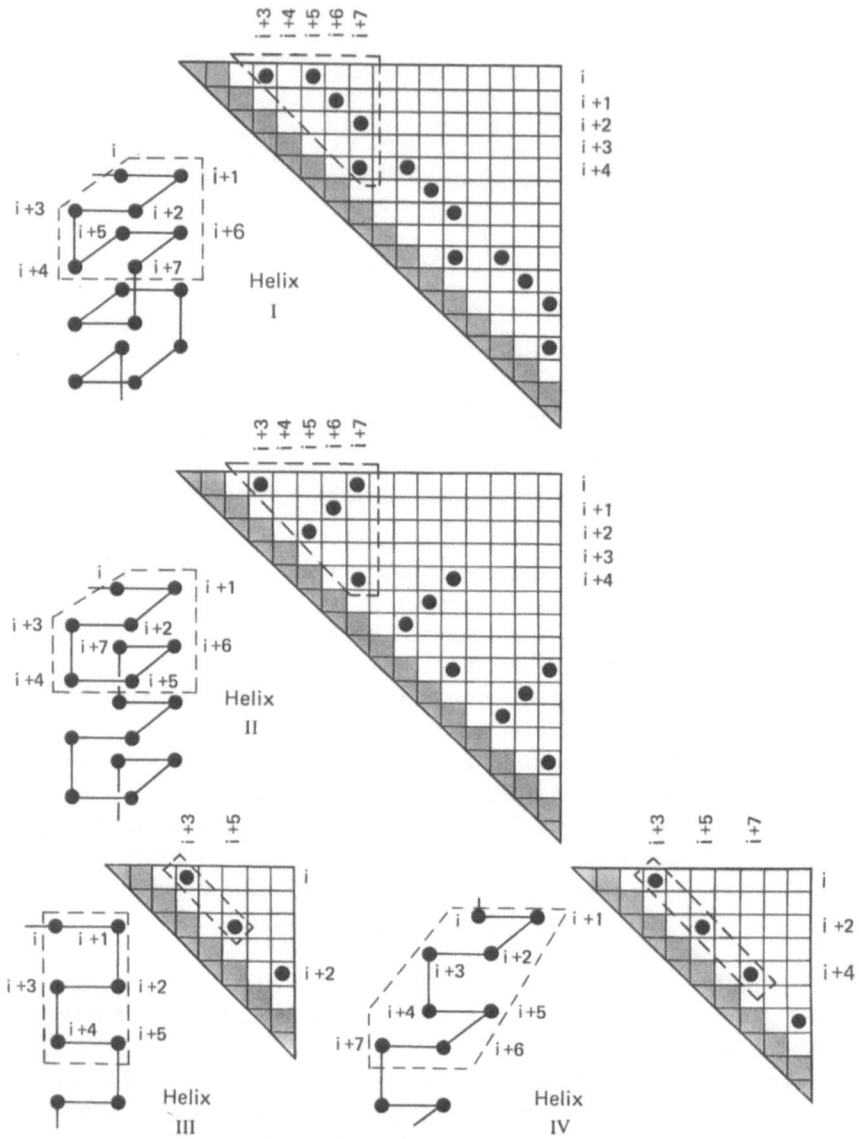

Fig. 5.11. Representation of helices on a cubic lattice [5.44]

Computer calculations are made to estimate changes in entropy upon formation of topological contacts. If the macromolecule has a topological contact (i, j), what succeeding contacts (i', j') are the most probable? The number of conformations with (i, j) and (i', j') contact is denoted by $Q(i, j; i', j')$; the total number of all possible conformations is designated as Q_0. A decrease in entropy upon formation of these two contacts is given by

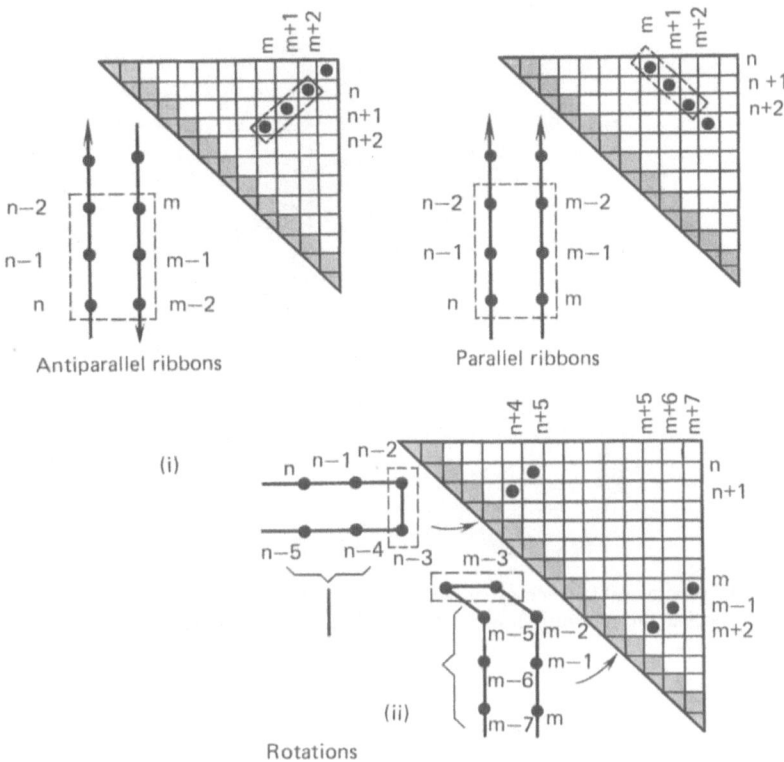

Fig. 5.12. Representation of antiparallel and parallel β-sheets and rotations on a cubic lattice [5.44]

$$\Delta S = -k \ln \left[\frac{Q(i,j;i',j')}{Q_0} \right] . \qquad (5.19)$$

Calculations performed for all possible contacts in the chain of a certain length give the distribution of free energies on the contact map, the "topological" free energy surface. Figure 5.13 shows the results of calculations for a three-dimensional chain consisting of 12 segments. The anti-parallel sheets (4,9) and/or helical contacts (3,6) (7,12) gives the largest decrease in free energy, since with them one can observe the least additional restrictions on conformational freedom. Hence, these conformations are preferred. Calculations show that the number of available conformations falls off rapidly with increasing ϱ, but the fraction of conformations corresponding to secondary structure increases significantly. Thus, any force that creates numerous contacts, i.e. makes the globule more compact, causes the formation of many secondary structures. For proteins in the cytoplasm this force is provided by hydrophobic interactions.

What is the probability of appearance of a certain fraction of secondary structure in an arbitrary chain conformation? Most conformations contain considerable fractions of secondary structures; there are a few ways in which con-

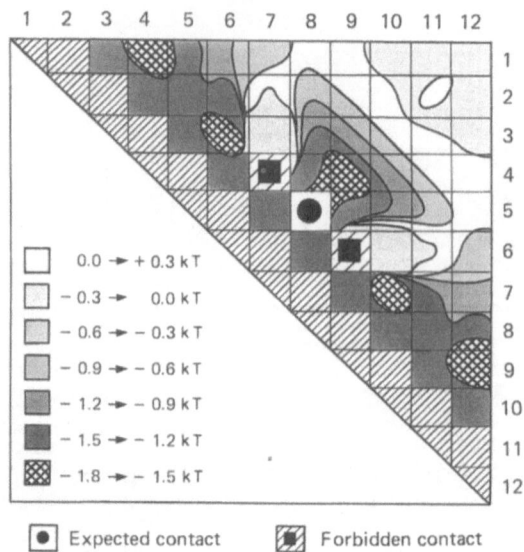

Fig. 5.13. The topological free-energy surface. The contact $(5, 8)$ in the three-dimensional chain of 12 units is specified, the contours show relative free energy upon formation of any second contact (i', j') [5.45]

Table 5.3. Fractions of secondary structures in chains of differing length on cubic lattice

Structure	Number of residues		
	12	18	27
Helices I	0.105	0.057	0.029
Helices II	0.160	0.072	0.042
Helices III	0.096	0.158	0.165
Helices IV	0.009	0.007	0.006
All helices	0.315	0.260	0.225
Antiparallel sheets	0.126	0.186	0.198
Parallel sheets	0.027	0.047	0.069
Turns	0.046	0.061	0.066
All sheets and turns	0.180	0.268	0.291
All structures	0.482	0.512	0.498

formations with maximum compactness and devoid of secondary structure can be formed. The dependence of the structure on the chain length is shown in Table 5.3.

The total fractions are less than the sum of individual fractions since the residues may be involved in several secondary structures.

The model considered above takes account only of conformational freedom and excluded volume (the chain has no self-crossings), ignoring chain heterogeneity and specific interactions. Accordingly, the resultant structures are not strictly localized in either the chain or the globule. The main principle is that the organization of secondary structures is the result of steric hindrances. It should be noted, however, that the predicted values in the region of 23–32 percent α-helix and 18–29 percent β-form are consistent with the data of protein crystallography, which give 21–40% aplha-helix and 14–28% β-form. The distribution of the lengths of alpha- and beta-forms is also in agreement with experimental data and in this sense the results obtained by Dill and Chan [5.44] are similar to the results obtained by Ptitsyn [5.42].

The dependence of molecular organization on primary structure is determined by particular interactions of topological neighbours and by interactions between bound neighbours. It has been shown on the whole that the principal structural features of globular proteins are governed by the very fact that they are macromolecular; they must arise in any polymers being compacted by any forces. The model explains why short helices, which are themselves unstable in solution, must be stable in the core of the globule. The model explains the origin of anti-parallel and parallel β-forms in proteins. And, finally, the model explains why some proteins which are unfolded by denaturation retain secondary structures.

A number of important inferences about the nature of the globule can be obtained with the aid of a simpler two-dimensional lattice model. Square lattices were used by Dill and his associates [5.45–47]. A detailed computer enumeration of the conformations of short chains has been made, but the model allows one to study long chains as well. Moreover, it turns out to be possible to study heterogeneous chains consisting, as in the Ptitsyn model, of two types of amino-acid residue: hydrophobic (H) and polar (P). Such a chain collapses to a relatively small number (often only one) of maximally compacted conformations with a hydrophobic core. The number of available conformations is found to be dependent on two factors: the conformational freedom and the entropy of the shape of the compact object. The formation of loops in compact polymers, and topological correlations between pairs of loops, have also been studied. There is an internal driving force which converts loops into helices and into anti-parallel sheets. It is very difficult to form a compact chain with less than 50% of secondary structures. In this sense, the conclusions made in studies of two-dimensional models coincide with the results of calculations carried out for three-dimensional models. Little is thus changed by two-dimensionality but it makes possible the solution of a number of important problems that are practically inaccessible for the three-dimensional model.

In the two-dimensional model the maximum compactness of the globule is characterized by a minimum perimeter, which depends, of course, on the number of monomeric units. Figure 5.14 shows compact structures of minimum perimeter with the number of monomeric units being $N + 1$ (N is the number of bonds between the residues). Calculations show that only one maximally compact structure a is possible with $N + 1 = (m + 1)^2$, ($m = 4$), two structures b with $N + 1 = m(m + 1)$, ($m = 4$), and three structures c out of 238 possible

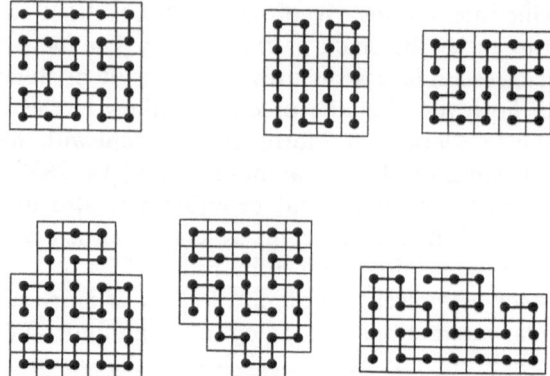

Fig. 5.14. Compact structures with a minimum perimeter on a square lattice [5.46]

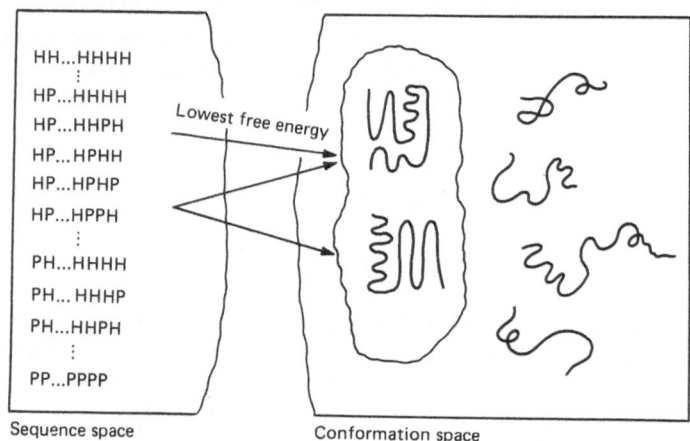

Fig. 5.15. The link between sequence space and conformation space [5.47]

ones for $N + 1 = 26$. A chain of $N + 1 = 21$ residues cannot be fitted into the compact structure composed of 21 units shown in Fig. 5.14 [5.46].

The key problem of the evolution of proteins is how mutational changes in the sequence of residues lead to changes in three-dimensional structure and stability, and hence to a change in the function of the protein. The HR copolymer is characterized by the presence of topological HH contacts, which lead to a decrease in energy by $\varepsilon < 0$. The two-dimensional model allows one to study a considerable fraction of the "sequence space". Figure 5.15 presents the corresponding scheme. The sequence space is a set of all primary structures with a given content of H and P, and the conformation space is a set of all possible conformations. Let us presume that native conformations are those which are maximally compact and contain a core consisting predominantly of H units and having a considerable content of two-dimensional analogues of helices and anti-parallel sheets.

Lau and Dill [5.47] have studied the effect of point mutations, (i.e. $H \to P$ substitutions) on the chain structure. Such substitutions naturally appear to be destabilizing. The units present on the surface are less susceptible to substitutions than those in the core. However, single substitutions of this kind do not usually affect the native structure and are neutral. The sequence space is relatively smooth; slight changes in the sequence seldom have an effect on he structure and stability of the native state. Calculations show that the number of "convergent" sequences (i.e. sequences that code for the same native structure) is very large, and many sequences fold up into relatively globular structures as a result of hydrophobic interactions. This means that the probability of formation of a globular, catalytically active protein from a random sequence is not low. An active site in this model arises as a result of the topological approach of quite definite residues. Lau and Dill have estimated the corresponding probabilities (see also [5.48]).

Figure 5.16 shows schematically the results of single $P \to H$ and $H \to P$ substitutions.

Thus, the study of the cubic and square-lattice models of proteins leads to

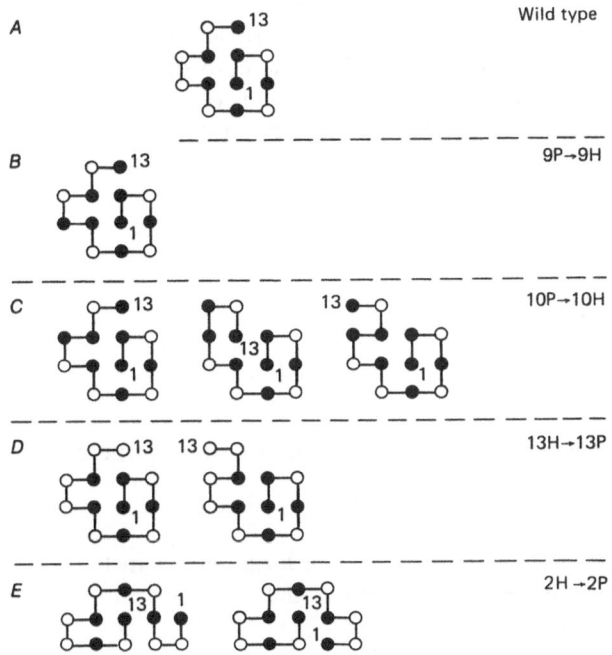

Fig. 5.16. Scheme explaining the effect of point substitutions on protein structure [5.47]. (*A*) The wild type has only one maximally compact conformation. (*B*) The replacement $9P \to H$ does not alter the native structure. (*C* and *D*) Mutations increase the degeneracy of the native state; the wild type is present. (*E*) The single mutation $2H \to 2P$ destroys the native structure

results that are in full agreement with the concept of the "edited" random co-polymer. The spatial structure of protein, which is responsible for its biological function, arises from its macromolecularity, conformational flexibility and hydrophobic interactions. This structure is only slightly sensitive to many unit substitutions; in other words, a large variety of sequences can form the same spatial structure. Thus, the correspondence between the primary and the three-dimensional structure must be degenerate.

Let us consider the theoretical works of Shakhnovich and Gutin, which are devoted to the same problem of protein folding [5.49–53].

These authors have applied the mathematical apparatus of so-called replicas, which had been worked out in the theory of spin glasses. It should be noted immediately that this by no means implies the vitrified state of proteins. We are speaking here only of the mathematical apparatus. The mathematical theory requires special treatment. Here we will confine ourselves to the main results.

Shakhnovich and Gutin have demonstrated, through the use of theoretical models of a heteropolymeric chain, that the fundamental possibility of protein self-organization stems from thermodynamics, i.e. at sufficiently low temperatures the protein globule is an equilibrium system. At such temperatures quite definite structures dominate. This allows one to treat the folded state of protein as corresponding to the global minimum of free energy, which coincides with the global energy minimum since the entropy is very low.

Each structure is characterized by the co-ordinates of atoms in the chain $\{r_i^{(m)}\}$, where i is the number of atom and m is the index of the spatial structure, to which the energy E_m corresponds. The statistical weight of the structure according to Boltzmann is given by

$$p_m = \frac{1}{Z} \exp\left(-\frac{E_m}{kT}\right) , \qquad (5.20)$$

where

$$Z = \sum_m \exp\left(-\frac{E_m}{kT}\right) \qquad (5.21)$$

is the partition function. The domination of a single structure m_0 means that

$$p_{m_0} = 1 - \varepsilon , \quad \varepsilon \ll 1 . \qquad (5.22)$$

This condition is observed if there is a large gap Δ in the energy "spectrum" between the "ground state" m_0 and the first "excited state" m_1, so that

$$\Delta = E_{m_1} - E_{m_0} \gg kT . \qquad (5.23)$$

For a random heteropolymer with a large variety of monomer units the interaction energies B_{ij} of units i and j, which are neighbours in the lattice, are random variables with a Gaussian distribution

$$P(B_{ij}) = \frac{1}{(2\pi B^2)^{1/2}} \exp\left(-\frac{(B_{ij} - B_0)^2}{2B^2}\right), \qquad (5.24)$$

where B_0 is the average value of B_{ij}; B is the standard variance, a measure of chain heterogeneity; $B = 0$ for a homopolymer.

A study of this microscopic model has shown that there are γ^N microstates, i.e. chain foldings, where γ is the number of possible conformations of the peptide bond, with the energy distribution given by

$$P(E_m) = \frac{1}{(2\pi N B^2)^{1/2}} = \exp\left(-\frac{(E_m - \bar{E})^2}{2\varrho N B^2}\right) \qquad (5.25)$$

where ϱ is the average number of contacts formed by the unit. We may introduce a quantity associated with the number of dominant microstates:

$$X = 1 - \sum_{m=1}^{\gamma^N} p_m^2 . \qquad (5.26)$$

When a single state dominates, $p_{m_0} \approx 1$ and $X = 0$. Theory shows that there is a critical temperature

$$T_c = \frac{B\varrho^{1/2}}{2k(\ln\gamma)^{1/2}} . \qquad (5.27)$$

With $T > T_c$, the average value of $\langle X \rangle = 1$ and with $T < T_c$, $\langle X \rangle = T/T_c$. The probability P_ε of a randomly chosen sequence being folded into a globule with the dominant value of p_{m_0} (5.22) is expressed by the formula

$$P_\varepsilon = \frac{\sin(\pi X_0)}{\pi X_0} \varepsilon^{x_0} \qquad (5.28)$$

where $X_0 \cong 2\varepsilon$. With $T \geqslant T_c$, $P_\varepsilon = 0$, but when $T < T_c$, P_ε rapidly increases. For example, $P_\varepsilon \cong 0.1$ with $T_c/2$ (or $X_0 = 0.5$). Under these conditions every tenth sequence is folded into a unique structure. The analytical result of expression (5.28) has been supported by a simple model of a 27-membered chain located in a cubic $3 \times 3 \times 3$ lattice (Fig. 5.17). The number of compact self-avoiding structures is 102,758 in this case. Calculations have given values of P_ε which agree well with the analytical formula.

Shakhnovich and Gutin [5.52] have estimated the probability of neutral mutations, i.e. random substitutions of units of one type by units of another type which do not alter the unique spatial structure of the globule, and thereby retain its minimum energy. Calculations show that this probability w depends directly on γ; its approximate value with $\gamma > 1.1$ is

$$w \approx \gamma^{-8} . \qquad (5.29)$$

The probability w is independent of the quantity B, the heterogeneity of interac-

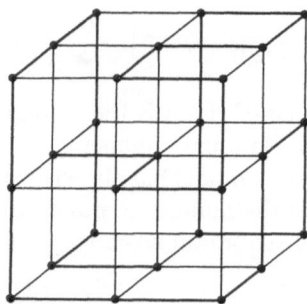

Fig. 5.17. The 27-membered chain on a cubic lattice (Shakhnovich and Gutin [5.53])

tions in the chain. The folded chain is the more stable towards point substitutions, the higher is its rigidity (i.e. the lower is the value of γ). Estimates for real proteins give $\gamma \cong 1.2$, whence $w \cong 0.2$.

The number of possible primary structures of a protein containing N residues is equal to 20^N, which is many times larger than the number of possible conformations γ^{N-1}. This means that to a single spatial structure $(20/\gamma)^N$ primary structures correspond. This ratio is actually lower, since the protein folding depends not on the 20-letter code but on the smaller number of the characteristics of amino-acid residues, such as hydrophobicity, charge, volume. A series of substitutions change nothing at all in this sense.

All the theoretical works discussed in this section point to a degenerate correspondence between the primary and the three-dimensional structure of protein. To a great degree of probability, point substitutions in the chain do not alter the globule structure. This is important evidence in favour of the neutral theory of molecular evolution.

5.4 The Primary and Spatial Structures of Proteins

Let us now turn to experimental data. Lesk and Chothia [5.54] have compared the spatial structures of a number of globins (proteins containing a prosthetic haem group). X-ray diffraction methods have been employed to establish the structure of the α- and β-sub-units of human haemoglobin, the α- and β-sub-units of horse metmyoglobin, seal myoglobin, sperm whale metmyoglobin, lamprey cyanohaemoglobin, carboxyhaemoglobin of the worm *Glycera dibraudata*, erythrocruorin of the larva *Chironomus thummi*, fetal human haemoglobin and rhizobium leghaemoglobin. All these structures have been obtained with a resolution of 2.0 and 2.5 Å, i.e. the spatial positions of all amino-acid residues have been determined.

The primary structures of all the globins mentioned above differ significantly. However, their spatial structures are very similar; they contain practically coinciding α-helical sections and "pockets", in which haem groups are

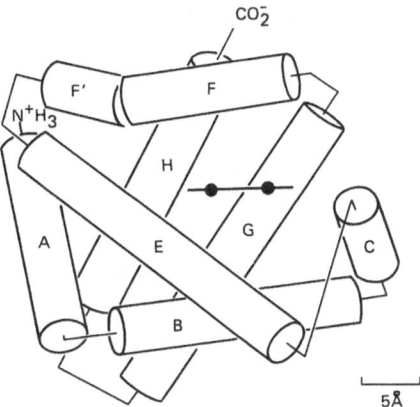

Fig. 5.18. The main three-dimensional structure of globins; alpha-helices are depicted as cylinders [5.55]

buried. Of special importance for keeping three-dimensional structure of globin intact are interhelical contacts and contacts between haem and α-helix. Nevertheless, in these regions numerous substitutions are observed, not to mention the non-helical portions and non-contacting units. The spatial differences between globins lie predominantly in small rotations and shifts of α-helices as solid entities. The situation in the region of the active site (the haem) remains constant, despite the considerable changes in the volumes of the amino-acid residues upon substitutions.

Figure 5.18 shows the basic three-dimensional structure of globins. The α-helices from A to G are sketched in the form of cylinders. It is worth comparing this figure with the structure of myoglobin shown in Fig. 5.18. Bashford and his colleagues [5.55] have compared 226 primary structures of globins that have a similar spatial structure. Only two residues (out of approximately 150) are absolutely conservative and the identites of the residues in any pair of sequences do not exceed 16%. There are more or less conservative regions, and the active part of the protein is naturally more intact than the passive part. In the conservative region there are 32 positions, which are almost always occupied by hydrophobic residues. In known structures these residues are located in the interior of the protein. There are also 32 other positions which are nearly always occupied by charged, polar or small non-polar (Gly or Ala) residues.

Spivak [5.56] has analyzed the data pertaining to mutants, the polymorphic types of human haemoglobin (Hb). Recall that Hb has a quaternary structure; the Hb molecule consists of four subglobules. In adult humans two subglobules belong to the alpha type, two to the beta type; such HbA is denoted as $\alpha_2\beta_2$ (do not confuse with the designations of the secondary structures of proteins). The foetal Hb of the newly-born contains, instead of β-chains, γ-chains, which are different from beta and alpha; the formula of foetal Hb F is $\alpha_2\gamma_2$. We also know of the less widespread Hb A$_2$ ($\alpha_2\delta_2$).

Table 5.4. Correlations between structural and functional changes in human haemoglobin

Intramolecular localization of substitution	Properties affected by substitution	Clinical and hematological consequences
Surface of molecule	None	None
Sub-unit interior	Stability of molecule	Haemolytic anemia
Haem crevice	Affinity for O_2	
	Oxidation of Fe	Cyanosis
	Stability of molecule	Haemolytic anemia
Interface between α1 and β1	Stability of molecule	Haemolytic anemia
Interface between α1 and β2	Affinity for O_2	Increase Erythrocytosis / Decrease Cyanosis
Portions of salt bridges and portion where diphosphoglycerate is bound	Affinity for O_2	Erythrocytosis
Incorporation of Pro into helical segment	Stability of molecule	Haemolytic anaemia

More than 400 kinds of human HbA have been identified, which differ from one another due to a single replacement of the amino-acid residue. Among these variants there are those which appreciably differ in function and therefore lead to pathologies, and also variants with a neutral character. The same problems were also examined in a later work by Imai et al. [5.57]. These authors compared more than 440 natural mutants of human Hb with single substitutions. More than half of these mutants are characterized by changed molecular properties (binding of O_2, structural stability, ability to undergo autoxidation, etc.). The changes depend significantly on which region of protein undergoes substitutions. Table 5.4 presents data characterizing the correlations between the structural and functional changes of Hb.

The methods of protein engineering, including directed mutagenesis, have been used to prepare artificial Hb mutants and to study the consequences of the substitutions that are not observed in nature.

Thus, the replacements of residues located at the surface of the protein molecule may have no consequences. Such replacements are neutral.

The neutrality of many mutations has been established not only for globins but also for a number of other proteins. In particular, Dickerson compared the spatial structures of four types of bacterial cytochrome c, and tuna cytochrome c [5.58]. Cytochrome c is a more ancient and universal protein than globins. The primary structures of cytochrome c of different species differ to a much lesser extent than the primary structures of globins. For example, hen cytochrome c differs from human cytochrome c in 11 amino-acid residues out of 104, whereas hen αHb differs from human αHb in 36 residues out of 141. In all known types of cytochrome c both the manner in which the haem is linked and its immediate environment are practically constant. There is no doubt that this constancy is essential for the electron-transfer function of cytochrome c in the respiratory chain (see [5.1]). At the same time, different types of bacterial cytochrome c have different chain lengths. They may be subdivided into three classes: short chains (in bacteria of the genus *Pseudomonas* cytochrome c_{551} contains 82 residues, and in *Chlorobium* there are 86 residues); long chains (*Paracoccus*, 134 residues) and medium-length chains (c_2, *Rhodospirillum*, 112 residues). The latter are also present in vertebrates (tuna, 103 residues). The elongation or contraction of the chain leads to the appearance of new loops in the non-haem region of the protein, with the haem portion remaining intact.

The possibility of neutral substitutions in polypeptide chains is also proved by results obtained in a study of peptide ligands that take part in regulatory systems [5.59]. Not all of peptide molecule is needed for its functioning. It is only the active part that is important. For example, out of 17 amino-acid residues of the hormone gastrin only 4 residues reside in the active site located at the C-terminus of the molecule. Moreover, it is also possible to modify the active-site structure without changing its functionality. Convincing models of peptide ligands have been synthesized.

In Crick's classical experiments, which have proved that the genetic code is a triplet and has no commas, the mutant forms of bacteriophage T4 reverted to wild type due to the compensation of a single base-pair deletion by a second mutation: the addition of a nucleotide at another site of the genome. With three single base-pair additions or three single base-pair deletions the wild type is also retained [5.1,60]. This means that the mutated proteins of the phage retain the properties of the original protein. Such frameshift mutations are also observed in the gene of bacteriophage T4, which is responsible for the synthesis of lysozyme. It turned out that the frameshifts really distort the protein text. The double mutant (revertant) of lysozyme differs from the original wild type by the sequences of 5 residues:

Wild type ···Thr—Lys—Ser—Pro—Ser—Leu—Asp—Ala···
Revertant ···Thr—Lys—Val—His—His—Leu—Met—Ala···

The properties of the wild type and revertant coincide. Two important propositions, which are rather general, follow from both theoretical models and experimental data.

The first one is the degenerate correlation between the primary and the three-dimensional structure of a protein. Different primary structures correspond to the same spatial structure. Hence, the correspondence between the primary structure and the biological function is also degenerate [5.61].

Secondly, the protein, which is justifiably treated as an intricate machine that performs a certain function, differs significantly from a machine made by man. In a protein, a cell and an organism there are a number of "clearances" and "tolerances", which alter many things in the design of the machine without affecting its operation. The presence of clearances and tolerances is a prerequisite for variability, for evolution at the molecular level. However, in artificial machines the trend is towards attaining minimal tolerances, and thus very precise fixation of the parts. The possibility of realization of such clearances and tolerances is determined by the fact that proteins and nucleic acids, cells and organisms are chemical, molecular machines. These machines are built up of molecules; signal sources, signals, signal transducers (enzymes) and signal receptors are molecules and supramolecular systems. Conversely, in artificial machines signalling is mechanical, electrical or magnetic.

At the first stage of development of the molecular theory of evolution it seemed that the correspondence between the primary and the spatial structure of protein is unambiguous, not degenerate. An argument in favour of this point of view was the discovery of protein renaturation, the recovery of the structure and properties of native protein from the denatured state. This argument was later shown to be insufficient. As early as 1959 Anfinsen pointed out that "... proteins can nevertheless be changed without loss of their function" [5.62].

In most cases the populations of organisms are characterized by the polymorphism of functional proteins, by the presence of so-called isoenzymes. These are proteins whose functions differ little or not at all, but the primary structures are different. Such polymorphism directly follows from what has been said above. The genetic factors that determine the polymorphism of proteins are considered by Kimura [5.63] and Altukhov [5.64]. We shall return to the genome in Chap. 6.

Associated with the above-described structural features of proteins is the solution of the problem that was put forward long ago: the prediction of the three-dimensional structure of protein on the basis of the known primary struc-

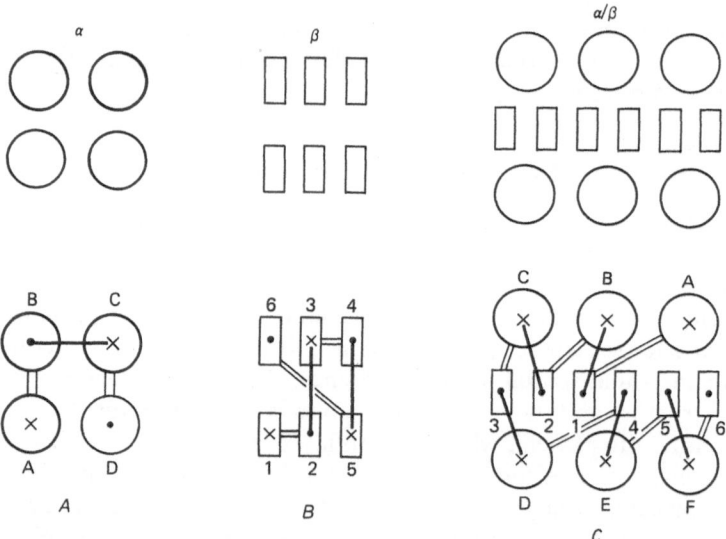

Fig. 5.19. Examples of thermodynamically preferred folds of alpha and beta forms. (*A*) A bundle of helices directed upwards and downwards. (*B*) A double antiparallel β-sheet with the topology of the "Greek key". (*C*) Rossman folding from [5.65]

ture. The finding of the corresponding algorithm would considerably facilitate
further development of the physics of proteins since the primary structure of
protein is much easier to determine than its spatial structure. A number of
works have been devoted to the solution of this problem, but here we will
confine ourselves to brief information about the works of Ptitsyn and his
colleagues.

The prediction of the three-dimensional structure is made in three stages:
(1) the prediction of the secondary structure on the basis of the primary structure; (2) the prediction of the manner in which secondary structures are folded
into a tertiary structure; (3) the prediction of the final folding result. Of course,
there is feedback between these stages but such a scheme does work to a first
approximation. The point is that there is only a limited set of thermodynamically favourable foldings of α- and β-portions into a compact globule, and this
set is independent of the details of the amino-acid sequence. Figure 5.19 shows
examples of such packing and the corresponding spatial structures for alpha,
beta and (α/β) proteins. Such an approach has made it possible to construct the
required algorithm and to attain in many cases a rather good agreement between the theoretically predicted structure and the structure found by X-ray
diffraction analysis. Figure 5.20 presents the results obtained for erythrocruorin
[5.10]. Mention should also be mode of the works of a number of other authors
[5.42,43,65]. This ideology may be applied in dealing with globule-folding
kinetics.

Scheraga and his coworkers have carried out detailed calculations of the
conformational energy of proteins based on the potential energy of the globule
[5.11,66–72]. Account is taken of the finite number of restrictions imposed on

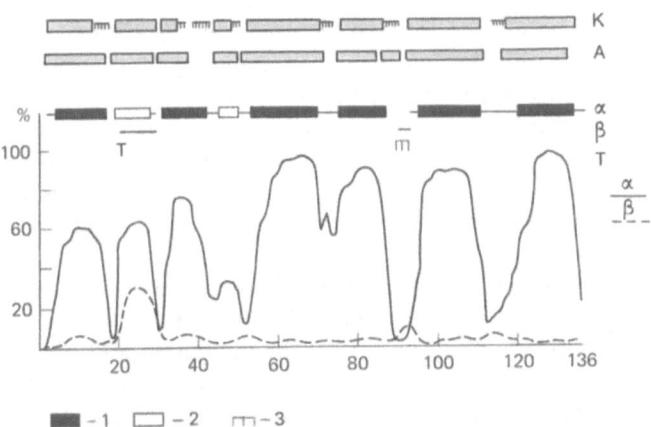

Fig. 5.20. Prediction of the secondary structure of erythrocruorin. The abscissa represents the numbers of residues in the sequence and the ordinate the probabilities of prediction of alpha-
(——) and beta-forms (------). *1* = reliably predicted forms; *2* = predicted forms; *3* = rotations.
Shown at the top is the localization of alpha- and beta-forms and rotations according to X-ray
diffraction data. From [5.65]

possible topological contacts, such as the covalent structure and disulphide bonds and also of statistical information about the known protein structures, the positions of the alpha and beta-forms, etc. Finding the global minimum by means of computer calculations leads to good agreement with experiment. The methodology of calculations has been described in detail by Dickerson and Geis [5.73]. It proves to be also applicable to enzyme-substrate complexes.

The works mentioned above are important; in a number of cases the authors have solved the problem of finding the three-dimensional structure of proteins on the basis of the knowledge of the amino-acid sequence. It has been shown that the observed structure of protein is an equilibrium one in the sense that it corresponds to the global minimum of potential energy. This is so at least in the cases studied.

It has also been found that the folding of the protein chain starts with the formation of thermodynamically stable local structures. Short-range interactions are found to be sufficient for the determination of local structures in the independent portions of the polypeptide chain.

5.5 Evolution of Proteins

Comparison of the amino-acid composition with the primary structures of homologous proteins shows that the more closely related the biological species, the smaller the differences between the corresponding proteins. This means that it is possible to construct an evolutionary tree on the basis of a comparison of homologous proteins. Figure 5.21 shows such a tree built on the basis of the primary structures of cytochrome c [5.73]. The primates differ from other mammals in 8–12 residues out of 104. The differences within the class are of the order of 5 residues. Mammals and birds differ, on an average, by 10 residues, all land vertebrates differ from fish (tuna), on an average, by 18 residues, all vertebrates differ from insects by 26 residues, and all animals differ from plants and microorganisms by 47 residues. Such trees are also constructed on the basis of data for a number of other proteins. The method of constructing the most reliable tree is an interesting problem, although we will not take it up here. The basic position has been established exactly: the more closely related the species, the closer the proteins.

Homologous proteins of the same kind are derived from one another in the course of evolution by way of point substitutions which are determined, of course, by the corresponding mutations in the genome DNA. The biological evolution of proteins, however, does not reduce to this (see the chapters that follow). The changes of amino-acid sequences may be very significant; so serious, for example, that the members of one protein family may contain the same residues in less than 20% of the chain links. Of course, such radical evolutionary changes modify to a certain extent the properties of proteins, but in such a case the active part of protein may be retained. The dependence of the accessible protein surface on molecular mass for 4 series of proteins, whose homology is

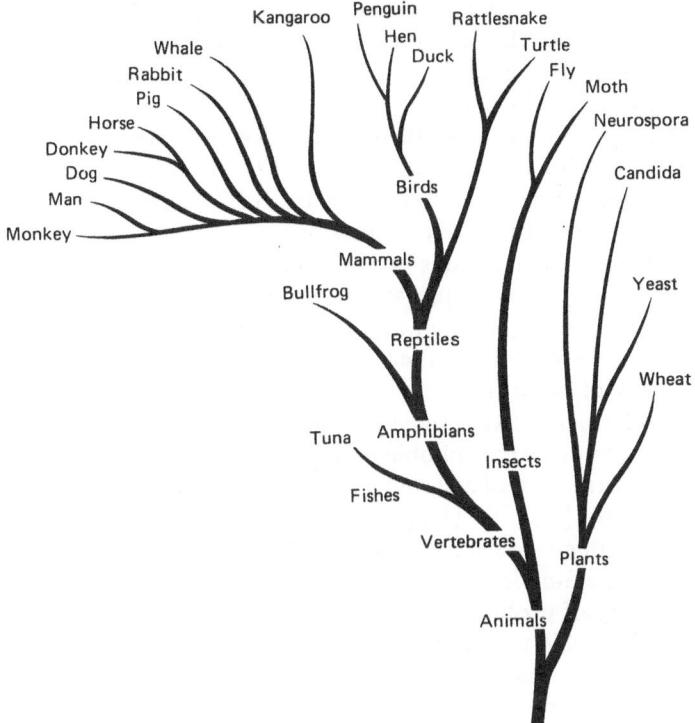

Fig. 5.21. The evolutionary tree constructed on the basis of data on cytochrome c (Dickerson [5.58])

far from obvious, is described by a simple expression. Lysozymes of phage T4 and of goose and hen eggs, cytochromes c of tuna and bacteria, protease B from *Streptomyces griseus* and bovine chymotrypsinogen, and plastocyanin from plants and bacterial azurin are all characterized by the dependence $A_s = 6.3 \cdot M^{0.73}$ [5.74,75]. The presence of a simple relationship indicates that the central core (the active part) of the globule remains intact in all these proteins. The active site geometry is retained. For example, in proteins that are involved in electron transport and in plastocyanin and azurin there are residues linked to the copper ion. Two of the four residues, which appear to be ligands of the Cu atom, Cys and Met, are located on the adjoining strands of the β-sheet. The third ligand, His, is in the loop of the linking strand. The amino-acid sequences in this region are as follows:

Plastocyanin	Cys—Ser—Pro	His-Gln-Gly-Ala-Gly-Met
Azurin	Cys—Thr—Phe—Pro—Gly—His	Ser-Ala-Leu-Met

A combination of insertions and deletions with changes in the loop conformation preserve the position of the ligand His. The active site geometry is retained, despite the considerable differences between the proteins. As Chothia wrote, 'the necessity of retention of the function makes evolutionary substitutions in proteins interdependent: possible changes are limited to their overall influence on the active sites geometry' [5.76].

Perutz has carried out a detailed analysis of the evolutionary features of amino-acid substitutions in haemoglobins [5.77]. In different species of vertebrates the allosteric properties of Hb differ (see [5.2]). Human hemoglobin differs from carp Hb by 140 residues. Most of the substitutions are functionally neutral, and the spatial structures of Hb remain practically unchanged during evolution, despite numerous replacements. Species adaptations, which lead to responses to new chemical stimuli are determined by only a few (from one to five) amino-acid substitutions at the key positions.

The majority of point mutations of codons, leading to the replacement of amino-acid residues, preserve the class of amino acids. The residues are replaced by similar ones and the hydrophobic nature of the residue (its most important property for the formation of a globule) changes little. This was demonstrated in 1966 [5.78], see also [5.1,72]. The genetic code, whose structure is the result of evolution, has in this sense high noise stability. Creighton and Darby [5.80] have introduced the functional divergence ratio (FDR) as a measure of changes in the functionality of protein as a result of mutations. FDR is the ratio of the frequency of amino-acid replacements in the active site of protein to that in the passive part. If the two regions are equally variable, the FDR is equal to unity. As a rule, FDR < 1, but the authors just cited have studied evolutionary hypervariability in a number of protease inhibitors. These proteins become specifically bound to the active sites of certain proteolytic enzymes, hindering their action. A number of families of such inhibitors have been established, each of which has a single progenitor. The spatial structures of inhibitors of different families differ, but the domains of inhibition are similar and are presumed to have arisen by way of convergent evolution. These domains are complementary to the active sites of the proteases inhibited [5.81].

It has been found that the values of FDR for a number of inhibitors are larger than unity; the probability of substitutions in the active site may exceed that in the passive part up to FDR $= 5.60$. The observed hypervariability of the functional domains of inhibitors may be accounted for by positive natural selection, by the co-evolution of inhibitors and proteases [5.80,82]. The proteases themselves are characterized by values of FDR ranging from 0.32 (cathepsin H) to 1.34 (renal callicreins). These findings are of special interest, but at present we still know little about the interactions of inhibitors with proteases *in vivo*.

The molecular evolution of proteins, which is realized by way of point substitutions, is gradual; the amino-acid residues are replaced one after another. Perutz [5.77] has carried out a stereochemical analysis of such substitutions in haemoglobin. During the evolution of vertebrates, the tertiary and quaternary structures of Hb remained practically unchanged and, as has been said above, only a few point substitutions are of adaptive value. One of a number of examples considered by Perutz is the comparison of the sequences in the Hb of fishes and mammals. Many species of fishes utilize Hb not only for breathing but also for the secretion of oxygen into the swimming bladder and the eye. This proves possible because the oxygen dissociation constants for the α- and β-sub-units of Hb in such fishes differ by two orders of magnitude at acidic pH. As a result, one pair of sub-units secrete O_2 into the swimming bladder and the eye even at hydrostatic pressure of a few hundered atmospheres. The primary structures of the α- and β-chains of Hb, say, in carp and in man differ by 140 residues. However, Perutz has shown that the indicated difference in function is provided by a single replacement of Cys-9 in the F of the α-helix of the β-chain of mammalian Hb by Ser in fish. This replacement gives rise to two extra hydrogen bonds which stabilize the "strained" structure of deoxyhaemoglobin and which lower the binding constant for O_2 in the β-chains. Moreover, another replacement, that of His by Lys, in the first position after the H-helix at the C-end of the β-chain, eliminates one hydrogen bond in oxyhaemoglobin, shifting the equilibrium further to the deoxy-form. The replacements of all the other residues (~ 140) presumably do not participate in the differentiation of the functions of the Hb of fishes and mammals and may be neutral.

We have so far discussed adaptive or randomly fixed mutations in proteins in connection with the evolution of existing proteins, neglecting their origin. Ptitsyn and Volkenstein [5.83] have shown that the "accidental choice remembered" (see Chap. 10) characterizes not only most mutations in existing proteins but also the origin of proteins. The hypothesis that the amino-acid sequences of many known globular proteins are just "memorized" random sequences, and that biological evolution reduces only to their "editing", has serious backing.

Indeed, the complete selection of a particular sequence is practically impossible during evolution since this sequence must be chosen out of 20^N possibilities, where 20 is the number of common amino-acids and N is of the order of 100 (the number of residues in the protein chain). Therefore, the time of existence of the Universe is insufficient for the evolutionary selection of all the residues in a single protein chain. The "accidental choice remembered" in proteins is evidently a prerequisite for the origin and existence of life.

As has been described above, the physical requirements for the formation of a functional globular protein are such that they do not include a strategic selection of sequences but are observed for random sequences. We have already seen that random sequences of polar and non-polar residues produce alpha and beta-domains, which can form three-dimensional structures similar to those in real proteins. Even a tight packing of side groups in the core of the globule does not require such an exacting selection of sequences. Biological selection is necessary mainly for the generation of active sites in proteins and for their stability in physiological conditions.

Of considerable interest is the question put in an interesting review by Doolittle [5.84]: Do "living fossils" contain old proteins? Living fossils are relict organisms that have survived in our days e.g. the Crossopterygian fish *Latimeria* or the gingko tree. This tree has not changed morphologically over 150 million years. The lamprey has also changed little – it is the most primitive species of all the existing vertebrates. The sequences of cytochrome c in these living fossils, like the cytochrome c of insects that appeared on Earth long before the flowering plants and higher vertebrates, have changed relatively little in the course of evolution. For example, the cytochrome c of the lamprey has undergone no more than one third of the changes experienced by mammals after their divergence, and the gingko tree has undergone less than half of the changes experienced by flowering plants.

Of course, the protein sequences observed today in living fossils carry the memory of numerous mutations that have occurred since the first appearance of these species in the palaeontological record. However, these sequences have changed to a considerably lesser extent than those in the more recent species. The retention of morphology correlates to a certain extent with the preservation of proteins.

An important inference from a consideration of the evolution of proteins is that there are only a relatively small number of fundamental structures that have given rise to most of the existing proteins [5.84]. The basis of this evolutionary limitation may be explained as follows. Once an amino-acid sequence appears and is capable of performing a certain basic function (say one of adding a nucleotide or a metal ion to another polypeptide chain) this implies the existence of mechanisms for the duplication of this sequence and for its utilization in other situations. It must also be noted that certain primary structures survive during evolution due to their inherent resistance to denaturation or proteolysis.

The evolution of proteins confirms on the whole the idea of Jacob about its similarity with the combination and reshaping of the existing structures and with their tinkering [5.85,86]. At all organismic levels, beginning with the molecular level, the memory of the preceding events is retained. We may say that this is the engineering principle of life. A similarity within a given taxon points to the fulfillment of this principle. All land vertebrates have four limbs modified to form paws, legs, arms, wings, flippers. A similar principle works in ordinary engineering: the first automobiles looked like carriages, the internal combustion engine incorporates a cylinder with a piston, just like a steam engine.

Let us return to theory. Shakhnovich and Gutin [5.52] have proposed a plausible explanation for the insensitivity of the spatial structure of protein to point substitutions. Let us consider, as an example, the phylogenetic tree of globins. Suppose that a point mutation occurs in the first branching of the tree. The mutation is lethal with a probability $1 - \gamma^{-8}$. This is followed by other mutational attempts which occur until a neutral mutation is encountered. The average number of such attempts is γ^8. The mutated structure is fixed. An analogous process takes place in the next branching of the tree, etc. As a result, from a single root arise Z species. Each of them differs from the original one by Z substitutions and the total number of attempts to obtain neutral mutations is $Z\gamma^8$, which is many times less than the number γ^{8Z} of independent neutral mutations. The authors emphasize that such a scenario of the neutral evolution of primary structures is possible due to an enormous degeneracy of the "physical code", i.e. the ratio of the possible number of physically differing primary structures to the number of conformations γ^N.

5.6 Protein Engineering

As has been pointed out above, the protein globule is a highly ordered, amorphous, compact solid entity, a kind of construction which performs a certain function, primarily enzymatic. The "protein engineering" of living nature initially involves a random combination of amino-acid residues, followed by "editing" of these combinations due to natural selection. In biology the accepted practice is to answer the "what-for" question. With respect to proteins, it may be said that their structures were built to perform a certain function. As a matter of fact, we have to answer the question "why" rather than "what for". A particular protein construction is chosen not because it is optimal for performing a particular function but because it is capable of fulfilling it. The polymorphism of proteins demonstrates once again that a given function can be performed by a set of proteins differing in primary structure. What we are speaking about here is ambiguous natural selection.

Due to big advances of genetic engineering it has become possible to create artificial proteins and to modify, through appropriate substitutions, the existing ones. The directed changes of proteins may be compared to the artificial selection of organisms, to the creation of new species (new species have not been produced so far). However, there is also a significant difference here.

Darwin paid special attention to artificial selection; he devoted one of his most important works to this subject. Artificial selection has provided a wealth of information for constructing the theory of natural selection, evolutionary theory. In protein engineering a reverse process occurs: the study of the natural selection of proteins and their structure and dynamics. This is a source of information required for artificial creation of new proteins.

Genetic engineering, i.e. the isolation and cloning of genes, and their incorporation into appropriate genomes, has become an area of technology rather than of science. It means important practical applications of molecular biology that do not require datailed theoretical investigations. The role of physics in genetic engineering is subsidiary. On the other hand, protein engineering is directly tied up with theoretical and experimental physics. The intricate protein constructions created by man for certain purposes require intensive work in the field of molecular biophysics. In the long run, such efforts will also be needed for further development of genetic engineering. The union of physics and biology is a necessity.

Here we shall restrict ourselves to a few examples illustrating the advances of protein engineering, an area which is rapidly and widely developing at present.

The task that was undertaken in the works of De Grado and his colleagues

[5.87–90], consisted of synthesizing multiple copies of secondary structures (alpha-helixes), which should fit in a certain way into the predetermined tertiary structure. An artificial protein was produced, which consisted of four α-helices. Such a protein is an idealized version of the class of natural proteins, including myohemerythrin, apoferritin, the protein of tobacco mosaic virus and cytochrome c. These proteins are composed of four helices linked by loops. The helices are packed in an anti-parallel fashion, forming an angle of $\sim 20°$ relative to one another. For such a structure to be produced, the artificial helices must be amphiphilic, i.e. they must have a non-polar, hydrophobic side contacting the neighbouring helices and a polar, hydrophilic side in contact with water. The packing is stabilized by hydrophobic interactions. Two amphiphilic polypeptides have been synthesized, each containing 16 residues, namely

$\alpha_1 A$ Ac Gly Lys Leu Glu Glu Leu Leu Lys Lys Leu Leu Glu Glu Leu Lys Gly—COOH ,

$\alpha_1 B$ Ac Gly Glu Leu Glu Glu Leu Leu Lys Lys Leu Lys Glu Leu Leu Lys Gly—CONH$_2$.

Each species of chain is tetramerized in solution, the energy of such a process being 20 kcal/mole. In the second stage of this work, short oligopeptides were introduced between the two $\alpha_1 B$ chains to produce a covalent cross-link between two helices in a four-helix bundle. $\alpha_1 B$-Pro-$\alpha_1 B$ and $\alpha_1 B$-ProArgArg-$\alpha_1 B$ chains were synthesized, and their ability to form dimeric aggregates, i.e., four-

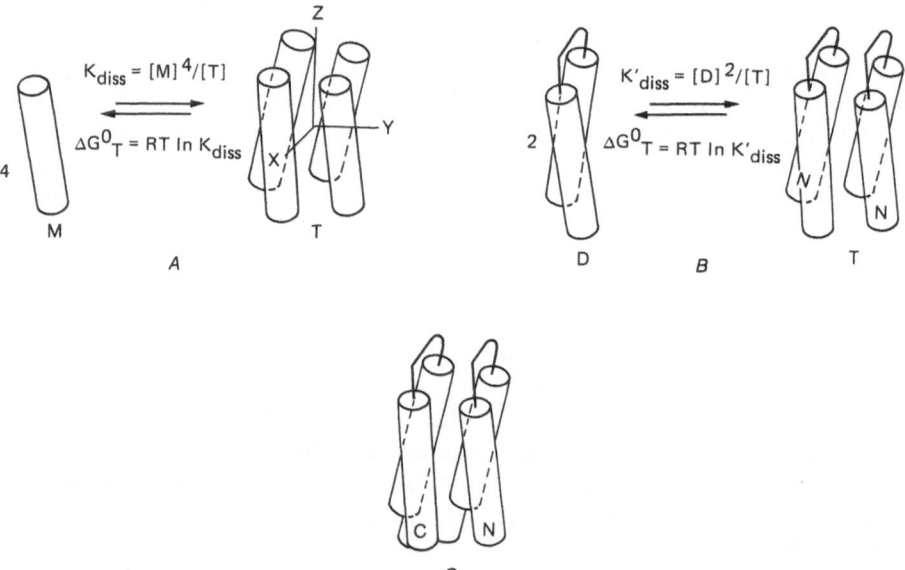

Fig. 5.22. The stages of formation of a four-helix protein. (*A*) Formation of tetrameric aggregates in solution. (*B*) Their union through loops. (*C*) Construction of the final four-helix bundle out of four optimized helices and three loops [5.88]

helix bundles, was explored. The peptide α_1B-Pro-α_1B forms trimers rather than dimers. The incorporation of extra charged Arg residues into a loop allows one to attain the desired result: the peptides α_1B-ProArgArg-α_1B form helical dimers in solution.

The scheme of these processes is shown in Fig. 5.22.

The synthesis of such proteins, which attained molecular masses of 8500–8700 Da, was carried out by the methods of protein engineering. A gene coding for the desired protein was synthesized; this was formed in cells of *Escherichia coli*, isolated and purified to homogeneity.

A special position is occupied by membrane proteins, in particular proteins forming ionic channels in membranes (channels through which small ions move to and from the cell (see [5.1])). In contrast to globular proteins, which function in an aqueous environment in cytoplasm, membrane proteins have non-polar residues on the external surface in contact with lipids. It has been found that such proteins are built up in the form of α-helices associated in such a manner that an ion-conducting channel is formed. Lear and his co-workers [5.91] have synthesized three model polypeptides containing only leucine and serine. A polyeptide composed of 21 residues with the composition H_2N—(Leu Ser Ser Leu Leu Ser Leu)$_3$—$CONH_2$ formed good ionic channels in membranes, which exhibited ionic permeability and a lifetime similar to those for an acetylcholine receptor. However, a polypeptide containing 14 residues appeared to be too short to penetrate the phospholipid bilayer in the form of an α-helix and did not form discrete, stable channels. The third polypeptide, H_2N—(Leu Ser Leu Leu Leu Ser Leu)$_5$—$CONH_2$, in which one Ser residue is replaced by Leu, produced proton-conducting channels. Energy minimization and computer graphics were used to construct molecular models compatible with the observed properties of the channels. The corresponding structure is shown in Fig. 5.23, where α-helices

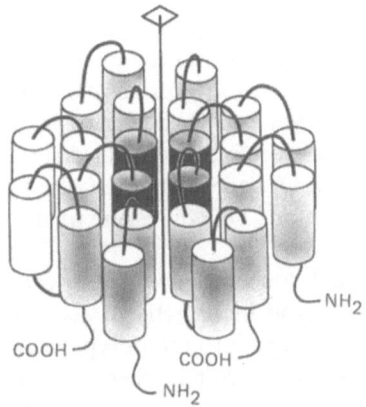

Fig. 5.23. Schematic representation of the structure of proteins forming ionic channels. See text for explanations [5.89]

are represented by cylinders, the four hatched cylinders in the center form an ionic channel, and B and C represent the projections of the α-helices [B = (LSSLLSL)$_2$ and (LSSLLSL)$_3$, C = (LSLLLSL)$_3$].

A striking example of the possibilities of protein engineering is the work of Matsumara and Matthews devoted to lysozyme [5.92]. Two residues were incorporated into the lysozyme of bacteriophage T4 (Thr-21 → Cys and Thr-142 → Cys) by way of mutagenesis directed by the corresponding oligo-nucleotide. These cysteines formed a disulphide linkage under oxidative condi-tions *in vitro* and the cross-link Cys—S—S—Cys was formed, which led to a complete loss of catalytic activity. However, under the action of a reducing agent the disulphide linkage was ruptured and the reduced non-crosslinked lysozyme regained its complete enzyme activity. Thus, an enzyme was artifi-cially built whose activity was controlled by a redox potential.

The works in the field of protein engineering provide a direct answer to the question of the relationship between the primary and the spatial structure of protein and confirm the concepts described above. It has been established that the proteins in which less than 10% of the same amino-acid residues is retained in the comparable positions may have very similar spatial structures. This de-generacy of the sequence-structure "code" implies that the forces that form the protein structure are thought to be dominant in the sense that they remain unchanged with strong variations of the amino-acid sequence [5.90,92].

Of great importance to protein engineering are theoretical investigations, as has already been pointed out. Warshell and co-workers [5.93] have described a method of estimating the catalytic free energies of genetically modified pro-teins. Use was made of the methods of quantum chemistry, a combination of the empirical valence-bond method and free-energy perturbation technique. It proves possible to semiquantitatively describe the observed effects of substitu-tions on the activity and free energy of binding of enzymes, such as trypsin and subtilisin. Analysis of the calculations performed points to an important role of electrostatic interactions. Of importance are the changes in the charges on the reacting systems and the corresponding changes in the energy of "solvation", i.e. the interactions between the charges and the microenvironment. Further devel-opment of this theory calls for more rigorous methods of quantum mechanics. It has, however, been shown that enzyme activity is associated with the fact that the protein provides a certain orientation of electric dipoles, which leads to a decrease in activation energy. Of course, such an orientation arises as a result of a conformational rearrangement. This takes the form of electron-conformation interactions.

A number of theoretical concepts have been tested in order to reveal the differences between proteins and incorrectly folded models that were obtained by introducing wrong side-groups into the polypeptide chain [5.94].

The importance of electrostatics has been directly proven experimentally [5.95]. The change in the charges on the surface of subtilisin brought about by directed mutagenesis produced enzymes with an altered dependence of the ac-tivity on the value of pH, increased activity and changed specificity. It is necessary to take into account the dielectric properties of water and the effect of small ions on catalysis.

5.7 The Role of Metals

The functioning and, hence, evolution of proteins and also of nucleic acids is inextricably tied up with metal ions. The ions of alkali metals (Na^+, K^+) and of alkaline-earth metals (Ca^{2+}, Mg^{2+}) are directly involved in the behaviour of (a) membrane proteins responsible for the transmission of the nerve impulse, and (b) of fibrillar proteins, which govern mechanochemical processes, beginning with the dynamics of the cytoskeleton and ending with muscle contraction. Such ions are always present in proteins, being bound by electronegative amino-acid residues. As has been said above, the electric conductivity of proteins is determined exactly by ionic impurities. Proteins function at a certain ionic strength.

Approximately one-third of all enzymes contain Zn ions and ions of transition metals: Fe, Cu, Mn, Mo, Co. Below are some data on the amounts by weight of metals per 70 kg of the human weight (in grams):

Ca	...1050	Zn	...1.9
K	...245	Cu	...0.15
Na	...105	Mn	...0.020
Mg	...35	Mo	...0.015
Fe	...4.24	Co	...0.003

Other transition elements (Cr, Se) are also present, though in much lesser quantities.

The general relationship regarding all complexes of biopolymers with metals is as follows: the metals are present in them as ions which form ionic and co-ordination bonds. Evidently, this relationship is accounted for by prebiotic evolution in an aqueous medium containing metals in dissolved ionized form. For the same reason, no organometallic compounds and no covelent C—Me bonds were formed in living nature. Organometallic compounds do not dissolve in water but are decomposed by it. The only exception is alkylcobalamines (derivatives of vitamin B_{12}) containing C—Co bonds.

An important role in nature is played by molybdenum, which is present as a co-factor in a number of oxidases, dehydrogenases, hydrolases and especially in nitrogenases. Nitrogenases participate in the fixation of atmospheric nitrogen by bacteria belonging to the genus *Rhizobium*, which are in symbiosis with legumes. At the same time, molybdenum is a rare element present in the earth's crust in minor quantities. On this basis Crick once put forward a half-serious hypothesis that molybdenum was brought to the earth for biological purposes by beings from other planets. This hypothesis is refuted by the fact of the pres-

ence of large amounts of molybdenum in sea water, which is comparable in
order of magnitude with the content of iron and zinc [5.96]. the important role
of molybdenum is one more argument in favour of the origination of life in sea
water.

The area of investigations of the role of metals in biology is the concern
of bioinorganic chemistry. The history of the relationships between chemistry
and biology is specific and very interesting. The modern concept of organic
chemistry emerged in connection with the production and study of substances
of plant and animal origin; the word "organic" is derived from the word "organism". Beginning with the work of Friedrich Wöhler, who synthesized urea
$(NH_2)_2CO$ from inorganic substances, organic chemistry divorced from biology
and transformed into the synthetic chemistry of carbon compounds, the number
of which is many times greater than that of all other elements. This was paral-
leled by the development of biochemistry. Over the last several decades the
efforts of organic chemists have again been directed towards natural and physiologically active substances and an intermediate area has developed, which is
called bio-organic chemistry.

Inorganic chemistry developed in the 19th century quite independently.
At that time the attention of the majority of chemists was focused on organic
chemistry; the theory of the chemical structure of organic compounds was inten-
sively worked out in the last century. We use this theory even today when we
present the structural formulae of amino-acids or nucleotides. The fact that
Mendeleev was denied admittance to the Imperial Academy of Sciences was
partly accounted for by the lack of attention to inorganic chemistry and by
non-understanding of its importance. Instead of Mendeleev, the other candi-
date, F.K. Beilstein, was elected, who could not be compared to his rival.

Theoretical inorganic chemistry is an area which is more complicated than
theoretical organic chemistry. Today the lion's share of inorganic chemistry
goes to the chemistry of co-ordination, complex compounds of transition
metals. A quantum-mechanical theory of such compounds, the ligand-field
theory, was worked out at a much later time than the quantum chemistry of
carbon compounds.

Zinc and transition metals, being present in the active site of proteins, form
co-ordination bonds with the amino-acid residues, which play the role of
ligands. A metal ion is charged positively; it is electrophilic. It may act as an
acid, according to Lewis, binding and activating the substrates. Transition
metals may exist in many states of oxidation. This allows such metals to take
part in electron transfers in oxidation-reduction processes. Such behaviour is
typical of the iron ion present in the prosthetic haem group in cytochromes.
The metal ion usually combines with four or more ligands. Binding many side
groups of protein, metals act as factors that organize cross-links in the active
site. The transition-metal ions function together with their ligands (amino-acid
residues) and the "symbiosis" of the functions of the metal and protein is realized [5.97]. The metal forms the general structure of the active site and its
conformational dynamics.

The active site of protein, which includes a metal ion or a prosthetic group, is similar to a co-ordination metal compound. The protein is a polydentate ligand, the co-ordination valences of the metal being occupied by amino-acid residues. The polydentate nature, the co-ordination of these ligands, determines the distinction of a protein co-ordination compound from the corresponding low-molecular analogs. In particular, sharp differences in absorption spectra arise. Vallee and Williams have proposed the concept of the stressed, so-called entatic state of the active site of a metalloenzyme [5.98,99].

As was said above, the metal ion acts as an acid according to Lewis. In the chemical reaction

$$HA + B \rightleftharpoons A^- + HB^+ \ .$$

HA and HB^+ are acids and B and A^- are bases. The Brönsted equations relate the rate constant of the reaction catalyzed by the acid, k_{HA}, to an ionization constant for the acid, K_{HA}:

$$k_{HA} = C_A(K_{HA})^\alpha \ , \quad 0 < \alpha < 1 \ .$$

For a reaction catalyzed by a base a similar relation is valid. Hence, $\log k$ is proportional to $\log K$. In low-molecular weight compounds the catalytic activity of a metal increases in the following sequence:

$$Mg < Ca < Mn < Fe < Co < Ni < Cu < Zn < Cd < Hg \ .$$

In metalloenzymes this sequence is upset. Anomalies in catalytic activity are determined not by differences in the binding of the metal but by its special state in the active site. For example, the addition of a halogen to Zn^{2+} in carboxypeptidase occurs in the following sequence:

$$I^- > Br^- > Cl^- > F^-$$

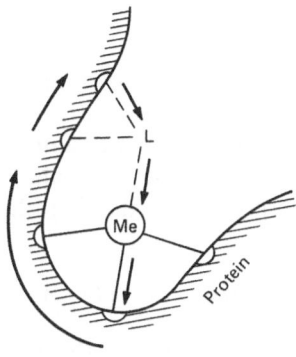

Fig. 5.24. Diagram explaining the role of the trans-effect in catalysis by means of an enzyme with a co-factor, a transition-metal ion

which is the reverse of the sequence for a compound with the free ion Zn^{2+} [5.99]. Vallee and Williams wrote: "Enzymes are catalysts not because they contain unusual chemical groups but rather because the specific three-dimensional structure of such proteins produces an environment which endows unusual properties on the groups" [5.98].

The transition-metal ions contain an unfilled d-electronic shell, which is "softer" than the s- and p-shells of first-period elements. This means that for the electronic state of a transition metal to be changed, perturbed, a relatively low energy is required, which is comparable to the energy of conformational changes. The electron-conformation interactions occurring in metalloenzymes are especially strongly pronounced. In co-ordination compounds a trans-effect takes place, which was discovered by Chernyaev in square complexes of platinum of the following type (X_i are ligands):

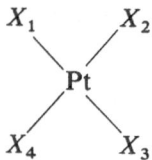

When one of the ligands, say X_1, is replaced by another, Y, the ligand X_3, which is in the trans-position with respect to X_1, assumes increased mobility. The stress of the d-shell of the metal leads to a change in the conformation of the polydentate ligand, the set of amino-acid residues in the active site co-ordinated with the metal. If a low-molecular ligand, i.e. a substrate, is co-ordinated with the metal, then its state is directly influenced by these conformational events. Such a situation is shown in Fig. 5.24. Through the trans-effect the substrate, i.e. an additional extra metal ligand L, acts on the metal-protein bonds, which brings about a change in the bonds between L and protein due to electron-conformation interactions. The electronic state of the substrate L is changed and the enzymatic reaction is facilitated. Of course, this is only a qualitative scheme; the quantum chemistry as applied to such complex systems has not yet been worked out.

The role of metals in proteins is varied. According to Berg [5.100], this role may be systematized by a scheme shown in Fig. 5.25. The metalloproteins are divided here into four classes; examples are shown in the scheme. Oxidoreductases and oxygen carriers function in such a manner that both the co-ordination sphere and the oxidation state undergo change. In cytochrome P450 the iron ion in the haem undergoes the following changes in the catalytic cycle [5.101]:

$$Fe^{3+} \to Fe^{2+} \to (Fe^{2+}[O_2] \leftrightarrow Fe^{3+}[O_2^-]) \to (Fe{-}O)^{3+} \to Fe^{3+} \ .$$

In a number of Zn-containing peptidases the ion Zn^{2+} does not change its valent state during functioning, but the co-ordination sphere is changed. On the other hand, in cytochrome c the iron ion changes its state:

Coordination sphere

	Static	Dynamic
	Binding protein	Hydrolases
Static	Aspartate transcarbamylase, regulatory subunit (Zn)	Carboxypeptidase A (Zn)
	Calmodulin (Ca)	Urease (Ni)
Dynamic	Energy transfer agents	Oxidoreductases and oxygen carriers
	Cytochrome c (Fe) Plastocyanin (Cu)	Cytochrome P450 Hemocyanin (Cu)

Oxidation state

Fig. 5.25. Classfication of the functions of metal ions in proteins [5.100]

$$Fe^{3+} + e^- \rightleftharpoons Fe^{2+}$$

but the structure of the co-ordination sphere and the conformation of the protein remain practically unchanged. X-Ray diffraction studies have shown that the conformations of the oxidized and reduced forms of cytochrome c differ very little.

Finally, in enzymes such as transcarbamylase aspartate, both the co-ordination sphere and the oxidation state are retained.

Of considerable interest are proteins that bind to nucleic acids. Metal-binding domains have been discovered, in particular in the protein transcription factor IIIA from the frog *Xenopus*. The protein consists of 9 tandem repetitions of the sequences containing about 30 residues. Each repeat has a tetrahedrally bonded ion Zn^{2+} attached to two Cys residues and two His residues; a structural domain is formed, which has come to be called "the zinc finger". Figure 5.26 presents the expected structure of this domain [5.100].

It can be shown that metal ions stabilize small domains, which do not fold up in the absence of such ions. This stabilization is based on two entropy contributions: the entropy of water molecules liberated upon transfer of the hydrated metal ion to protein, and the decreased conformational entropy of the polypeptide with the metal ion, which is linked to four side groups. The corresponding scheme is given in Fig. 5.27 [5.100]. To this may be added the consid-

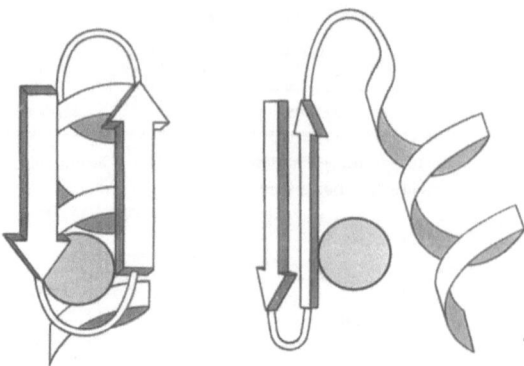

Fig. 5.26. Two types of expected structure for metal-binding domains in proteins ("zinc fingers"). The structure is $\beta\beta\alpha$ with a Zn^{2+} ion co-ordinated tetrahedrically by means of two Cys and two His residues [5.100]

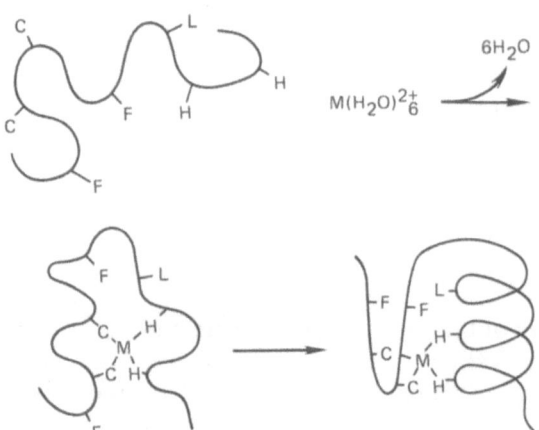

Fig. 5.27. The hypothetical path of folding of a peptide with a metal ion [5.100]

erable energy of electrostatic interactions induced by the ion. Structures of the "zinc finger" type play an important role in the regulation of gene action.

Evolution has led to the specialization of metals in living organisms. There are striking examples of a very narrow choice. For example, the vanadium ion is present only in Tunicata. Superoxide dimutase contains ions of various metals as co-factors in various organisms. Eukaryotes contain copper or zinc; pro-karyotes, iron or manganese; and mitochondria, manganese [5.102]. As we have seen, the important role of a metal in the determination of the spatial structure and dynamics of protein has been established. A metal is a powerful factor in the evolutionary "editing" of the protein structure, the function of the protein being determined by a few residues interacting with metals. We have already mentioned the evolutionary intactness of residues that interact with the haem group in haemoglobin. In plastocyanin and azurin, copper-containing proteins, the amino-acid residues that surround a metal undergo the least change during evolution [5.103].

In well-known experiments carried out by Fox and devoted to the problems of the origin of life, mixtures of 18 amino acids were subjected to thermal polycondensation in the presence of dehydrating substances. This led to the formation of proteinoids, polypeptides with a random arrangement of the chain links. Some of these possessed catalytic activity similar to enzyme activity; they were used to conduct proteolysis, decarboxylation, amination and deamination reactions [5.104]. This fact by itself confirms the main result of the theoretical and experimental investigations described above; in a number of cases the accuracy of the amino-acid text is not too important to the function of protein.

It would be of interest to study the interaction of proteinoids with metal ions and also to carry out the synthesis of proteinoids in the presence of such ions. It is possible that this will lead to the formation of polypeptides with a more specific catalytic activity: polymerase (Zn) or redox (Fe, Cu) activity.

An extensive literature has been devoted to complexes of metals with biopolymers and bioinorganic chemistry (see [5.105–108]). However, the role of metals in biological evolution has been explored insufficiently so far.

6. Neutrality of Point Mutations

6.1 The Neutral Theory

In 1968 the journal *Nature* published a brief article written by the Japanese geneticist Kimura with the title "Evolutionary Rate at the Molecular Level" [6.1]. Kimura laid the foundations of neutral theory, one of the most important theories in modern evolutionary biology.

Kimura compared the rates of substitution of amino-acid residues in homologous proteins (haemoglobin, cytochrome c and triosophosphate dehydrogenase) of various organisms, and came to the conclusion that the rate of substitution in different proteins is about the same. He estimated this rate to be 2.8×10^7 years in a polypeptide chain 100 residues long. In application to the entire genome this means a very high rate of point evolution. Estimation of the time of substitution of one base pair in a genome consisting of 4×10^9 pairs gives about two years. Such a rate is evidence that most of the substitution is almost neutral and is not subjected to the pressure of natural selection. Kimura calculated the corresponding "load of substitution" and showed that it does not lead to any limit for the rate of substitution. The probability of fixation of such mutations is approximately equal to their initial frequency. This leads to an unexpected conclusion, as Kimura writes, that in mammals the neutral (or nearly neutral) mutations occur at an annual rate of 0.5 per gamete. The arguments based on relevant factual data are also given for *Drosophila*.

Kimura and Crow had shown earlier [6.2] that for neutral mutations the probability of appearance of a homozygote individual is equal to $1/(4N_e v + 1)$, where N_e is the effective size of population and v is the rate of mutations per gamete per generation. Hence, the probability for a heterozygote is equal to

$$H_e = \frac{4N_e v}{4N_e v + 1} . \tag{6.1}$$

The derivation of this formula is given on page 190. The effective population size N_e may be much less than its actual value N. The population size N_e is equal to the number of crossbreeding individuals in a single generation. Each zygote is the result of the fusion of male and female gametes. The probability of the origin of two genes contained in different individuals of a generation t from males of a generation $t - 1$ is $1/2 \times 1/2 = 1/4$ and the probability of their origin from a single male is lower by a factor N_m; N_m is the number of males in a population. Thus, this probability is equal to $1/4 N_m$. The probability of the origin of two genes from a single female is $1/4 N_f$. The effective population size is thus equal to

$$N_e = \left(\frac{1}{4N_m} + \frac{1}{4N_f}\right)^{-1} = \frac{4N_m N_f}{N_m + N_f} \ . \tag{6.2}$$

If $N_m = N_f = N/2$, then

$$N_e = N^2/N \equiv N \ .$$

But with $N_m \neq N_f = N - N_m$ we have $N_e < N$. Let us assume that $N_m = 1$ and $N_f = 100$. We then obtain $N_e = 400/101 \cong 4$. According to the estimate made by Kimura for *Drosophila*, $v = 1.5 \times 10^{-5}$ per generation. This estimate is probably valid for neutral mutations. Indeed, if we assume that the frequency of such mutations is of the order of unity per genome per generation, then for a gene (cystron) 3000 nucleotide pairs long we have the value of v given above. According to the formula for heterozygosity $H_e = 0.35$, i.e., the fraction of heterozygous individuals is 35% if $N_e = 9000$. This value of N_e may be too high, but if the migration between the subgroups is taken into account, $H_e = 0.35$ will also be obtained for lower N_e in each subgroup.

Kimura came to the following conclusion: "If neutral or nearly neutral mutations are produced in each generation at a much higher rate than the rate adopted earlier, then we must recognize the great importance of a random genetic drift, which is determined by a finite population size, for building up the genetic structure of biological populations" [6.1].

Such is briefly the content of the first article by Kimura. Similar conclusions have been reached independently by King and Jukes, who treated evolution by random drift as being non-Darwinian [6.4]. This gave rise to many misunderstandings. We must say at once that the neutrality of point mutations and, hence, molecular drift does not contradict Darwinism at all. Darwin studied the evolution of phenotypes and not of molecules. Kimura, who has devoted about 25 years to the development of his neutral theory, is a staunch follower of Darwin.

The neutralist or neutral theory boils down to the statement that numerous point mutations in proteins and, accordingly, in DNA, are independent of natural selection. This theory has been expounded in detail by Kimura in his monograph [6.5] and in a number of his review and popular articles [6.6–9].

Before the publication of his article [6.1], it had been assumed in the synthetic theory of evolution that at any level, beginning with the molecular level, the characteristics of an organism are adaptive and are fixed by natural selection. Ernst Mayr wrote in 1963: "I think ... that it is hardly probable that any gene remained selectively neutral for any period of time."

In Chap. 4 we pointed out the limitedness of panselectionism. It should be added that in the sixties any reasonings about the nature of genes and genomes and their changes in populations were based on the structure and properties of phenotypes. Even today we are still far from a complete understanding of the relationship between genotype and phenotype. The process of solving this problem has only just begun. However, the methods of population genetics used in conjunction with molecular-biological investigations have made it possible to work out neutral theory.

Let us first outline the content and main arguments of neutral theory.

Kimura formulates five principles of the theory. The first principle is as follows: for each protein the rate of evolution expressed by the number of amino-acid replacements per residue per annum is approximately constant for different lines unless the tertiary structure and function of the protein molecule undergo substantial change.

The experimental facts and the results of theoretical investigations that support this principle were given in Chap. 5. They will be considered further at a later time.

The second principle is: the functionally less important molecules or their constituent parts evolve (the rate of evolution being expressed by the number of mutational replacements) faster than the more important ones.

The third principle states that mutational substitutions that are less dangerous to the existing structure and function of the molecule (conservative substitutions) occur in evolution more frequently than the more dangerous ones.

The fourth principle is: the duplication of genes must always precede the appearance of a gene with a new function.

The problem of new genes will be considered in the next chapter, which is devoted to the drive and not to the drift.

The fifth (theoretical) principle is as follows: the elimination of definitely harmful mutations as a result of the selection and the random fixation of selectively neutral mutants, or mutants that are only slightly harmful, occur in evolution much more frequently than the positive Darwinian selection of definitely useful mutations.

The first principle is associated with the "molecular clock of evolution", which will be considered in more detail in Sect. 6.3. At this point we give a table of substitution rates for a number of proteins borrowed from the monograph by Nei [6.10] (see Table 6.1).

These values were obtained by comparing the primary structures of homologous proteins of different species. The palaeontological findings allow us to determine the times of divergence of kingdoms through species. For example, the divergence of animals and plants occurred about 10^9 years ago and the divergence of mammals and arthropods dates as back as 7×10^8 years; the horse diverged from the cow and pig 54 million years ago and from the donkey 2 million years ago; man separated from the orangutan $13-16 \times 10^6$ years ago and wheat separated from barley 50×10^6 years ago. Figure 6.1, which was borrowed from Kimura [6.7] gives the relevant data for the α-chains of vertebrate haemoglobin. The numbers in the matrix on the right indicate the corresponding numbers of substitutions. The approximate constancy of the numbers along the diagonals running from the left lower part to the right upper part

Table 6.1. Rate of substitution in paulings (one pauling is the substitution rate equal to 10^{-9} per unit per year)

Fibrinopeptides	9.0	Parathyrin	0.73
Growth hormone	3.7	Parvalbumin	0.70
Ig κ chain, region C	3.7	Protease inhibitors	0.62
κ Casein	3.3	Trypsin	0.59
Ig γ chain, region C	3.1	Melanotropin β	0.56
Lutotropin, β-chain	3.0	α Crystallin, chain A	0.50
Ig λ chain, region C	2.7	Endorphin	0.48
Complement	2.7	Cytochrome b_5	0.45
Lactalbumin	2.7	Insulin	0.44
Growth factor of epidermis	2.6	Calcitonin	0.43
Somatotropin	2.5	Neurophysin 2	0.36
Pancreatic ribonuclease	2.1	Plastocyanin	0.35
Haptoglobin, α-chain	2.0	Lactate dehydrogenase	0.34
Plasma albumin	1.9	Adenylate kinase	0.32
Phospholipase A_2	1.9	Triosophosphate isomerase	0.28
Protein inhibitor PSTl	1.8	Vasoactive peptide	0.26
Prolactin	1.7	Corticotropin	0.25
Pancreatic hormone	1.7	Glyceraldehyde dehydrogenase	0.22
Carboanhydrase C	1.6	Cytochrome c	0.22
Haemoglobin, α-chain	1.6	Ferrodoxin	0.19
Haemoglobin, β-chain	1.2	Collagen	0.17
Lipid AII-binding protein	1.0	Troponin c	0.15
Gastrin	0.98	α Crystallin, β-chain	0.12
Animal lysozyme	0.98	Glucogen	0.12
Myoglobin	0.89	Glutamate dehydrogenase	0.09
Amyloid AA	0.87	Histone H2B	0.09
Nerve growth factor	0.85	Histone H2A	0.05
Acidic proteases	0.84	Histone H3	0.014
Basic myelin protein	0.74	Ubiquitine	0.010
Tyrotropin, β-chain	0.74	Histone H4	0.010

indicate the approximate constancy of the rates of point mutations in protein and the constant running of the molecular clock of evolution.

From Table 6.1 it follows that the operation of the clock varies within wide limits for different proteins. The most rapid change is shown by fibrinopeptides and the slowest by histone H4. The difference in rates for these proteins is three orders of magnitude. Histone H4 is surprisingly constant – this protein in the cow differs from its counterpart in the pea only by two residues out of 105.

The physical significance of the differences indicated above is obvious. Roughly speaking, a protein is composed of an active and a passive part. The active part is the active site and its nearest environment; the remainder is the passive moiety of protein. It is obvious that mutations in the passive moiety have a greater probability of being neutral than those in the active site. Indeed, the fraction of neutral mutations in the passive moiety of haemoglobin is an order of magnitude higher than their fraction in the active part. The active part is the site of electronic-conformational events and the passive part is the place

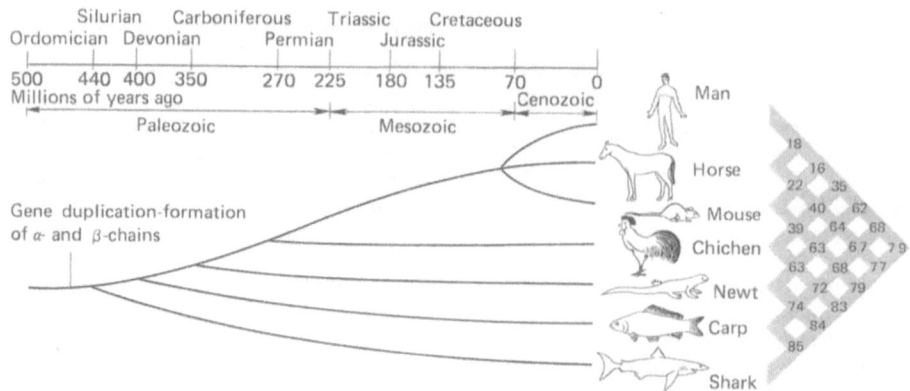

Fig. 6.1. The phylogenetic tree of vertebrate haemoglobin and the numbers of differing residues for seven species [6.7]

where only conformational events take place. Accordingly, the rate of substitutions in the passive part of protein is significantly higher than that in the active part and the figures given in Table 6.1 characterize roughly the fraction of the active site; they are the lower, the greater is this fraction. It may be said that the entire molecule of histone H4 is active, this being determined by its incorporation into the chromosome where histone H4 interacts with DNA, other histones and with non-histone proteins. As a rule, more ancient and more universal proteins change more slowly; it is sufficient to compare the rates of substitutions in haemoglobin chains and in cytochromes. Researchers have noted the presence of an appreciable negative correlation between the content of Gly in proteins and the rate of amino-acid substitutions, irrespective of the function of protein [6.10]. The amino acid Gly occupies a special position in protein, while being the smallest of the residues, which is why its substitution may have a greater effect on the spatial structure.

Thus, the presence of the molecular clock of evolution is an argument in favour of neutral theory. A considerable proportion of point substitutions in proteins are really neutral; these substitutions have no appreciable effect on the function of protein. However, the problem of the quantitative relationship between the molecular clock and neutral theory is not simple. This topic will be discussed in Sect. 6.3.

The second important argument in favour of neutralism is provided by the phenomena of polymorphism of numerous proteins established by electrophoretic techniques, i.e. their presence in various forms in the sense of the primary structure within a species. In many cases, the polymorphism of proteins does not manifest itself phenotypically and does not correlate with environmental conditions. It follows from this that such polymorphs are neutral.

The molecular clock and polymorphisms are both expressions of very important and general features of the molecular structure and dynamics of living systems. The physical meaning of neutral theory boils down to the degeneracy of two correlations, one between the primary and the three-dimensional structure of protein, and the other between structure and biological function [6.11]. This degeneracy implies the fundamental feature of the protein-machine. In contrast to man-made machines, in which the tolerances and clearances are reduced to a minimum (say, in Japanese automobiles or electronic devices); tolerances and non-rigid construction are also necessary for a protein and an organism for their life activity and evolution. The constructional non-rigidity and functional flexibility are in turn the result of the chemical nature of living machines. In solid-state machines built by man, signalling and control are accomplished by mechanical, electrical or magnetic means. In chemical machines, e.g. our bodies, the signals are molecules or ions, and the signal sources and receptors are molecular devices. The enzyme molecule may be looked upon as a signal transducer, as a converter of substrate into product. The non-rigidity of a living system or of its part is manifested not only in construction but also in dynamics, time-dependent behaviour.

The non-rigidity manifests itself in all specific features of living organisms. The molecular clock of evolution does not run with perfect regularity; the data of Table 6.1 must be taken *cum grano salis* (see Sect. 6.3).

After examining Fig. 6.1 and Table 6.1, we come to important conclusions. Neutral mutations are not the cause of appearance of new proteins with novel functions. Hence, molecular drift has no direct relation to speciation and, the more so, to macroevolution. In determining the course of the molecular clock, we proceed from the time of divergence established from palaeontological data pertaining to phenotypes alone. In other words, it is assumed that (e.g.) the time elapsed between the appearance of amphibians and the reptiles branching off from them is the time during which those substitutions occurred that distinguished snake haemoglobin from frog haemoglobin. But these substitutions, while being neutral, occurred both before and after the divergence. Molecular drift is not so much a factor as a fact of evolution. These considerations, however, cannot affect the averaged estimate of the course of the molecular clock for a given protein.

We give here one more table borrowed from Kimura [6.7] (see Table 6.2). Sixty-two and, accordingly, 61, amino-acid residues are not different in the

Table 6.2. Comparison of haemoglobin (Hb) chains

Type of change	Human alpha v. human beta	Carp alpha v. human beta
No change	62	61
Replacement of one nucleotide	55	49
Replacement of two nucleotides	21	29
Addition or deletion	9	10
Total:	147	149

chains being compared. What is essential is the closeness of the figures in both columns, one of which compares one species, while the other compares two species; the two species have developed separately for 400 million years. From Table 6.2 it follows that the alpha-chains of Hb accumulated mutations at a constant rate in both lines.

It must be stressed that neutral theory does not imply that neutral genes, i.e. genes in which neutral mutations are accumulated, have no biological functions. It only implies that alleles that differ in neutral substitutions are equally efficient for survival and reproduction.

In the preceding chapter we gave the results of theoretical model calculations for proteins, which were performed by Ptitsyn, Dill, Shakhnovich and Gutin (Sect. 5.3). From these calculations it follows that point substitution in model chains may have no effect on the spatial structure of a condensed globule and, hence, on its function. This is an independent argument in favour of neutral theory. As we have seen, the degenerate correlation between the primary and the spatial structure of protein is also confirmed by experimental data.

In connection with what has just been said, let us consider the problem of the origin of protein [6.12]. The general requirements of the three-dimensional structure of a functional protein are: the presence of an active site; its incorporation into the extended structure (backbone), built up of alpha-helices and/or beta-ribbons, in a tight packing of side groups in the "core" of the structure; and the stability of such a system while it is functioning. In a number of cases proteins that have different three-dimensional structures possess similar functions [6.13]. It may be said that on the whole the three-dimensional structure of protein is not specified by its function.

In their theoretical works Ptitsyn and Finkelstein [6.14–16] show that α-helices and β-ribbons are formed in regions enriched with non-polar residues. Localization of the elements of the secondary structure can be predicted on the basis of information about the primary structure. The theory tallies well with X-ray diffraction data. Based on this theory, Ptitsyn and Volkenstein [6.12] have determined secondary structures for a number of random sequences. Comparison of the results obtained with the structure of real globules shows once again that the primary structures of many globular proteins are the result of the "accidental choice remembered." Biological evolution reduces to the "editing" of these random sequences. It has been shown that the physical requirements of a functional globular protein do not include a strategic selection of sequences but are also met by typical random sequences. The evolutionary selection of sequences is necessary not for the formation of the three-dimensional structures of globular proteins but for the formation of active sites in these structures and the attainment of appropriate stability. These tasks do not require the selection of the entire amino-acid sequence; they can be realized by "editing" the drift, i.e. selecting random mutations, which lead to the appearance and refinement of active sites and to the stabilization of the protein structure. Of course, this can be done more easily than choosing the entire primary structure of protein from an infinite number of variants.

Indeed, if we suppose that in the α-chain of haemoglobin about 150 amino-acid residues long, and that each of the residues is chosen at a rate of 1.6×10^{-3} per year (Table 6.1), then the formation of the chain will require $150 \times (1.6 \times 10^{-9})^{-1} = 94 \times 10^9$ years, which is 6 times longer than the time of existence of the Universe.

"Editing" is done at the genetic level and in the protein structure. In the first case codons are selected, and in the second an important role is played by cofactors (metal ions).

The argumentation of neutral theory in the form of the concept of the molecular clock of evolution has been criticized. Lewontin writes: "Typically we are dealing with times of 200 million years, putting the common ancestor 100 million years ago, in the middle of the Cretaceous. Since that time every phyletic line has undergone numerous episodes of rapid and slow evolution, so that the substitution rates shown in the table are averages over vast periods of time and many cycles of speciation, extinction, and phyletic evolution. [Table 44 of Lewontin's book presents estimates of the number of years of evolution separating two forms – M.V.] The claimed "constancy" [of the molecular clock – M.V.] is simply a confusion between an average and a constant. It is like claiming that the temperature never varies in Chicago because the total number of degree days measured there was the same in the last fifty years of the nineteenth century as it was in the first fifty years of the twentieth" [6.17]. These considerations were repeated in an excellent book by Grant [6.18].

Kimura gave the following answer to Lewontin [6.7]: "The remarks [made by Lewontin – M.V.] reveal a misunderstanding of the nature of molecular evolution. One is attempting here to compare intrinsic rates of evolution in different lineages. The death rates characteristic of man and of an insect do not become equal by merely being averaged over a long period of time or over a large number of individuals; there is no reason to expect two averages to converge on each other unless the intrinsic factors shaping them are the same. My point is that intrinsic evolutionary rates are essentially determined by the structure and function of molecules and not by environmental conditions."

I think that Kimura is right here. The example cited by Lewontin is erroneous. Since Chicago does not change its geographical location, it is natural that its average number of degree-days should remain constant, but also that this should appreciably differ from the average for Los Angeles, for example. Speaking figuratively, it may be said that Kimura establishes the closeness of certain properties of Chicago and Los Angeles that are independent of their positions, such as, for example, the number of automobiles per inhabitant.

The groundlessness of the reasonings put forward by Lewontin does not mean, however, that the molecular clock is fully consistent with neutral theory (see Sect. 6.3). There are unsolved problems here. Let us consider the polymorphism of proteins. In his time, one of the most eminent geneticists and evolutionists of our century, Dobzhansky, maintained that the heterozygote advantage observed in many cases is accounted for by the so-called balance hypothesis, according to which adaptive polymorphism exists in a population maintained by the heterozygosity of the environment [6.19]. In other words, natural selection fixes heterozygosity and polymorphism, which correspond to the optimal fitness of a population. However, Kimura has shown that the level of heterozygosity is, as a rule, too high for selective balance, assuming that selection is not too weak. The subsequent study of the polymorphism of proteins and DNA has confirmed the neutral or nearly neutral character of corresponding substitutions. The polymorphism of proteins is also observed in prokaryotes, in particular in *Escherichia coli*. In such a case we cannot speak of heterozygosity, since prokaryotes are haploid organisms. Detailed investigations of the fitness of various alleles corresponding to isozymes of a number of proteins in *Drosophila melanogaster* and *D. pseudoobscura* have demonstrated the absence of selective differences in these polymorphs. In *Drosophila* a certain correlation between the polymorphism of proteins and the molecular size of their sub-units is observed. This is easily explained by neutral theory: the larger the size of the sub-unit, the more frequently there arise mutations. Enzymes that have a quaternary structure and form hybrid molecules with enzymes coded for by other genes are less polymorphic. This is determined by large functional constraints. Polymorphism is an indication of neutralism.

The relationship between the polymorphism of proteins and their thermal stability is fundamental [6.20]. One of the most significant environmental factors that determine the conditions of existence of a species is temperature. Changes in climate undoubtedly had an effect on both speciation and species extinction. As was mentioned above, Aleksandrov has carried out a detailed study of the problems associated with the thermal stability of organisms at the cellular and molecular levels [6.21,22]. He has found that the thermal stability of proteins is tied up with the level of conformational mobility which is of greatest importance to the functionality of proteins. Point substitutions exert an appreciable influence on conformational mobility, which has been shown by Yutani et al. [6.23], amongst others. This determined Aleksandrov's negative attitude toward neutral theory [6.22].

Slight changes in protein structure caused by most single substitutions of codons (the code is noise resistant) produce slight changes in conformational mobility. Changes in thermal stability brought about by neutral mutations are not significant, and are overlapped by polymorphism. Therefore, the differences in conformational behaviour of proteins of related species might result from molecular drift and adaptive evolution from negative selection. The polymorphism of proteins caused by neutral mutations makes possible adaptive evolution with changes in the conditions. This eliminates proteins that are extremely or insufficiently thermally stable.

The data obtained by Yutani et al. [6.23], which are also cited by Aleksandrov [6.22], are indirect. What was determined was the residual enzymatic activity of tryptophan synthesase after heating up to 58°C. The degree of denaturation of a protein after thermal treatment is not a direct manifestation of its conformational mobility under normal conditions. Moreover, 8 out of the 12 mutants studied are definitely not neutral since replacements occurred in

Table 6.3. Estimation of evolutionary rate

Protein	k_2	k_s
Pressomatotropin (man and rat)	1.13 ± 0.19	4.13 ± 0.66
β-Globin (man and mouse, man and rabbit)	0.59 ± 0.12	2.48 ± 0.48
α-Tubulin (hen and rat)	0.008 ± 0.005	1.18 ± 0.13

them in which a given residue was replaced by another residue which did not differ strongly from the first one. The mutations Glu → Met and Tyr → Ile cannot arise as a result of a single substitution of a nucleotide in the codon.

A very large number of data pertaining to the nucleotide sequences of genomes have accumulated since neutral theory was enunciated. These data have confirmed the neutral theory. It has been established that the synonymous silent substitutions in codons (in the third nucleotides, Z in XYZ) occur at an especially high rate, which is nearly equal in different molecules. Table 6.3 presents estimates of the rates of such substitutions in structural genes coding for pressomatotropin, β-globin and α-tubulin ([6.24], see also [6.5]), where k_2 is the rate of substitution in 10^{-9} per nucleotide per year in the second position of the codon and k_s is the corresponding rate for a silent mutation. Since all substitutions of the second nucleotide and most of the substitutions of the first one change the kind of amino-acid residue, $2k_2$ is the crude value of the rate of amino-acid replacement.

The rates were determined on the basis of a comparison of the DNAs of the organisms indicated. It is remarkable that k_2 changes by two orders of magnitude in going from pressomatotropin to α-tubulin, but k_s changes only fourfold.

Figure 6.2 compares values of k_s with those of k_A, which change the kind of amino-acid residues for a number of proteins [6.24]. The values of k_s are always substantially higher than those of k_A and are close to one another.

In accordance with the neutral theory of evolution, the highest rates of point substitution of nucleotides are characteristic of non-functional codons. Especially high are the rates for pseudogenes, the former genes which have lost their function, and for introns. For pseudogenes equal rates of substitution of all the three nucleotides of the codon have been established [6.25,26]. For example, for globin pseudogenes $k_0 = 4.6 \times 10^{-9}$ per nucleotide per year. In normal genes of globin the rates of substitution of the first, second and third nucleotides are as follows: $k_1 = 0.71 \times 10^{-9}$, $k_2 = 0.62 \times 10^{-9}$, and $k_3 = 2.63 \times 10^{-9}$. In pseudogenes there are no constraints associated with functionality.

The unequal use of synonymous codons in various organisms occurs at all times. As has been shown by Ikemura [6.26], this is a consequence of selective constraints caused by the unequal content of the corresponding tRNAs; more often than not the codon used corresponds to the isoacceptor tRNA present in the largest amount. We are dealing here with negative selection. As has been

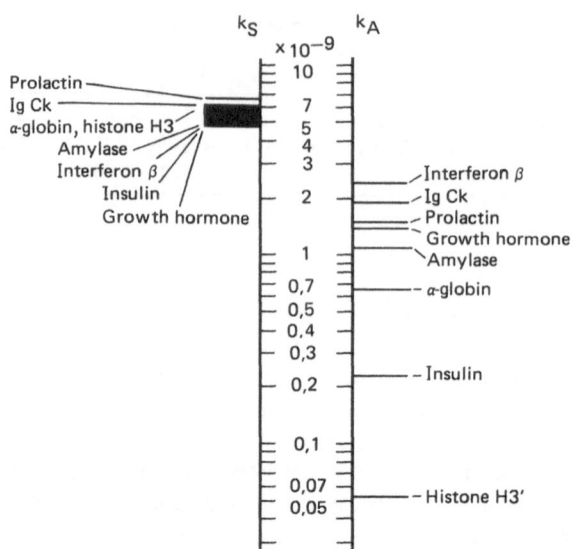

Fig. 6.2. Distribution of synonymous substitution rates (k_S) contrasted with that of amino acid-altering substitution rates (k_A) for several protein genes [6.24]

Table 6.4. Selection of codon and rate of evolution

Protein	Frequency of use of optimal codon	Number of synonymous substitutions per nucleotide	Number of amino-acid substitutions per codon
Main protein of external membrane	0.92	0.18	0.07
Tryptophan synthesase α	0.61	1.34	0.15

emphasized by Kimura, more stringent selection of a synonymous codon leads to a slowing-down of such substitutions. This is illustrated in Table 6.4 [6.24].

Attention has long been drawn to the noise resistance of the code in the sense of the increased probability of preservation of the hydrophobicity of an amino-acid residue upon single substitution of a nucleotide in the codon ([6.27], see also [6.28]). One of the factors that determine the noise resistance of the code is the synonymy of the substitutions of a nucleotide in the codon. Despite the neutrality of these silent mutations, they play a role in evolution. The results of the subsequent mutations in a codon that has arisen by way of a silent mutation are different in a number of cases from those of the preceding mutations. Silent mutations that are absolutely neutral at the protein level are thus the raw material for adaptive evolution.

The reliability of the code is not limited to its resistance to hydrophobicity. The greater proportion of single substitutions in codons do not change the charge on an amino-acid residue and do not significantly alter the structure of the radical. For example, for a series of substitutions

$$GGZ \to GCZ \to GUZ \to CUZ \to UUPu—UAPy$$

where Z is any nucleotide, Pu is A or G, and Py is C or U, there is a corresponding series:

$$Gly \to Ala \to Val \to Leu \to Phe \to Tyr \ ,$$

in which the side radicals of the neighbouring residues differ by one or two carbon atoms or by the replacement of H with OH. A similar character is displayed by the series

$$Gly \to Asp \to Glu \to Gln \to His \to Arg \ ,$$

which corresponds to the substitutions

$$GGZ \to GAPy \to GAPu \to CAPu \to CAPy \to CGPy \ .$$

The noise resistance of the code is directly associated with neutralism. The replacement of a residue by a residue with similar properties does not alter the structure or function of a protein [6.20].

Molecular evolution governed by drift is substantially different from phenotypic evolution. Molecular drift is characterized by a constant rate and by the retention of the function, while phenotypic evolution occurs at a non-regular rate and with adaptive changes in functions.

There is no direct correlation between molecular evolution and the evolution of phenotypes. A striking example is frogs, whose phenotypes have changed relatively little, whereas their biopolymers have undergone substantial changes. On the other hand, the phenotypic evolution of mammals is very sharply pronounced and their biopolymers have changed much less. As is known, the most important proteins of man and chimpanzee differ very little.

In answer to the question of the causes of predominantly neutral mutations, Kimura writes [6.24] that the most ordinary type of natural selection at the phenotypic level is stabilizing selection. This type of selection has been studied in detail by Schmalhausen [6.29] (see below).

While discussing the foundations of neutral theory, Crow points out its advantages [6.30]:

1. The theory provides at present the best explanation for significant differences in rates of molecular and morphological evolution.
2. The theory accounts for high rates of substitution in such dissimilar objects as the third nucleotide of the codon (silent mutations), pseudogenes and introns.
3. The theory has predicted high rates of evolution of pseudogenes, and also that molecular differentiation and polymorphism in "living fossils" are of the same order as in rapidly changing species.

4.   The theory has stimulated the development of rigorous methods of calculation and a systematic study of molecular polymorphisms.
5.   From the theory it follows that the rate of molecular evolution in a population and the mutation rate per individual are remarkably similar (see the next section).

We conclude this section with the words from *The Origin of Species* by Darwin, which were cited by Crow [6.30]:

"I am inclined to suspect that we see, at least in some of these polymorphic genera, variations which are of no service or disservice to the species and which consequently have not been seized on and rendered definite by natural selection ..."

"A structure, originally formed by the aid of natural selection, when rendered useless to a species, may well be variable, as we see in rudimentary organs, for their variations can no longer be checked by natural selection."

One can find everything in Darwin's great book, just as in the Bible.

6.2 Quantitative Relationships

Let us first estimate the probabilities of homozygote and heterozygote in the case of selectively neutral isoalleles; we introduce formula (6.1), first given by Kimura and Crow [6.2]. If v is the average rate of mutation in alleles of a diploid population, then in a population of size N there are $2N$ genes and there arise $2Nv$ mutants per generation. In a randomly mating population with an effective size N_e the probability of fusion of two gametes with identical alleles (i.e. gametes that originate from the same allele of a common ancestor) is equal in generation t to

$$H_t = \frac{1}{2N_e} + \left(\frac{2N_e - 1}{2N_e}\right)H_{t-1} \; . \tag{6.3}$$

Two alleles are identical if neither has mutated. The probability of this event is $(1 - v)^2$. Formula (6.3) is generalized as follows:

$$H_t = \left[\frac{1}{2N_e} + \left(\frac{2N_e - 1}{2N_e}\right)H_{t-1}\right](1 - v)^2 \; . \tag{6.4}$$

In equilibrium the loss of alleles as a result of random drift is equal to the appearance of new alleles due to mutations $H_t = H_{t-1} = H$. Dropping small terms of the order of v^2, we obtain from (6.4)

$$H \cong \frac{1 - 2v}{4N_ev - 2v + 1} \cong \frac{1}{4N_ev + 1} \tag{6.5}$$

and the probability of a heterozygote is

$$H_e = 1 - H = \frac{4N_e v}{4N_e v + 1} \; . \tag{6.1a}$$

Any population or species develops discretely in the course of evolution; half of an individual cannot appear or disappear. Systems of this kind must be treated on the basis of the theory of Markov chains, i.e. the probabilities of the subsequent events depend on the realization of the preceding ones. These are stochastic processes. However, as has been shown by Kimura, in population genetics use may be made of the continuum approximation, which considerably facilitates the solution of the corresponding problems [6.1,31,32]. One of the problems consists of the calculation of the average time, i.e. the number of generations required for a mutation to be fixed.

Let us assume that there is a randomly mating diploid population with an effective size N_e. There is a wild-type A and a mutant allele A'. The relative fitnesses of wild-type homozygote AA, mutant homozygote $A'A'$ and heterozygote AA' are respectively equal to 1, 1 + and 1 + h. The frequency of A' in such a Mendelian population is p, and the frequency of A is $1 - p$ (Table 6.5).

Kimura applied the diffusive approximation, which is represented by the inverse Kolmogorov equation. If $u(p, t)$ is the probability of fixation of allele A' in population t, i.e. in time t, then it satisfies an equation in partial derivatives:

$$\frac{\partial u}{\partial t} = \frac{1}{2} V_{\delta p} \frac{\partial^2 u}{\partial p^2} + M_{\delta p} \frac{\partial u}{\partial p} \tag{6.6}$$

where

$$M_{\delta p} = p(1 - p)[sp + h(1 - 2p)] + v(1 - p) \; . \tag{6.7}$$

This is an average change of p per generation due to mutations and natural selection; v is the probability, i.e. the rate of mutation $A \rightarrow A'$; there are no reverse mutations. Thus

$$V_{\delta p} = \frac{p(1 - p)}{2N_e} \; . \tag{6.8}$$

The average time of fixation of allele A' is given by

Table 6.5. Fitness and frequencies of three genotypes in a single locus

Genotype	AA	AA'	A'A'
Fitness	1	$1 + h$	$1 + s$
Frequency	$(1 - p)^2$	$2p(1 - p)$	p^2

$$\overline{T}(p) = \int_0^\infty t \frac{\partial u}{\partial t} \, dt \, . \tag{6.9}$$

It can be shown that $\overline{T}(p)$ satisfies an ordinary differential equation [6.31]:

$$\frac{1}{2} V_{\delta p} \frac{d^2 \overline{T}(p)}{dp^2} + M_{\delta p} \frac{d\overline{T}(p)}{dp} + 1 = 0 \tag{6.10}$$

with the boundary conditions

$$\overline{T}(0) \text{ is finite}, \quad \overline{T}(1) = 0 \, . \tag{6.11}$$

If the population has a finite size, the mutant allele is fixed, i.e.

$$u(p, \infty) = 1 \, . \tag{6.12}$$

The time of fixation of a deleterious mutant may be very long, but it is fixed. The solution of Eq. (6.10) has the form

$$\overline{T}(p) = 4N_e y(p) \tag{6.13}$$

where

$$y(p) = \int_p^1 e^{-B(\eta)} \eta^{-V} d\eta \int_0^\eta \frac{e^{B(\xi)} \xi^{V-1}}{1 - \xi} d\xi$$

$$B(\xi) = 2N_e s \xi^2 + 4N_e h \xi (1 - \xi) \tag{6.14}$$

$$V = 4N_e v \, .$$

For mutations that are only slightly harmful we may put

$$h = -s', \quad s = -2s' \, , \tag{6.15}$$

where $s' \geqq 0$ is the selective coefficient against A'; the fitness of heterozygote AA' is $1 - s'$, and that of homozygote $A'A'$ is $1 - 2s'$. In this case

$$B(\xi) = -4N_e s' \, . \tag{6.16}$$

If the initial probability of a mutant allele is $p = 0$, formula (6.13) transforms to

$$\overline{T}(0) = 4N_e \int_0^1 e^{4N_e s' \eta} \eta^{-4N_e v} \, d\eta \int_0^\eta e^{-4N_e s' \xi} \xi^{4N_e v - 1} \frac{d\xi}{1 - \xi} \, . \tag{6.17}$$

For a neutral mutant $s' = 0$ and the formula simplifies. If $4N_e v \neq 1$, we obtain

$$\overline{T}(0) = \frac{4N_e}{4N_e v - 1} \int_0^1 \frac{1 - \xi^{4N_e v - 1}}{1 - \xi} d\xi \, . \tag{6.18}$$

With $4N_e v = 1$ we have

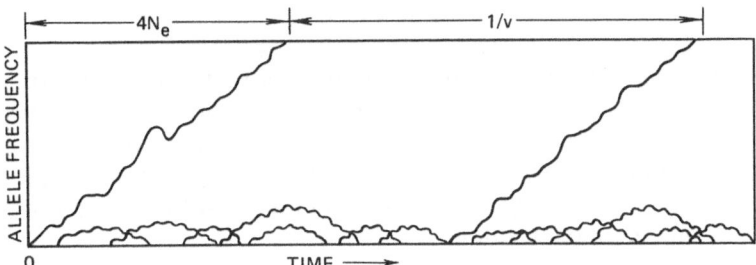

Fig. 6.3. The fate of mutant genes after their appearance in a population of limited size. The "trajectories" of frequency fluctuation for mutant alleles that have attained fixation are shown by black lines [6.7]

$$\bar{T}(0) = 4N_e \frac{\pi^2}{6} \cong 6.58\, N_e \ .$$

At $2N_e v = 1$

$$\bar{T}(0) = 4N_e \ .$$

On average, $4N_e$ generations are required for a neutral mutant to be fixed with its zero initial frequency. This is shown schematically in Fig. 6.3. Mutant alleles arise randomly in a population, their frequencies fluctuate and a greater part of them disappear with time, but some of them spread over the entire population and attain one-hundred-percent probability. The average time, i.e. the number of generations between two successive fixations is equal to the inverse probability of finite fixation of mutations. For a neutral mutant any of the $2N$ genes in a population has an equal probability of being fixed; the probability that this will occur with any of the genes is given by

$$v = \frac{1}{2}N \ .$$

Here the process is assumed to involve a very large number of generations since the average time required for a neutral gene to be fixed is equal to $4N_e$.

Let us consider a diploid population with a total size N. Thus, there are $2N$ sets of chromosomes and at a mutation rate per gamete per generation of v, we obtain in each generation $2Nv$ different new mutations. The rate of accumulation of mutational substitutions per generation k is equal to the product $2Nv$ and v is the probability of fixation of a single mutation. Thus,

$$k = 2Nvu \ , \tag{6.19}$$

where k is the rate of evolution expressed in terms of the number of mutational substitutions.

If a mutation is neutral, $u = \frac{1}{2}N$ and, hence,

$$k = v .\tag{6.20}$$

The rate of evolution expressed by the number of mutational substitutions per generation is simply equal to the mutation rate per gamete. Crow points out the elegance and unexpectedness of this proposition [6.30]. Indeed, k is the rate of molecular evolution in a population and v is the mutation rate for an individual.

This proposition is valid only for neutral alleles. If the mutant has a small selective advantage s, the result is

$$u \cong 2s \tag{6.21}$$

$$k = 4Nsv .\tag{6.22}$$

In this case the rate of evolution depends on the population size. Relation (6.22) is obtained from the selectionist rather than from the neutralist model.

If a certain fraction of mutants f is neutral, then the fraction of deleterious mutations

$$k = vf .\tag{6.23}$$

The fraction of deleterious mutations is $1 - f$; the fraction of especially favourable mutations may be neglected. Formula (6.23) refers to mutations per generation and per unit time, which appears in the diffusion equation (6.6). In order to find the mutation rate per year we have to divide v by the number of generations per year g. The thus-modified equation (6.23) has the form

$$k = \frac{v}{g}f .\tag{6.24}$$

Let us consider the intraspecies variability. We assume that for each locus there are n possible selectively neutral alleles and that the rates of all mutations are the same. At equilibrium, when a balance is attained between mutational variability and the random extinction of alleles, the mean heterozygosity in a given locus is given by

$$\bar{H}_e = \frac{4N_e v}{4N_e v \dfrac{n}{n-1} + 1} .\tag{6.25}$$

If the locus is a single nucleotide in DNA, then $n = 4$. But if we consider a gene, then n is very large and

$$\bar{H}_e = \frac{4N_e v}{4N_e v + 1} .$$

We again obtain formula (6.1).

In 1973 Ohta showed that mutations that are only slightly harmful behave

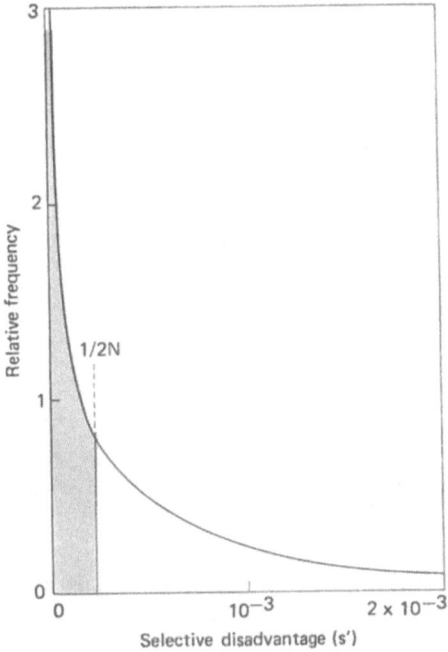

Fig. 6.4. Dependence of the relative mutation frequency on the coefficient of negative selection [6.7]

as neutral mutations under certain conditions [6.33]. Kimura has carried out a quantitative study of a model in which the relative frequency of mutations depends on the absolute value of negative selection coefficient $|s|$. The corresponding distribution is shown in Fig. 6.4. The mutation rate for variants, whose negative coefficient, s', is less than $\frac{1}{2}N$, may be treated as an effectively neutral rate. Thus, the condition of effective neutrality is

$$\frac{1}{2N|s|} > 1 \ . \tag{6.26}$$

As the population size increases this condition is progressively less fulfilled, the fraction of effectively neutral mutations falls off and the rate of evolution decreases. At the same time, heterozygosity slowly increases with increasing N.

If we sketch the hyperbola $N_e|s| = 1$ (Fig. 6.5), then the region below the curve (b) corresponds to the domination of the drift and the region above the curve (a) to the domination of selection [6.30].

Quantitative approaches to neutral theory are based on population genetics. Genetic investigations allow us to test the theory directly. For example, Mukai [6.34] has studied three kinds of selection: overdominance, frequency-dependent selection, and disruptive selection, the latter being determined by the interaction between the genotype and the environment. His objective was to

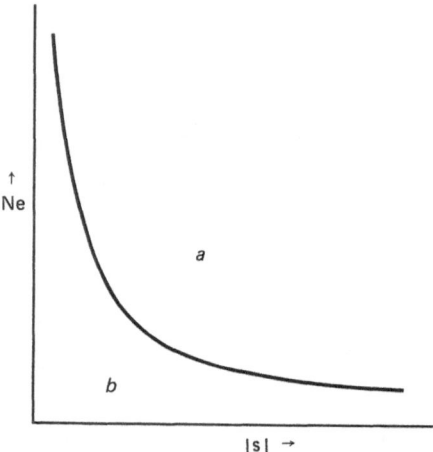

Fig. 6.5. Domains of the dominance of selection (*a*) and of mutation drift (*b*) [6.30]

elucidate the factors responsible for the maintenance of polymorphism for a number of enzymes in populations of *Drosophila melanogaster*. The results of his study have not confirmed the significant role of these kinds of balancing selection. On the contrary, direct evidence in favour of neutral theory has been obtained. As we have seen, independent but no less weighty evidence comes directly from experimental and theoretical studies of the structure and function of proteins (see Chap. 5).

It is necessary to emphasize that the population-genetics quantitative theory of neutralism, as developed in the works of Kimura, is based on a consideration of the Mendelian inheritance of a multi-allele system. From what has already been said it follows that the molecular drift of neutral mutations cannot have a direct effect on speciation and macroevolution. It is exactly for these reasons that the mutations under consideration are neutral. An important question arises regarding the compatibility of Mendelian inheritance and speciation. This question will be dealt with in the next chapter. The second question, of no less importance, concerns the subsequent fate of neutral mutations. Do they always remain neutral or can changes in environmental conditions and in genome structure convert neutral mutations into objects of selection? Darwin wrote (see [6.30]): "... The indirectly arisen structures, which did not at first confer advantages on a species, can be used later by their modified progeny in newly acquired habitats."

In an article called "Accelerated evolution following gene duplication and its consequences for the neutralist-selectionist controversy", Li [6.35] points out that neutral theory predicts a constant rate of gene evolution only if the selective constraints remain the same over the entire time. If a gene is duplicated, then one of the two resultant genes can freely undergo change and evolve rapidly. Therefore, accelerated evolution, which follows duplication, does not contradict the neutral theory. Gene duplication and its consequences will be discussed in the next chapter.

At this point it is appropriate to cite the words of Kimura [6.8]: "Even if the so-called neutral alleles are selectively neutral under a dominating set of environmental conditions in which a species exists, it is possible that some of them will be selected if new environmental conditions arise. Neutral mutations may serve as raw material for adaptive selection."

As will be pointed out in the next chapter, neutrality can disappear as a result of the interaction between molecular drift and molecular drive.

In Chap. 4 we considered certain problems associated with adaptation and fitness. So-called pre-adaptation is discussed on page 190. This term has been given the saltationist meaning. However, according to Mayr [6.36], an organism is pre-adapted if it is capable of moving to a new ecological niche; a structure is pre-adapted if it can acquire a new function without destroying the old one. Hartl and Dykhuizen [6.37] have examined enzymes of *E. coli*. Alleles that result in electrophoretic polymorphism were transferred by way of transduction to a genome of a standard laboratory strain K12. The resultant strains were placed in pairs into a chemostat, in which the limiting nutrient was the substrate of an enzyme of interest. The objective was to establish differences in rate of growth of the competing strains. Naturally-occurring alleles, whose products take part in the metabolism of carbohydrates, were studied. They were all found to be neutral or nearly neutral if the limiting substrate was glucose. However, if competition occurred under changed conditions, appreciable selective effects in the electrophoretic variants were often observed. The authors maintain that much selection occurs seldom in natural populations. However, the potential that does not manifest itself in selection provides a molecular basis for pre-adaptation (exaptation) of organisms to changing conditions.

We may consider the important role of neutral evolution to have been established at the molecular level. How is this associated with the Darwinian evolution of phenotypes?

As has already been pointed out, the most widespread type of phenotypic natural selection is stabilizing selection, which preserves and protects a certain norm. Neutral evolution can occur upon stabilizing selection provided that the quantitative trait is determined by a large number of loci. Kimura [6.24] believes that one may speak of a random drift only with stabilizing selection. As a matter of fact, this removes the selectionist-neutralist controversy, the consequences of which correlate well with stabilizing selection. Only a small fraction of nucleotide or amino-acid substitutions are favourable for a species that adapts itself to a new environment. In other words, only a small fraction of point mutations are subject to positive natural selection.

The neutral theory assumes the survival of the luckiest rather than of the fittest. The physical meaning of the theory boils down to a degenerate correspondence between the primary and the spatial structure of proteins, i.e. to a degenerate correspondence between the genotype and the phenotype (in the long run).

6.3 The Molecular Clock of Evolution and Neutral Theory

It is obvious that the molecular clock of evolution, which was first mentioned by Zuckerkandl and Pauling [6.38], can tick with precision only if the mutation rate for neutral alleles per annum is the same for all organisms at all times. Violations of the precise ticking are always possible: they can be caused by a change in mutation rate per year as a result of changes in v/g. Alternatively, they can be brought about by changes in selection constraints induced by changes in the internal molecular environment, i.e. by changes in f. The molecular clock requires special consideration. It is not by chance that a double issue of *Journal of Molecular Evolution* (Vol. 26, 1–2, 1987) was devoted entirely to this problem.

A striking example of the violation of the ticking of the molecular clock is the so-called enigma of superoxide dismutase [6.39]. The number of substitutions in this enzyme for man, horse and cow is 30.9 per 100 residues for 10^8 years; it is equal to 10.6 if these animals are compared with *Drosophila*, and to 5.8 if the yeast *Saccharomyces cerevisiae* is compared with the animals. Evidently, the structure and functionality of superoxide dismutase differ significantly in the organisms of mammals, insects and yeast. It has already been noted that co-factors (transition-metal ions) are different in superoxide dismutases of different organisms. In fact, we are speaking of different enzymes and, hence, there is no enigma at all.

Li and coworkers [6.40] state that mammals have no global molecular clocks. The rates of synonymous substitution in DNA are equal to 6.5×10^{-9} per nucleotide per year in rodents, 3×10^{-9} in primates and the even-toed, and to 1×10^{-9} in human beings and anthropoids. The rates of non-synonymous substitutions are 1.1×10^{-9} for rodents and 0.6×10^{-9} for primates and the even-toed. These data have been obtained, in particular, for the gene prolactin, which changes in rodents more rapidly than in other orders of mammals that have been studied.

The authors cited above maintain that the rates of neutral mutations must be estimated in terms of the number of changes not per year but per generation. The number of substitutions per year is larger for rodents than for other mammals since rodents multiply more rapidly.

One might think that the functionality of the gene prolactin, a protein hormone which plays an important role in lactation, is correlated with lactation and is significant specifically to generation. An actively operating gene is more strongly subject to mutational substitutions than a gene that does not take part in duplication and transcription. Point mutations occur exactly in these processes, i.e. upon convariant reduplication of DNA. Li and coworkers [6.40], in fact, reject molecular clocks, whose presence is confirmed by a large number of facts. The amplification of most genes is not tied up at all with the appearance of new generations. The data obtained by Li and his colleagues are insufficient for the molecular clock to be discarded.

The behaviour of superoxide dismutase contradicts the results obtained by Li et al. [6.40]. The rate of substitution in this enzyme does not increase, it slows down with increasing number of generations per year, e.g. if going from mammals to *Drosophila* to yeast.

The work of Li and his coworkers [6.40] is typical. Occasionally attempts are made to refute the existence of the molecular clock. These attempts are accounted for, eventually, by the fact that a reliable establishment of the "ticking of the clock" is not a simple matter. As will be seen at a later time, the subsequent studies have not confirmed the propositions made by Li and his colleagues.

A detailed analysis of the operation of the molecular clock has been carried out by the authors of a series of papers in the number of *Journal of Molecular Evolution* indicated above. Let us dwell on some of them.

Ohta [6.41] was concerned with slightly deleterious mutations which behave as neutral ones. It is stressed that amino-acid substitutions are more strongly subject to natural selection than the replacements of nucleotides. The latter may be neutral, the former nearly neutral. In the model proposed, the distribution of the selection coefficients of mutants (amino-acid substitutions) is presumed to be continuous at a point near to zero, with the average value of s being negative. This distribution is shown in Fig. 6.6 for a large population under heterogeneous conditions and for a small population in homogeneous conditions. There exists a negative correlation between the generation time and the population size N. Presumably, two effects are mutually compensated: the effect of the population size due to weak selection, and the effect of the generation time on mutation rate. This will be discussed in more detail at a later time.

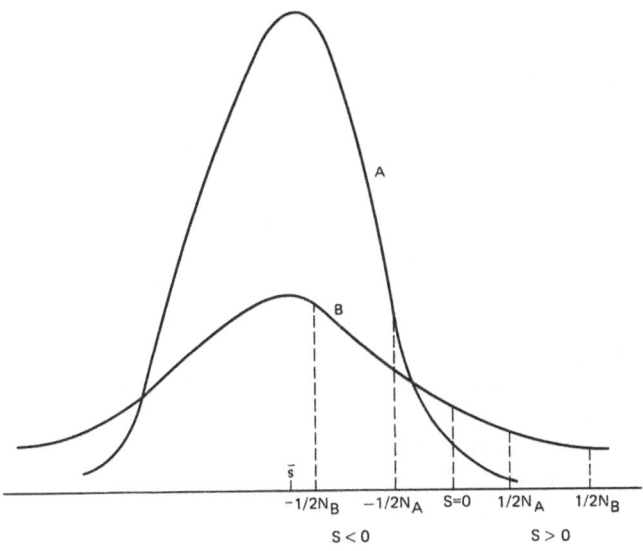

Fig. 6.6. Distribution of the probabilities of values of selection coefficients for a large population in a heterogeneous environment (*A*) and a small population in a homogeneous environment (*B*) [6.41]

Preparata and Saccone [6.42] have proposed a simple quantitative model of the molecular clock. These authors base their model on the theory of Markov chains. The evolution of nucleic acids is treated as a stationary stochastic Markov process. Any nucleotide appears in DNA as a result of such a process. The probability of mutation and of the subsequent fixation does not depend on the previous history of the nucleotide. It depends only on the molecular environment and on biological constraints that operate at the moment of a substitution. A stationary process corresponds to the ticking of the molecular (Markov) clock. Departures from the stationary state depend on the evolutionary distance between the species being compared. From the calculations performed it follows that as the evolutionary distance between species increases, certain critical value can be found at which the nucleotide frequencies and the dynamics of evolutionary changes undergo significant changes. The clock ticks correctly for a certain set of species. It is possible that the enigma of superoxide dismutase is due to precisely these critical values.

A discrete, stochastic approach to evolutionary problems seems to be promising. The work of Preparata and Saccone [6.42] deserves to be continued and developed further.

Syvanen [6.43] has suggested that deviations from the correct ticking of the molecular clock be accounted for by the horizontal transfer of genes from some organisms to other distantly related organisms. This is not convincing. Phenomena of such transfer are rare and they may be neglected in studies of evolution.

Kimura [6.44] has analyzed the data pertaining to a number of proteins and has shown that the two aforementioned factors responsible for violation of the ticking of the clock are really dominant. Three other factors that have an effect on the precision of the clock may also be mentioned, namely:

1. Fixation of slightly deleterious mutants.
2. The participation of compensatory neutral mutants in evolution.
3. Fluctuations of the neutrality region during mutational substitutions.

Compensatory pseudoneutral mutations will be considered in the next section.

The work of Zuckerkandl [6.45] is mostly methodological. Three concepts of neutralism are discussed: neutrality as the non-functionality of mutations; neutrality as the equifunctionality of mutations; and neutrality as a mode of mutation fixation. As we have seen, only the second concept has a true sense. While speaking of phenotypic and molecular evolution, the author associates the first notion primarily with regulatory differences at the genetic level. Genes are changed with time uniformly; the morphological characters have no relation to the molecular clock.

Jukes points out the different behaviour of transitions and transversions in molecular evolution [6.46]. Transitions are the replacements of one purine by another purine, or of one pyrimidine by another pyrimidine in DNA or RNA, i.e. for RNA the replacements are adenine $\rightleftharpoons$ guanine and uracil $\rightleftharpoons$ cytosine.

Transversions are the substitutions of a pyrimidine for purine or of a purine for a pyrimidine, $A, G \rightleftharpoons U, C$. Transitions occur much more often (see [6.28]).

Ochman and Wilson [6.47] deal with the interesting and important problem of the evolution of bacteria. The time-scale of evolution is established by way of tying ecological events that occurred during certain periods in the geological past to the geneological tree for 16S ribosomal RNAs of eubacteria, mitochondria and chloroplasts. This allows one to estimate approximately divergence times for other branches of the bacterial tree. According to this estimate, *Salmonella typhimurium* and *Escherichia coli* diverged between 120 and 160 million years ago, this being consistent with the time of appearance of the regions of existence of these bacteria. The distinction between the DNAs of these species is expressed by the rate of synonymous substitutions, 0.7–0.8% for 10^6 years, a rate close to that observed for the genes of mammals, invertebrates and flowering plants. The similarity between prokaryotes and eukaryotes is not limited to synonymous substitutions in genes coding for proteins. The average rate of substitution for 16S rRNA in eubacteria is approximately equal to 1% for 50×10^6 years, which is close to the average rate for 18S rRNA in vertebrates and flowering plants. The average rate of about 1% for 25×10^6 years for 5S rRNA is the same for eubacteria and eukaryotes. The rate of amino-acid substitutions in eubacteria is less (by a factor of 20) than the rate of synonymous substitutions. In mammals this difference is less by a factor of 4–5.

We give here the table borrowed from Ochman and Wilson [6.47] (see Table 6.6).

Let us consider some other works devoted to the molecular clock of evolution.

Easteal has compared the rates of nucleotide substitutions in the globin genes of mice, cows, rabbits and man [6.48]. This comparison is based on the order of branching of genes and species established as a result of the analysis of nucleotide sequences. Figure 6.7 shows the most economical cladogram (a rootless evolutionary tree) for genes similar to the gene of β-globin. The lengths of

Table 6.6. Events used for calibration of data on divergence of bacteria and rate of evolution of 16S rRNA

Event	Time ($\times 10^6$ years ago)
Diversification of *Cyanobacteria*	> 1300
Appearance of photosynthetic eukaryotes	> 800
Appearance of oxygen	< 2000
Appearance of oxidative eukaryotes	> 800
High oxygen concentrations	< 800
Appearance of luminous organs	> 50
Appearance of eyes	< 500
Appearance of land plants	< 400
Appearance of mammals	< 150
Appearance of legumes	> 100

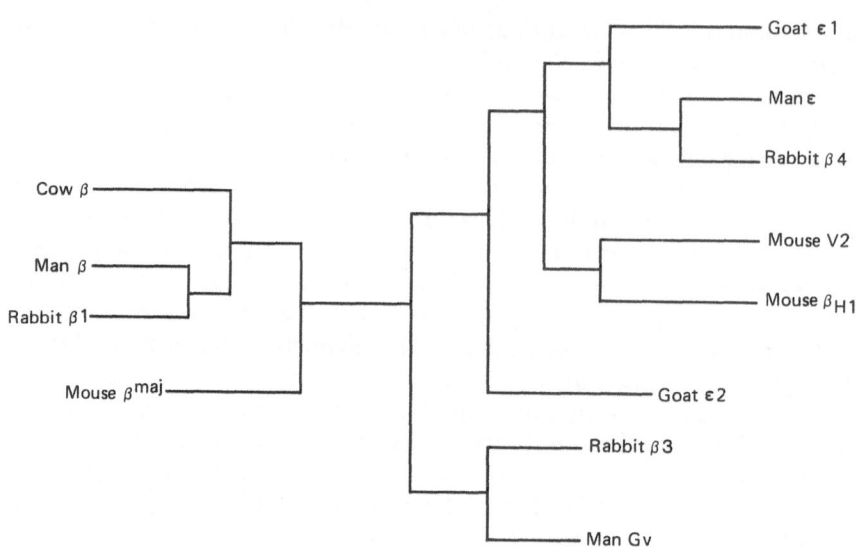

Fig. 6.7. The rootless evolutionary tree for vertebrate genes similar to the β-globin gene [6.48]

horizontal branches are proportional to the numbers of corresponding substitutions. The relative rates are not evidence of any heterogeneity of species. For different mammals the rates of gene evolution per year coincide. This result, which refutes the data of Li and coworkers [6.40], was obtained on the basis of reasonable phylogeny and an analysis of a large number of genes and species. Easteal's work clearly demonstrates that the rate of molecular evolution does not depend on the generation time. This is one of the most direct pieces of evidence for the molecular clock of evolution.

Palmer and Herbon [6.49] have studied the tempo and character of the evolution of mitochondrial DNA in 6 species of cruciferous plants. These DNAs differ strongly in the size of subgenome ring-shaped chromosomes. In these, 3 to 14 inversions occur. At the same time, the primary structures of 6 mitochondrial DNAs are very similar, and the rates of point mutations are about the same. This rate is less by a factor of 4 than in the DNA of chloroplasts of land plants, and less by a factor of 100 than in the mitochondrial DNAs of animals. On the other hand, the rate of rearrangement of the structure in mitochondrial DNA is especially high.

The work of Palmer and Herbon is evidence of the complexity and specificity of the evolution of DNA. The results obtained by these authors are in agreement with the concept of the molecular clock.

Sharp and Li [6.50] have studied the rate of evolution of DNA sequences in *Drosophila*. The study of synonymous substitutions has shown that their rates vary in different genes and are associated with the degree of use of synonymous codons. The selection among such codons limits the rate of such substitutions of certain genes. Additionally, the genes of mitochondria developed in *Drosophila* only at the rate of nuclear genes with the weak directedness of the selection of synonymous codons. These flies differ from mammals, in which the mitochondrial genes evolve 5–10 times faster than the nuclear genes. The absolute rate of silent substitutions in *Drosophila* is about 3 times higher than in mammals.

Dover [6.51] has explored the effect of molecular drive, i.e. the turnover of DNA, on the operation of the molecular clock. Molecular drive will be considered in the next chapter. Here we limit ourselves to a brief consideration of Dover's work.

The DNA turnover, or molecular drive, includes unequal crossing-over, gene conversion, slippage of genes and their transposition and also certain transcription events. All these phenomena have an effect on the ticking of the molecular clock of DNA. The mechanisms of divergence of nuclear genomes of eukaryotes are directly associated with the drive. The greater part of the observed DNA sequences did not arise by way of passive accumulation of base-pair substitutions, which is entirely determined by diffusion laws at the population level. Variations of the rates, units, directionality and gradients of many mechanisms of DNA turnover make contributions to the divergence of DNA. These mechanisms are capable of slowing down or increasing the rate of DNA differentiation between populations or maintaining it at a constant level. Molecular drive therefore bears a direct relation to speciation. We know examples of the joint action of various drive mechanisms on coding and non-coding DNAs. As a result, various ways of changing the structure and divergence appear. One cannot consider selection and neutral drift separately in such cases. The constancy in the rate of divergence for a certain period of time may be a reflection of constancy in the rates and units of turnover. However, the significant disparity between the rates of turnover and point mutations shows that the DNA clock, controlled independently by drive, is seldom encountered. Of course, the use of a given DNA sequence for measuring the time and relationships between species depends on the extent to which all contributing forces to the evolution of the sequence are understood.

Figure 6.8, which was borrowed from Fitch [6.52] and Ayala and Kiger [6.53], shows the accumulation of nucleotide pairs during evolution. The minimal values of the number of nucleotide substitutions have been found for all possible pairs of 17 mammalian species from the data on the primary structure of 7 proteins. The location of the points near a straight line corresponding to 0.41 of the substitution of nucleotides for 1 million years is an indication of the approximately constant rate of tick of the molecular clock of evolution of coding DNA.

A conclusion may be made as to the actual existence of the molecular clock, which, of course, runs non-uniformly but sufficiently well on average. Let us now consider the important problem of the correspondence between the molecular clock running constantly per year, and neutral theory, which is based on a constant rate per generation (formulae (6.20) and (6.24)). There is some difficulty here, which was clearly indicated by Kimura [6.5]. A number of authors maintain that the molecular clock contradicts the neutral theory and offer a selec-

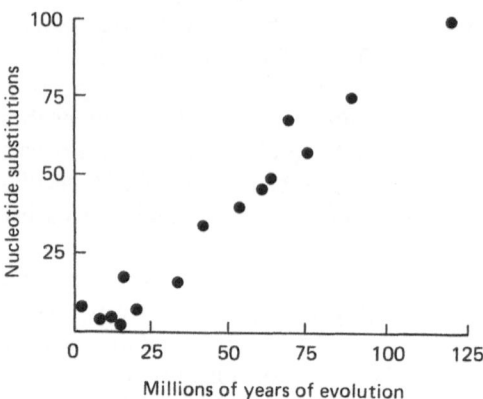

Fig. 6.8. Accumulation of nucloeotide substitutions during evolution [6.53]

tionist explanation (see, for example, [6.54]). Kimura puts forward arguments against such explanations and proposes a solution of the problem based on the neutral behaviour of mutations that are only slightly harmful (Ohta). Such mutations behave as neutral ones in small populatioins; in very large populations they are subject to selection. A general regularity for the existence of animals is known: the duration of a generation g and the small effective population size N_e are inherent in large animals (e.g. elephant); in small animals (e.g. mouse) an inverse relation is valid. Thus, for animals with a long generation time g the values of N_e are lower and the rate of appearance of effectively neutral mutations is higher. It thus follows that the rate of such mutations is higher. It thus follows that the rate of such mutations may prove approximately equal for animals with different generation times.

While developing the model of effectively neutral mutations proposed by Ohta, Kimura constructed a mathematical model based on the proposition that true selective neutrality is an extreme case of selective disadvantage of mutations when selective disadvantage is infinitely small. It can be shown that

$$k_g \approx v_e \; , \tag{6.27}$$

that is, the rate of evolution per generation is equal to the rate of appearance of effectively neutral mutations. This condition is a generalization of the condition $k = v$ (6.20) for truly neutral mutations. With a reasonable value of the parameter of the quantitative theory, the rate of evolution is inversely proportional to $N_e^{1/2}$:

$$k_g \sim 1/N_e^{1/2} \; . \tag{6.28}$$

If the same relation connects the generation time with the population number as shown by the equation

$$g \sim 1/N_e^{1/2} \; , \tag{6.29}$$

then the rate of evolution per year

$$k_T = k_g/g \qquad (6.30)$$

appears to be constant.

Of course, there are no grounds for thinking that relation (6.29) is strictly satisfied. We are speaking here of the values averaged out over sufficiently large periods of time. This is valid for both the molecular clock of evolution and the constancy of the rate of neutral evolution.

One might think that the constancy of the rate of point mutations per year (and not per generation) is particularly associated with the preservation of thermal stability. The rate of mutation fixation is the rate of deviation of protein structure from the corresponding optimal functioning. This rate is an average over a cycle of environmental conditions, particularly over the ambient temperature, i.e. presumably over a yearly cycle and, hence, over large periods of time [6.20].

The difficulties indicated cannot be considered to have been overcome. However, there are no serious grounds for believing that the molecular clock contradicts the neutral theory. If we proceed from the physical significance of the theory, which boils down to a degenerate correlation between genotype and phenotype, it becomes clear that the molecular clock and the neutral theory point to the same thing.

6.4 Pseudoneutral Mutations

As has already been said, mutations that are slightly deleterious behave as neutral mutations provided that the condition $1/2N_e|s| > 1$ is satisfied, i.e. at small values of N_e and/or $|s|$. We called such mutations effectively neutral. However, pseudoneutral mutations also exist, whose harmfulness does not manifest itself, due to compensation at the genetic or metabolic level [6.53].

Suppose that a certain mutation has somewhat impaired the function of a protein-enzyme. A mutant enzyme catalyzes the formation of a product at a lower rate than does a wildtype enzyme. This decrease can be compensated for by an increase in the amout of the enzyme synthesized. Such a compensation can be determined by a change in the regulation of the operation of the corresponding gene due to a change in the content of the regulatory protein synthesized by another gene. In this connection let us cite the words of Mayr [6.36]: "The selection pressure is directed for the generation of all kinds of feedback, regulators of development and its canalization with a view to providing the uniformity of the phenotype, despite the changes in the genic composition. All mechanisms of this kind are to a considerable extent stabilizing, and tend to protect changes of the genotype against natural selection ... a number of well-defined developmental channels provide a standard final result." In other words, the regulatory mechanisms provide a stationary state of the system: a

sort of homeostasis or (according to Waddington) homeorhesis, i.e. the maintenance of constant flow. We are speaking here of stabilizing selection.

The regulation of genomes (repressed or inducible genes) is realized in such
processes. The synthesis of anabolic enzymes is usually repressed by the products of these enzymes (anabolites). The anabolite level can be kept constant due
to the feedback between the product and the enzyme. If the catalytic properties
of the enzyme being synthesized are impaired because of a deleterious mutation,
a decrease in the amout of the product leads to a decrease of the amount of the
repressor bound to it, which in turn brings about an enhancement of protein
synthesis. As a result, the amount of the product remains unchanged. The impairment of the catalytic properties of the enzyme is compensated for in this
case by the weakening of the repression and by an increase in the amount of
the enzyme. The synthesis of catabolic enzyme is activated by their substrates
(catabolites). Here a positive feedback operates between the substrate and the
synthesis of the enzyme. The substrate eliminates the action of the repressor,
which leads to the induction of enzyme synthesis. If the catalytic properties of
the enzyme are impaired because of a deleterious mutation, the amount of the
substrate is increased. This in turn will lead to the removal of repression and to
induced synthesis of a large amount of the enzyme. As a result, the substrate
level will be kept constant. Phenomena of this kind have been explored in
bacteria on the basis of the operon model proposed by Jacob and Monod.

Let us assume that a mutant enzyme participates in a certain metabolic
chain, providing a rate v' of formation of an intermediate product. The rate v' is
lower than the rate v specified by a wild-type enzyme. Consider the simplest
form of stationary kinetics described by the Michaelis-Menten formula. We
have

$$v = \frac{k_1 k_2 ES}{k_{-1} + k_2 + k_1 S} \tag{6.31}$$

where k_1, k_{-1} and k_2 are the rate constants and E and S are the enzyme and
substrate concentrations. If the decrease of the rate is determined by a decrease
of the constant k_2 to $k_2 - \Delta k_2$, with $\Delta k_2 \ll k_2$, the new value of the constant will
be given by

$$v' \cong v - \Delta k_2 \frac{k_1 ES(k_{-1} + k_1 S)}{(k_{-1} + k_2 + k_1 S)^2} . \tag{6.32}$$

The feedback, however, provides the constancy of the rate v due to an increase
in the amount of the enzyme from E to $E + \Delta E$. In such a case we have

$$v'' = v + \Delta E \frac{k_1 k_2 S}{k_{-1} + k_2 + k_1 S} . \tag{6.33}$$

Full compensation is obtained if

$$v'' - v = v - v' \tag{6.34}$$

or

$$\Delta E \frac{k_1 k_2 S}{k_{-1} + k_2 + k_1 S} \cong k_2 \frac{k_1 ES(k_{-1} + k_1 S)}{(k_{-1} + k_2 + k_1 S)^2}$$

that is,

$$\frac{\Delta E}{E} \cong \frac{\Delta k_2}{k_2}\left(1 + \frac{k_2}{k_{-1} + k_1 S}\right)^{-1} . \tag{6.35}$$

If the functionality of the enzyme is impaired because of a decrease of the rate constant k_1, we have

$$\frac{\Delta E}{E} \cong \frac{\Delta k_1}{k_1}\left(1 + \frac{k_1 S}{k_{-1} + k_2}\right)^{-1} . \tag{6.36}$$

We are dealing with related processes in the case of adaptation of phenotypes (cells and organisms) to changing conditions. For example, in the bodies of humans and animals transferred to a high-altitude region, i.e. conditions of decreased oxygen content in the atmosphere, the haemoglobin content of the blood increases. The production of erythrocytes is regulated by a special globulin, erythropoietin. The regulation also involves hormones that stimulate erythropoiesis. The mutations of haemoglobin itself will be described at a later time.

Compensatory mechanisms are inherent in living systems just because they are chemical machines with considerable tolerances and clearances. Hochachka and Somero [6.56] consider three main types of biochemical adaptation:

1. Quantitative adaptations, in which the concentrations of macromolecular components are adaptively altered.
2. Qualitative adaptations, in which new types of macromolecules are incorporated into the system in response to an altered environment.
3. Modulation-type adaptations, in which the activities of existing macromolecules are adaptively adjusted.

The first type of adaptation has already been encountered. Evidently, the compensation of point mutations may be looked upon as a sort of adaptation to conditions of the internal environment. The substrates and products of enzymatic reactions may exert either an indirect (through synthesis of an enzyme) or a direct influence on enzymes, correcting the properties that have been impaired by deleterious mutations. There are various possibilities for maintaining the homeostasis of metabolites; some of these possibilities have been discussed by Volkenstein and Goldstein [6.57].

Let us consider the simplest cycle

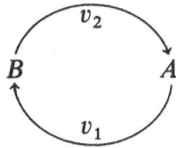

in which the substrates A and B are interconnected by the balance relation $A + B = C$. Analysis of this simple system shows that its only stationary state is stable with respect to a perturbation of variable substrate concentrations and/or of its parameters. Let the rates be given by the Michaelis–Menten equations:

$$v_1 = \frac{kA}{K + A} , \quad v_2 = \frac{k'B}{K' + B} = \frac{k'(C - A)}{K' + C - A} . \tag{6.37}$$

In particular, v_1 may correspond to the consumption of adenosine triphosphate ATP and v_2 to oxidative phosphorylation. Under stationary conditions, $v_1 = v_2$. These values correspond to the points of intersection of the $v_1(A)$ and $v_2(A)$ curves shown in Fig. 6.9. In the region of small values of v_2, when B is small, the variation of the parameter k only slightly changes the concentration of A (see [6.58]). This implies homeostasis, stabilization of the level of metabolite A with a change in k. As k increases, the homeostasis disappears, since the concentration of A is strongly changed; in contrast to $v_1^{(I)}$ and $v_1^{(II)}$, the values $v_1^{(III)}$ and $v_1^{(IV)}$ correspond to different A values.

In more complicated cases there are systems with several stationary states. The behaviour of such systems is of the trigger nature. The study of these systems is conveniently carried out by means of the graph theory [6.59]. Let us consider an open system with substrate inhibition [6.60]. The influx of the substrate can be realized by way of diffusion to compartmentalized or membrane-bound enzymes. The transport of the substrate is expressed by the equation

$$-\frac{dS}{dt} = K(S' - S) , \tag{6.38}$$

where S' and S are the substrate concentrations in a macro- and a micro-

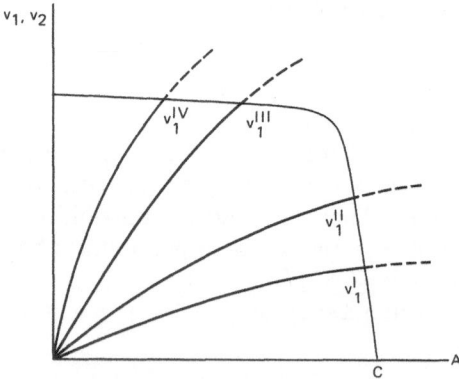

Fig. 6.9. Stationary values of the rates of two oppositely directed enzymic reactions corresponding to $v_1 = v_2$. With a high affinity of enzyme for substrate B there is a region of concentrations $A \approx C$ in which the variation of the parameter v_1 causes little change to the stationary value of A. v_1^{I} and v_1^{II} are in the region of homeostasis of metabolite A [6.57]

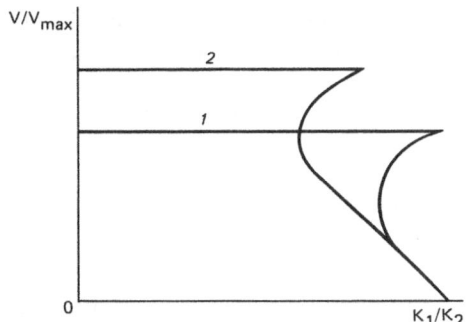

Fig. 6.10. Dependence of the rate of enzymic reaction suppressed by substrate on the dimensionless inhibition constant. The plateau region corresponds to the stable stationary state. Curves *1* and *2* correspond to different rates of diffusion of substrate [6.57]

environment. Let us assume that substrate inhibition occurs in the system. The conversion of the substrate is then described by the equation

$$v(S) = \frac{v_{max}}{K_1(1 + S^{-1} + SK_1K_2^{-1})} \ . \tag{6.39}$$

In the stationary state $v(S) = K(S' - S)$. The dependence $\bar{v}(\bar{S})$ (the bar indicates the stationary state) has a hysteresis character at not-too-high K values. The dependence $\bar{v}(K_1/K_2)$ is shown in Fig. 6.10. Within a certain region of variation of the inhibitior properties of the substrate, characterized by the constant K_2, the rate $\bar{v}$ remains constant. The curves 1 and 2 correspond to the lower and higher values of the constant K. Sailfullin and Goldstein [6.60] have examined the effect of the substrate structure (tautomerization, hydration) on the kinetics of reactions catalyzed by lactate dehydrogenase and glyceral-dehyde-3-phosphate dehydrogenase. Conditions for the appearance of hysteresis in a two-substrate reaction with a change in the structure of one of the substrates have been found. The kinetic behaviour remains unchanged with a mutational change of the constant K_2.

Selkov and co-workers [6.58,61] have studied the stabilization of the energy metabolism of a cell. These are important examples of a possible neutral variation of the catalytic properties of enzymes during their mutational changes.

In a large set of allele mutations and in a corresponding set of proteins with substitutions of various amino-acid residues, partially compensated mutations are often encountered [6.61]. It has been shown that compensation occurs upon formation of proteins from several sub-units as a result of the compensation of conformational distortions of their tertiary structures in the quaternary structure. Such compensation is possible only in heterozygous mutants upon formation of heteromultimer proteins, in which the function lost in homomultimers built up of identical sub-units is partially recovered (see [6.57]).

The mechanism of conformational compensation of defective enzymes becomes especially efficient in aggregates made up of different proteins, i.e. in multi-enzyme complexes. There are compensation variants at the level of various genes. On the other hand, the union of different interlinked active sites in a single protein (allosteric proteins, etc.) makes possible such a compensation by means of conformational mechanisms. This will be discussed at a later time.

The efficiency of conformational co-ordination and regulation by means of conformational changes is especially significant when these changes are associated with a particular protein modification. Special modifying enzymes govern the behaviour of an enzyme system, causing conformational changes (see, for example, [6.63]). A change in conformation makes possible the fit of a defective (because of mutation) enzyme to its substrate. Moreover, a forced change of conformation that occurs with a corresponding energy consumption can keep the enzyme far from equilibrium in such a conformation. The confinement in non-equilibrium states and, as a consequence, the approximate unidirectedness of the stages, are prerequisites for high specificity of an enzyme towards the substrate. Nature creates various kinetic mechanisms for the formation of a non-equilibrium population of conformers which retain the memory of a correct substrate [6.64].

The functioning of multi-enzyme complexes and multi-functional enzymes involves a change in the conformation of such a molecular machine. Presumably, at each moment of time a part of the complex performs the first step of the process, while the other part is involved in the realization of the subsequent steps. Upon completion of the slowest step a conformational change may occur, as a result of which the roles of the sub-units or domains are changed. The entire system can then function, performing cycles of conformational "swinging". Such a mechanism has come to be known as the flip-flop mechanism for two conformational states [6.65]. In such a case, one conformer determines the rate of the process and the activity of the entire system. Thus, mutations that involve only the "fast" conformer will not manifest themselves in the kinetics limited by the "slow" conformer. Both thermodynamic and kinetic mechanisms govern organized enzyme systems. This provides their compensatory abilities.

We see that the non-rigidity of biological structures at all levels creates diverse possibilities for the compensation of deleterious mutations. One might think that most of the neutral mutations observed are really pseudoneutral in the sense that these mutations are harmful but are compensated.

Spivak has studied the polymorphism of haemoglobins (Hb) and compared the role of neutrality and pseudoneutrality with natural selection [6.66]. The total number of possible mutants of human Hb with a single amino-acid substitution is about 2200. More than 400 Hb A mutants have been identified. There are mutants that are lethal in the homozygous state, such as, for example, Hb S, a haemoglobin responsible for sickle-cell anemia, in which the amino-acid residue 26Glu in the chain is replaced by a non-polar Val. At the same time, there are a number of less dangerous substitutions, in particular the moderately widespread anomalous Hb. Table 6.7, which was borrowed from Spivak [6.66], presents 8 such variants.

Table 6.7. Moderately widespread anomalous human haemoglobins

Haemoglobin	Carriers	Remarks
1. J Oxford α 15 Gly → Asp	Italians, the English	Only heterozygotes have been described
2. Hasharon α 47 Asp → His	Jews, Italians	Only heterozygotes have been described; decreased stability of Hb
3. Porto Alegre β 9 Ser → Cys	Spaniards, Latin-Americans	Heterozygotes and homozygote have been described
4. J Baltimore β 16 Gly → Asp	Europeans	Only heterozygotes have been described
5. G Coushatta β 2 Gly → Ala	Indians of Northern America, Chinese	Heterozygotes and homozygotes have been described
6. Korle Bu β 73 Asp → Asn	Negroes of Western Africa	Heterozygotes and homozygotes have been described
7. N Baltimore β 95 Lys → Glu	Negroes of the United States	Only heterozygotes have been described
8. New York β 113 Val → Glu	Chinese, Vietnamese	Heterozygotes and homozygote have been described. Decreased stability of Hb

In all the cases given there are no pathological symptoms. One has to presume that in homozygotes the haemoglobins 1, 2, 4, and 7 may be dangerous because substitutions have taken place in them which sharply change the properties of the residue.

Presumably, homeostasis at various levels determines the stability of an organism with respect to haemoglobins that are present in heterozygotes. As Spivak has pointed out, anomalous haemoglobins that cause the premature destruction of erythrocytes, haemolysis, stimulate erythropoiesis and perform other functions that are useful to the organism. In Spivak's opinion, most rare Hb mutants that are in the heterozygous state are pseudoneutral rather than neutral. When the environmental conditions and the state of the genome are changed, such mutants may cease to be neutral and become harmful. However, as we have seen, true neutrality is not absolute either, and cannot always be distinguished from pseudoneutrality.

Evidently, most of the anomalous human haemoglobins are pseudoneutral. In some cases such mutants may appear to be advantageous. Spivak cites, as an example, the fitness of llamas to high-altitude areas. The llama haemoglobin has a higher affinity for oxygen than the haemoglobin of mammals living on plains. An increase in the affinity of Hb for O_2 is associated with a decrease in the affinity of llama Hb for 2,3-diphosphoglycerate caused by the fact that Asn is located at the position N2 of the β-subunits. In a camel, a close kin of the llama, this position is occupied by His. The replacement His → Asn, which could have been pseudoneutral under flatland conditions for the llama ancestry, made it possible for the llama to occupy a new ecological niche.

It must be stressed that a number of truly neutral human Hb variants are probably unknown at present. Only those variants which manifest themselves to any degrees in pathology have been identified and studied. About 80% of possible mutant forms of human Hb have not yet been identified.

A quantitative theory of the role of compensatory pseudoneutral mutations in a single gene has been worked out by Kimura [6.7,32]. Let us describe this theory.

There is a diploid population with an effective number N_e. Two linked genes are considered, whose wild-types are A and B, and whose mutant forms are A' and B'. For simplicity, the average rates of $A \to A'$ and $B \to B'$ mutations are assumed to be identical and equal to v. We assume that the mutants A' and B' are harmful if they occur seperately, but the double mutant $A'B'$ displays the same fitness as AB. The intermediate states $A'B$ and AB' are also harmful. Table 6.8 presents the fitness coefficients and frequencies for four genotypes.

The diffusion model described earlier is used. There are two independent variables, p_1 and p_2; the process is a kind of two-dimensional diffusion. The probability of fixation of the double mutant $A'B'$ during t generations $u = u(p_1,p_2,t)$ satisfies the equation

$$\frac{\partial u}{\partial t} = \frac{p_2(1-p_2)}{4N_e}\frac{\partial^2 u}{\partial p_2^2} - \frac{p_1 p_2}{2N_e}\frac{\partial^2 u}{\partial p_2 \partial p_1} + \frac{p_1(1-p_1)}{4N_e}\frac{\partial^2 u}{\partial p_1^2}$$

$$+ M_{\delta p_2}\frac{\partial u}{\partial p_2} + M_{\delta p_1}\frac{\partial u}{\partial p_1} , \tag{6.40}$$

where

$$M_{\delta p_2} = vp_1 + s'p_1 p_2 \tag{6.41}$$

$$M_{\delta p_1} = 2vp_0 - s'p_1(1-p_1) - vp_1 \tag{6.42}$$

express the average changes in p_2 and p_1 per generation which are caused by mutations and natural selection. The analytic solution of Eqs. (6.40) is rather difficult. Kimura limits his consideration to an approximation in which

$$s' \gg v > 0 .$$

Table 6.8. Fitness coefficients and frequencies for four genotypes

Genotype	AB	$A'B$ or AB'	$A'B'$
Fitness coefficient	1	$1-s'$	1
Frequency	p_0	p_1	p_2

This means that the selection coefficient for a single mutant is much higher than the rate of its mutational product. Its frequency is thus very low during the entire process and a quasi-equilibrium is approximately realized, in which the mutational change in each gene is compensated for by negative selection. In such a case

$$M_{\delta p_1} = 0 \ . \tag{6.43}$$

From Eqs. (6.42) and (6.43) it follows that with p_1^2 being neglected, and with the condition $p_0 + p_1 + p_2 = 1$ being taken into account, we have

$$M_{\delta p_2} = (v + s'p_2)\frac{2v}{s' + 3v}(1 - p_2) \ . \tag{6.44}$$

It will thus suffice to consider the change of p_2, which is denoted henceforth as p. The probability of fixation of the double mutant $A'B'$ in a population with its initial frequency p is $u(p, t)$, which satisfies a much simpler equation:

$$\frac{\partial u}{\partial t} = \frac{1}{2}V_{\delta p}\frac{\partial^2 u}{\partial p^2} + M_{\delta p}\frac{\partial u}{\partial p} \ , \tag{6.45}$$

where

$$V_{\delta p} = \frac{p(1 - p)}{2N_e}$$

$$M_{\delta p} = s_1 p(1 - p) + v_1(1 - p) \ .$$

$V_{\delta p}$ is the average value of p and $M_{\delta p}$ is the variance of p per generation:

$$s_1 = \frac{2sv}{s' + 3v} \ , \quad v_1 = \frac{2v^2}{s' + 3v} \ .$$

The average time of fixation of $A'B'$ is given by the following equation (see Sect. 6.9):

$$\overline{T}(p) = \int_0^\infty t\frac{\partial u(p, t)}{\partial t}dt \ . \tag{6.46}$$

The quantity $\overline{T}(p)$, as before, satisfies the equation

$$\frac{1}{2}V_{\delta p}\frac{d^2\overline{T}(p)}{dp^2} + M_{\delta p}\frac{d\overline{T}(p)}{dp} + 1 = 0 \tag{6.47}$$

with the boundary conditions

$$\overline{T}(0) \text{ is finite} \ , \quad \overline{T}(1) = 0 \ .$$

The solution of (6.10) has the form

$$\bar{T}(p) = 4N_e \int_0^1 e^{-B(\eta)}\eta^{-V}\,d\eta \int_0^\eta \frac{e^{B(\xi)}\xi^{V-1}}{1-\xi}\,d\xi\;; \tag{6.48}$$

$$B(\xi) = 2N_e s\xi^2 + 4N_e h\xi(1-\xi)\,,$$

$$V = 4N_e v\,.$$

In this case $s = 2s$, $h = s_1$ and $v = v_1$.

The selection coefficient s_1 represents very weak selection, being of the order of the mutation rate $v \ll s'$. Approximate calculations show that at $2N_e v = 1$, the average fixation time is $7.4N_e$ generations with $4N_e s' = 20$; $8.7N_e$ with $4N_e s' = 30$; and $54N_e$ with $4N_e s' = 400$. This time is found to be lower than the time of fixation of one deleterious mutant A'.

Wykoff compared the pancreatic ribonucleases of rats and cows and noted that "many of the changes are pairwise". For example, cow ribonuclease contains the residues 57Val and 79Met, whereas in rats these residues are replaced by 57Ile and 79Leu [6.67]. In the three-dimensional structure these amino-acid residues are close to each other. Kimura points out that it is precisely the spatial proximity of the residues in the globule molecule that determines the compensation. Yanofsky and co-workers [6.68] have studied triptophane synthesase in the wild-type and found that it contains 210Gly. With Gly being replaced by Glu the functionality of the enzyme is lost. The same occurs upon replacement of 174Tyr by Cys. But if the two mutations are simultaneously present, the functionality is retained. Figure 6.11 shows schematically the location of two interacting amino-acid residues, A and B, in the globule.

The compensation studied by Kimura differs from the situations described above. Obviously, the possibilities of such a compensation, just as the other facts and considerations that support the neutral theory, are evidence of a limited sensitivity of the structure and dynamics of protein to point substitutions.

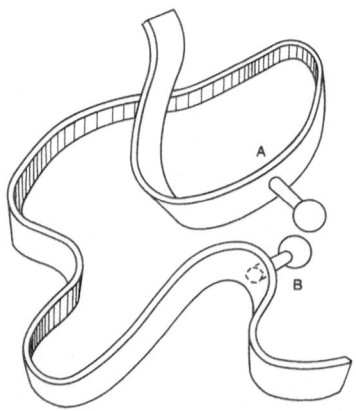

Fig. 6.11. Schematic representation of a protein with interacting residues A and B [6.32]

6.5 Extinction of Species

Neutral and pseudoneutral mutations are not directly related to speciation. Molecular drift may play a significant role in this most important phenomenon in conjunction with molecular drive, which will be described in the next chapter. On the other hand, molecular drift may appear to be a substantial factor for extinction, disappearance of species [6.69].

Extinction of species is inseparably associated with speciation. It is known from palaeontology that about 99.9% of all species that have ever existed on earth have become extinct. What are the causes for the occurrence of species extinction?

Darwinism and synthetic theory of evolution rightly state that the main causes for the extinction of species are the struggle for existence, competition and natural selection, all of which discard the species that appear to be insufficiently adapted to particular conditions. Apart from the biotic factors, there also operate abiotic factors determined by environmental conditions. The situations that are usually considered are as follows.

1. A species is changed due to changes in environmental conditions and is replaced by a new species. This is an event which is similar to a first-order phase transition (see Chap. 9).
2. A species disappears as a result of changes in environmental conditions and diverges, being replaced by two new species. This is an event which implies a decrease in symmetry and which is similar to a second-order phase transition (see Chap. 9).
3. A species becomes totally extinct as a result of changes in environmental conditions.
4. A species is an impasse; it exists for a limited period of time since its original characters impede its adaptation.

As an example of a dead-end species usually cited is the extinct giant deer *Megaloceros giganteos* with antlers weighing 25 kg. However, the weight of the antlers of the currently living argali (*Ovis ammon*) is up to 32 kg and there is no evidence that this species is in danger of extinction. As a matter of fact, the very existence of impasse species has not been proved. We do not know why the giant deer became extinct. An evolutionary cul-de-sac is presumed, on the basis of attempts to ascribe some significance, either adaptive or adaptation-hindering, to any characters. Of course, it is difficult for a biologist, who has never seen the argali, to understand how such heavy horns could be carried on the head. But he forgets that the deer was a giant.

In contrast to situations pertaining to other species, the self-destruction of *Homo sapiens*, if it ever takes place, will be caused by internal rather than external factors, i.e. by the fact that nature has endowed this species with a brain capable of producing nuclear weapons and anti-ecological industry.

Mention should be made here of the fifth situation:

5. A species disappears as a result of the action of anthropogenic factors, that
    is direct extermination (examples include Steller's sea cow, the passenger
    pigeon, and the marsupial wolf), or changes in the environment caused by
    man.

In all the cases indicated (except for the fourth case) we are talking about
extinction due to external factors. A species is decimated as a result of the effect
of both biotic and abiotic events. The latter include catastrophes, which appear
in hypotheses concerning events such as the disappearance of dinosaurs. There
is a well-substantiated hypothesis that these events were caused by the collision
of the earth with a sizeable meteorite, which induced drastic changes of climate.

Do internal causes of species extinction exist? Davitashvili [6.70] thinks
not. The author criticizes the work of some biologists, according to whom a
species, like an organism, irrespective of the environmental conditions, passes
through the stages of birth, maturation, old age and death. In point of fact, no
real mechanisms of aging and death of well-adapted species under invariable
conditions are put forward in these and other works. Davitashvili regard the
conceptions of the internal causes of species extinction as being idealistic. This
is nonsense, of course, but the question of internal causes still remains open.

A species is a real open system, in which various internal processes are
taking place. The belief in the absence of the internal causes of aging and death
of species reminds one of the notion of the primitive nations that the death of
humans is always the result of witchcraft or pure murder.

In palaeontological literature there are indications as to the aging of
species, which precedes their death and which manifests itself in the accumulation of various monstrosities. This refers, in particular, to the ammonites [6.71–
73]. Rejecting the internal causes of extinction, Davitashvili casts doubts on the
facts reported in these works.

One can cite arguments, both general phenomenological and molecular-
biological, in favour of the aging and death of species caused by internal factors.
Modern theory asserts the biological reality of a species and stresses an analogy
between a species and an organism. The emergence and disappearance of a
species are analogous to the birth and death of an organism [6.74,75]. This has
already been discussed in the book.

This statement is formal and does not disclose the causes of species extinction. The phenomenological similarity between a species and an organism is
determined by the fact that both are far-from-equilibrium open systems, i.e.
dissipative systems. The existence of such systems is supported by the export of
entropy into the environment. A dissipative system, however, cannot exist for an
indefinitely long period of time; sooner or later it moves on to a state of equilibrium, i.e. it dies out. The transition from the stationary to the equilibrium state
occurs due to the instability of the stationary state caused by random, stochastic
factors.

The internal mechanisms of species extinction must be associated one way
or another with deleterious mutations, i.e. they must be governed by natural

selection. These are either macroscopic events in chromosomes (molecular drive, discussed in the next chapter), or point mutations of genes. Neutral and pseudoneutral mutations are not subject to the direct pressure of natural selection.

Deleterious mutations, which directly upset adaptation, are eliminated by selection and, as a rule, do not appear in a species as a whole. They may be lethal to an individual. As regards neutral and pseudoneutral mutations, they may be very important in the extinction of species, while playing no determining role in speciation.

Three mechanisms of species extinction associated with such mutations may be presumed [6.69].

1. Mutations which are neutral in isolation and therefore accumulating may become harmful in certain combinations. A protein whose structure has been somewhat changed due to the accumulation of neutral mutations, but which still retains its function, may lose it as a result of a new mutation. On its own this new mutation is neutral, but it exerts a harmful trigger action on a system that has already accumulated neutral mutations. We have already seen that neutralism is independent of the interactions between the elements of a structure.

2. Pseudoneutral mutations can accumulate as long as they are compensated. However, it is always possible that new mutations will appear which will lead to violation of the compensation. In particular, the replacement of an amino-acid residue adjacent to interacting residues, shown in Fig. 6.11, may have a significant effect on compensation if changes thus occur in the charge, polarity and even in the volume of the residue. Such mutations will lead to impairment of the fitness and, eventually, to the death of the species.

The mechanisms indicated above can operate at the level of populations and species if neutral and pseudoneutral mutations occur prior to meiosis.

3. As pointed out earlier, with a small effective population size, mutations that are only slightly harmful behave as neutral mutations. Accordingly, as the population size increases these mutations again become deleterious. It is possible that these transitions from neutrality to harmfulness and vice versa play a role in so-called life waves. This term, introduced by Chetverikov (see [6.76]), implies oscillations in the population number induced by various factors; these oscillations may be periodic (more often, seasonal) or aperiodic (for example, a catastrophic increase in the population of locusts or lemmings). One might envisage, as a result of an increase in population number, neutral mutations becoming harmful and the population number falling off. There are also other factors that bring about an increase in population size, and the cycle repeats itself. At the same time, the extinction of a species caused by an increase in its numbers is always a possibility, however paradoxical it may seem.

These are the internal causes of species extinction connected with point neutral, effectively neutral and pseudoneutral mutations, and with molecular drive. The internal factors operate at the chromosome level, in phenomena of molecular drive.

The extinction of a species is an autocatalytic process. A decrease in the number of species self-accelerates since, in Darwin's words, the "chance for extinction" increases with each generation. This is evidenced, for example, by The Red Data Book.

One might think that the general proposition of the similarity between speciation and phase transition is also applicable to species extinction.

Internal causes of extinction which are independent of changes in biotic and abiotic factors must therefore exist. These causes are also ultimately determined by natural selection.

Of course, an interaction between the internal and external factors of species extinction occurs in nature.

Schmalhausen [6.77] states that since neutral or slightly detrimental mutations appear much more often than useful ones, such characters must also predominate in the first stages of speciation. He continues, "... degenerate phenomena are even possible due to an uncontrolled accumulation of mutations. Such degenerate phenomena sometimes preceded the extinction of animals in the geological past." Schmalhausen wrote these words before the enunciation of the theory of molecular evolution. There is every ground for believing that the accumulation of neutral and pseudoneutral mutations in a species characterized by polymorphism is similar to the accumulation of somatic mutations in an organism.

7. The Molecular Drive

7.1 Genomes

When we say that DNA is the genetic substance, we mean it is the substance of structural genes which are responsible for protein synthesis. As the neutral theory shows, the point mutations in such genes usually do not produce material for speciation and macroevolution. At the same time, biological evolution is, of course, realized on the basis of the variability of genomes. The changes that the genomes undergo are varied, especially in eukaryotes.

As is known, the eukaryotic genome contained in the nucleoprotein systems of chromosomes is a complex one. Structural genes are composed of a number of fragments, known as exons, which are interrupted by non-coding "spacer" or "intron" sequences, which carry no information about the structure of proteins, i.e. about the amino-acid sequence. The genes that code for the polypeptide chains of proteins are made up of exons (coding sequences), the number of which may reach a few dozen. The transcription of DNA, i.e. the formation of the corresponding RNA, is followed by processing, the formation of mature RNA molecules from the precursor by snipping-off the terminal sequences and splicing. Splicing consists of removing the introns and joining the coding sequences (exons) into a whole gene. Both the transcription and the processing occur with the participation of a number of enzymes [7.1]. The scheme of these events is shown in Fig. 7.1.

The prokaryotic genome is a cyclic double-helical DNA molecule, which, unlike eukaryotic chromosomes, is not linked to proteins. There are no introns or exons in this cyclic structure, and the structural genes are divided into parts. The number of nucleotide pairs in eukaryotic introns ranges from 100 to 10,000. Their functions are still unknown, just like those of a number of other sequences that form chromosomes. The portions of DNA, such as the one shown in Fig. 7.1, occupy 5 to 30% of the eukaryotic genome [7.2].

In the last forty years two spectacular discoveries have been made in the field of molecular genetics. Firstly, the DNA that does not directly participate in protein synthesis has been discovered and explored to a certain extent. It is the "excess" or non-coding DNA (ncDNA). Secondly, the instability and mobility of the genome have been detected, topics to be discussed in the next section.

The non-coding DNA consists of comparable numbers of intergenic regions of interspersed repetitive sequences and repeats that form tandems. This non-

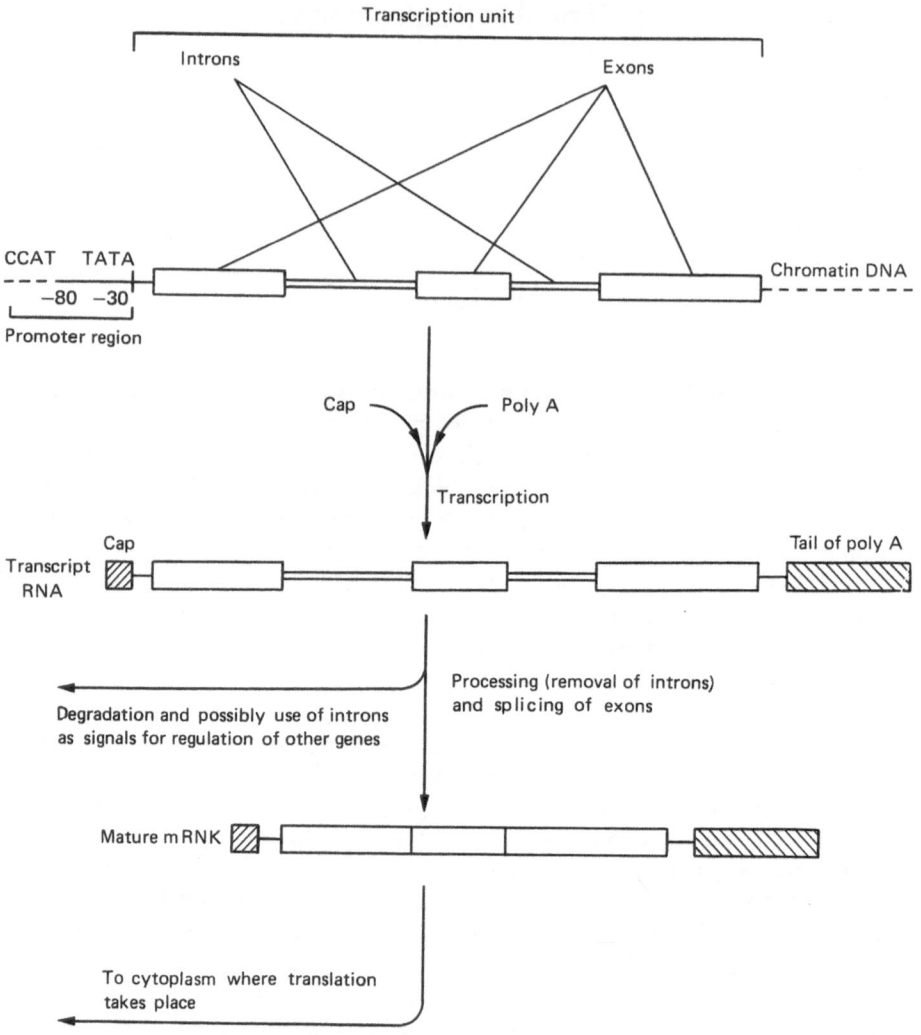

Fig. 7.1. Schematic representation of processing and splicing

coding DNA, which constitutes up to 95% of the genome, does not code for protein synthesis and (it is commonly believed) does not affect the phenotype. This is the reason for the appearance in the literature of various terms for the non-coding DNA: "junk DNA", "egotistic DNA" [7.3–6], "ignorant DNA" [7.7].

In other words, no functional biological significance is attached to this part of DNA. But is this the case in reality?

Before we try to answer this very important question, it should be noted that the "reading" of even the coding nucleotide sequence may be ambiguous;

one and the same nucleotide text can code for various overlapping amino-acid sequences [7.8]. Such situations are also possible in any ordinary language. Let us cite an example from the Russian language: "genotip uma" (mind genotype). If we begin reading this phrase, omitting the first letter and changing the spaces, we obtain the names of two mammals: enot i puma (racoon and puma).

Elegant cases of overlapping are possible in the French language. The following question is put:

"Que est ce qu'il y a de commun entre Paris, Virginie, l'ours blanc et Amundsen? ("What is common to Paris, Virginie, a polar bear and Amundsen?") (Virginie is the heroine of the novel *Paul and Virginie* by Bernardine de Saint-Pierre). The answer looks as follows:

1. "Paris est mètropole." (Paris is the capital.)
2. "Virginie aimait trop Paul." (Virginie loves Paul very much.)
3. "L'ours blanc est maitre au pol." (A polar bear is the master of the Pole.)
4. "Amundsen aime être au pol." (Amundsen likes to be at the Pole.)

As we see, the common feature between these sentences is that they all end with something which is pronounced "emetropol".

Let us consider the problem of the non-coding DNA (ncDNA). This problem has been discussed in detail in the literature [7.3–6]. Doolittle and Sapienza [7.4] maintain that the sole function of ncDNA is to survive inside genomes. According to Dawkins [7.3], organisms are only a means of multiplication for DNA and natural selection acts on genes and not on species or populations. It is, of course, difficult to agree with this; no genes can exist and undergo evolution outside an organism. However, there is no direct phenotypic substantiation for the presence of ncDNA. Therefore, one may speak of the survival and selection of ncDNA at a purely molecular level within the cells.

Orgel and Crick [7.5] also believe that the non-specific nature of ncDNA is evidence that at least the greater part of it originated through the multiplication of the sequences that either have a slight effect on the phenotype or do not affect it at all. The non-coding DNA appears in numerous copies, whereas the genes coding for proteins are represented in a single copy. Accordingly, the total DNA content in eukaryotes is much larger than in prokaryotes. At the same time, among eukaryotes we also find strongly differing amounts of DNA. For example, the amount of DNA in lilies and salamanders is 20 times greater than that in man. Is this connected in any way with evolution?

Orgel and Crick compare the non-coding DNA with not-too-harmful parasites that are converted into symbionts. The inheritance of ncDNA is evidently non-Mendelian since it involves the processes of DNA multiplication and redistribution among the chromosomes. There is a certain correlation between the DNA content and the time of generation, i.e. the rate of meiosis. Orgel and Crick note that the essential point here is not so much the total content of DNA, as the ratio of the amount of coding DNA to non-coding DNA. In eukaryotes this ratio is small.

Naturally, one cannot regard any DNA whose function is not obvious as "egotistic" or "junk". This notion of non-coding DNA is, however, useful for the simple reason that it is also unreasonable to seek a special function for each portion of DNA that has multiplied and is migrating in the chromosome.

Dover, in his brief article "Ignorant DNA?" [7.7], points out that the adjective "ignorant", which he suggests for DNA, is better because it implies a chance for DNA to learn something. The behaviour of ncDNA is governed by random processes, and it is quite capable of generating a number of sequences, including those which may prove to be functional. The "egotistic" process is teleological in character and undoubtedly one should take into account the existing randomness of the behaviour of non-coding DNA. It is not only non-coding DNA which is "ignorant"; we ourselves are "ignorant" too because we are still trying to understand its significance. Dover notes a number of points. Firstly, the overall growth of the genome is not quite accidental; it finds its expression in chromosomal sizes. This could possibly be associated with their specific behaviour in meiosis. Secondly, there are data indicating the effect of differences in the organization of long transcripts on splicing and transfer of short transcripts. Thirdly, there is the correlation pointed out by Orgel and Crick [7.5] between the total DNA content on the one hand, and the duration of the cell cycle and the time of generation on the other. Fourthly, certain DNA families exert some influence on the state of condensation of chromatin by their very presence and position. Finally, in a number of animal species the non-coding DNA forms rather specific structures in chromosomes. We may thus presume that ncDNA is not just "junk" DNA with no important function.

Zuckerkandl has written an extensive paper on the "polite" DNA [7.9]. He offers arguments in favour of the idea that DNA, while not being directly functional, is subject to limitations of the nucleotide composition and sequence since it interacts with the functional segments of DNA in chromatin. The "polite" DNA "respects" its neighbours.

Trifonov and his colleagues [7.10] have advanced a hypothesis which imparts a functional diversity to ncDNA other than the coding of proteins. Apart from the protein triplet code, other codes are also possible: (1) the code of the reading frames for translation, which is responsible for the correct counting of nucleotide triplets by ribosomes during protein synthesis; (2) the chromatin code which provides nucleosomes with appropriate locations along the DNA molecules, and their spatial orientation; (3) the "loop code" which determines the interaction of a single-chain RNA with protein. The key argument in favour of the participation of "egotistic" DNA in the coding of reading frames is the periodicity of mRNA-rRNA contact sites. They have an (N—N—C) structure, where N is any nucleotide and C is cytosine.

The chromatin code is also determined by the periodicity of the dinucleotides AA and TT, which recur along the sequence every 10.5 nucleotides on average. The formation of loops is associated with the presence of palindromes (inverted repeats) in ncDNA. One can encounter palindromes in any language. (By standard definition, a palindrome is a word or sentence that reads the same forward and backward.) Here are some examples from the English language:

"Madam", "A man, a plan, a canal, Panama", "Madam I'm Adam", "AND MADAM DNA".

The work of Holmquist [7.11] is devoted to the "molecular ecology" of ncDNA. He asserts that positional restrictions are imposed on ncDNA, which are determined by its participation in the structure of chromosomes. Many of the specific features of coding sequences (exons) are in turn determined to a larger extent by their positions; where the gene "lives" is more important than what it does. The dynamic organization of non-coding DNA leads to the occurrence of a feedback with the use of codons and stabilizes the structure of chromatin. The DNA sequences determine the affinity for proteins that form chromatin, which modulates the rate constants of DNA modification. The hypotheses suggested by Holmquist [7.11] concerning the hierarchial selection and molecular ecology show how selection can influence the Darwinian units of ncDNA at the genome level by producing a positionally restricted DNA.

There is, so far, no direct convincing evidence for the functionality (in particular, the "politeness") of the non-coding DNA. The conceptions outlined above are probable but still speculative hypotheses.

Undoubtedly, further research in this area is of great interest. But even at present, certain general considerations can be formulated which seem to be indisputable.

The problem of the non-coding DNA is tied up with the questions "what for?", "for what purpose?". The authors of numerous works proceed from the explicit or hidden concept of the fitness of any character of an organism to conditions of existence at any level, beginning with the molecular level. This implies that any character is subject to natural selection. Since the adaptive properties of ncDNA are still unknown, the idea of "junk" DNA arises.

As has already been pointed out, one should ask not "what for?" but rather "why?". Why is so much DNA contained in eukaryotes? Why do prokaryotes contain no introns?

The answer to the first question is embedded in the structure and dynamics of DNA. "Why Nature Chose Phosphates" is the title of a paper written by Westheimer [7.12]. Phosphoric acid is especially well suited to its role in nucleic acids since it is capable of linking two nucleotides and of being ionized at the same time. The negative charge stabilizes diesters against hydrolysis and provides the intactness of the molecules both inside the lipid membranes and within the cytoplasm. At the same time, the energies of phosphodiester linkages are such that only a small fraction of these linkages can be cleaved due to hydrolysis at acid pH. The nucleic acids can be cut and spliced together again under the action of enzymes. Moreover, they are capable of replication by way of template synthesis. Since the discovery of the ability of RNA to catalyze the splicing proper, the idea of the decisive role of RNA in the origin of life has been finding ever-increasing recognition [7.13]. To what has just been said we may add the variety of conformational possibilities of DNA and RNA, which plays an important role.

The first primitive cells probably contained the various DNA and RNA sequences. As has been shown in the Eigen theory [7.14–16], the totality of such macromolecules, which are built up of monomers and which break down in an open system, undergoes natural selection; prebiotic evolution is also Darwinian. Various templates compete in such a system. It is exactly the autocatalytic nature of template replication that makes possible an evolutionary optimization, which is missing in conventional chemistry. A detailed exposition of this theory, which is of prime importance to the theory of evolution as a whole, is given in the next chapter. For the moment we will confine ourselves to just certain points [7.16]:

1. The target of selection is not the single wild type, but rather the total mutant distribution, called quasi species.
2. Neutral mutants modify the quasi-species distribution to a considerable extent but, in sufficiently large populations, generally do not lead to random drift.
3. Despite the appearance of a unique sequence, the underlying wild type is usually present only in relatively small numbers. The sequence in the population appears to be uniform, since it is a defined consensus sequence. So-called "hot spots" in sequences are more likely to be due to the presence of nearly neutral mutants than to greatly increased mutation rates.
4. Nearly neutral mutants not only exist, but also form more frequently than deleterious ones, because fitness is clustered in connected mutant domains.
5. Evolutionary optimization is biased by the mutant population and proceeds along continuous routes of increasing fitness through intermediates that are nearly neutral.

These propositions have been confirmed by direct experiments on the phage $Q\beta$ RNA [7.16].

It might be thought that the molecules of ncDNA obey natural selection within a cell, i.e. that a kind of prebiotic evolution is realized in them. Of course, this process is more complicated than the process considered in the Eigen theory. Firstly, competing macromolecules appear in the open system of the cell not by chance but predominantly as a result of numerous duplications of genome fragments. Secondly, the eukaryotic genomes are localized in chromosomes, in complex nucleoprotein structures. From this it follows that not only the primary structure but also the conformational properties play an important role in the behaviour of DNA.

One may be surprised why the diverse possibilities of cleavage, splicing and template duplication of the polynucleotide chain do not lead to a complete random disorder in an assembly of DNA molecules. Evidently, it is precisely the presence of those chromosomes which determine the processes of mitosis and meiosis that creates the organization and directionality in the multiplication of ncDNA.

The statement to the effect that the non-coding DNA is non-functional hinges on the fact that only one function of DNA (the biosynthesis of proteins by way of transcription and translation) has been studied well. DNA is normally considered only from the viewpoint of adaptation to the familiar "ecological niche", i.e. to the protein biosynthesis system. However, the general proposition of biology is that the results of natural selection, be it species or molecules, not only adapt to existent ecological niches but create new ones; it is not they that adapt to the environment, but the environment itself that adapts to them (see Sect. 7.6). It may be believed that the numerous structural and dynamical possibilities of any DNA are such that the non-coding DNA is also utilized in biology one way or another. Of course, these considerations are purely speculative; we have no solid facts at our disposal to prove them. Junk is something to be discarded as worthless. The non-coding DNA may rather be compared to the raw material used in some technological process. We have already described the reasonable hypotheses of the functional importance of the non-coding DNA.

Introns are also regarded as "junk". In most cases, they break down after being spliced. Situations are known, however, in which the intron of one gene proves to be a coding sequence (exon) of another gene. An intron has been found in yeast mitochondria, which codes for a regulatory protein [7.1].

The complexity of the problem of ncDNA boils down to the interrelationship and competition of two types of selection: molecular, "prebiotic" selection of molecules in the cell; and natural selection of organisms and populations. It is perhaps in the course of such competition that two major kingdoms of living things arose: prokaryotes and eukaryotes. Two ways of realizing evolutionary diversity exist. The first is the utilization of a large reserve of ncDNA, in particular introns, for creating wide variability at the molecular level. During processing the formation of a set of various mature mRNAs is possible, whose translation produces various proteins. Of course, the doubling of DNA requires both time and energy and, naturally, eukaryotes are sufficiently energized and the division of their cells occurs slowly. The second pathway is the rapid production of material for evolution in a large amount in prokaryotes. Bacterial cells are small and undergo division often; they obey relatively simple physical laws that govern the division. The appearance of daughter cells of the prokaryote is directly associated with the doubling of the DNA of the mother cell. In this sense, the multiplication and competition of mutant prokaryotes and also of viruses and phages is similar to prebiotic evolution according to Eigen.

There is every ground for believing that the most ancient prokaryotes, archaebacteria, were capable of splicing. The intron-exon structure is inherent even in adenoviruses. The subsequent development of the prokaryotic kingdom is associated with the loss of introns [7.1].

To conclude this section we note that the exon structure of a gene finds in many cases its direct expression in the domain structure of proteins. It is possible that the observed relationship between exons and protein structures is the degenerate state of the archaic correspondence between exons and structurally functional units in proteins [7.17].

7.2 Instability of the Genome

One of the fundamental premises of classical genetics, the foundations of which were laid by Mendel's works, is the constancy of the genome, whose variability boils down to the appearance of point mutations and to a crossing-over of chromosomes (see Sect. 7.4). Phenotypic changes in the course of evolution are associated only with the appearance of mutant alleles, and the inheritance of traits obeys Mendel's laws. These ideas have become an inseparable part of the synthetic theory of evolution.

Today, now that we have become aware of the enormous potential for conversion of nucleic acids, of the ability of DNA to be cleaved and spliced, to duplicate through template synthesis and also to undergo various conformational changes, it seems to be quite natural that the genome must be a mobile, dynamic system. It is necessary to understand how the mobility of the genome can be combined with the constancy of genes over numerous generations; recall the drooping lower lip of representatives of the Habsburg dynasty mentioned by Schrödinger [7.18]. It means that the mobility is limited. And if this is so, then what is the essence of speciation and evolution at the molecular level?

In 1951, two years before the discovery of the DNA double helix, Barbara McClintock, who was concerned with the genetics of maize, published a pioneering work devoted to the instability of the genome: she had discovered transpositions, i.e. the movements of a portion of a chromosome either within the host chromosome or into another chromosome (see [7.19]). Much later McClintock was awarded the Nobel Prize for her work. Her Nobel lecture has been published (see [7.20]).

At present numerous facts are known which provide evidence for the instability of the genome. The events in which this instability, mobility and dynamics manifest themselves are of primary importance to biology in general, and evolutionary theory in particular.

Only a brief description of this important area, based on a fundamental monograph written by R.P. Khesin [7.21], will be given here.

Transposable (mobile) genetic elements (TGE) exist in the cells of both prokaryotes and eukaryotes. Prokaryotes contain the following types of TGE:

1. Insertion sequences (IS elements), which move from site to site in the genome, without assuming a free state.
2. Transposons, which are complex translocatable elements containing structural genes.
3. Episomes, which are plasmids capable of independent replication and of integration into a chromosome.
4. Temperate bacteriophages, in particular the lambda phage (or phage λ), which insert their DNA into the bacterial chromosome.

Transposable genetic elements (TGE) of the first two types, as well as plasmids and viruses, also exist in eukaroytic cells. A number of viruses which are capable of altering the genome have been found in *Drosophila*; the genetics of this organism has been studied more thoroughly than that of any other eukaryote. The so-called I-factors are capable of moving within the genome, like bacterial transposons.

Hogness, Georgiev and Gvozdev have discovered a new type of *Drosophila* TGE ([7.22], see also [7.23]). These elements are multiplicative, are scattered throughout the entire genome and capable of transposition. The TGE families are represented by a large number of copies. They are species-specific, that is to say, they are different even in closely related species such as *Drosophila melano-gaster* and *Drosophila virilis*. According to Khesin, these TGEs appeared in the species mentioned even after their differentiation. Species-specificity is deter-mined by the high evolutionary variability of TGE. The positions of copies of a single TGE differ appreciably in chromosomes of the different lines of *Dro-sophila*. The lengths of TGE run to thousands of nucleotide pairs.

We also know of families of short transposable DNA segments in the genomes of *Drosophila* and mammals, particularly palindromes.

The so-called Alu-family in the genome of man and monkeys has been studied well. This comprises short sequences of about 300 nucleotide pairs, and their numbers may amount to 500,000, which is up to 3 percent of the entire DNA.

A number of unstable genes in *Drosophila*, which are characterized by high mutability, have been studied. The unstable loci can spontanously move from site to site in chromosomes. These are transpositions of genes.

This brief information does not exhaust the large variety of factors that cause changes in the genome of *Drosophila*. In the light of what was said in Sect. 7.1, we should not be surprised by these phenomena. The lability of DNA, its ability to be cleaved and spliced together and to duplicate, points to the mobility and instability of the genome. However, eukaryotes also possess highly stable genes which determine heredity. It is difficult so say why some genes are stable and others are not. We can only assert that the changes of hereditary traits brought about by transpositions of the mobile elements occur rather sel-dom, though much more frequently than point mutations. The changes caused by the instability of the genome may not be fixed and may not manifest them-selves in heredity since the newly formed genes are not stable either.

Presumably, the mobile elements are representatives of the "egotistic" DNA. As we have seen, its egotism is probably relative. We have already put forward arguments in favour of the random nature of the non-coding DNA. Since the transposable elements are represented by a large number of copies, their behaviour in cells must obey the laws of population genetics. Calculations have been made which show that the repetitions are scattered throughout the genome and dissociate into original tandems [7.24]. The exchange of genetic material presumably preserves their average population. We have already pointed out that the properties of the non-coding DNA may be described in terms of the Eigen theory of pre-biotic evolution. In 1984 Khesin wrote that "intracellular evolution, which is a very intriguing problem, awaits a special study" [7.21].

Georgiev has suggested that transposable genetic elements be divided into two classes, active and passive [7.25]. Active transposons include genes which code for the enzymes required for transposition. The number of passive transposons is much larger. They can move only with the aid of enzymes that are coded for by other genes. There are two ways of transposition of TGEs. The first includes a so-called revertase, an enzyme coded for by retrovirus proviruses or by active TGEs. The revertase catalyzes the synthesis of DNA on a RNA template, which arises as a result of the transcription of TGE. This DNA is then incorporated into another site of the genome. In this way new copies of TGE can be created.

The second way is realized with the participation of transposases. These enzymes determine a direct excision of TGEs from the genome. This portion of DNA may then be lost or moved to another site.

The "egotism" of TGEs is so called because they either do not code for proteins at all, or code only for proteins that do not take part in the functioning of the cell. TGEs of the first type are passive and those of the second type are active. The latter are responsible only for proteins that participate in the transposition of TGEs (transposases). Transposable genetic elements may be looked upon as specific genome parasites since they utilize energy for their autonomous replication, transposition and, in a number of cases, for translation. Georgiev has carried out a detailed study of a number of transposable genetic elements of *Drosophila* and other eukaryotes and has described them. He has detected phenomena of TGE competition and has clearly shown that the effect of TGE on the expression of neighbouring genes is significant [7.25].

The role of TGEs in evolution is undoubtedly very great. Transposable genetic elements induce the increased mutability of neighbouring genes. The turning-on of TGEs is an important source of genetic variability. Though the transpositions of TGEs are rare phenomena, the rates of transpositions being less than 10^{-4} to 10^{-5}, they can be activated under special conditions. Here, the interactions of various mobile elements are essential. In certain lines there is observed an inherited genetic instability. Georgiev has described "transposition bursts" which are observed in some germ cells of *Drosophila* [7.25]. In these cells diverse independent transposition phenomena occur, in which various TGEs take part. "Transposition bursts" lead to multiple mutagenesis. Presumably, they play a significant role in evolution and can also serve as an important factor in somatic mutations. Thus, they probably have a role in carcinogenesis. In most tumour cells the transpositions are appreciably enhanced.

Further details are given in Khesin's monograph [7.21] and in a book by Georgiev [7.26] (see also [7.27]).

The instability of the genome implies the possibility of transfer of genes from some organisms to others ([7.21], Chap. 8). In a number of experiments translocations have been realized between various representatives of eukaryotes, including translocations between organisms of different classes. The transfer of genes can be realized with the aid of viruses and plasmids. Khesin maintains that the existence of the flow of genes from species to species is possible, although it would be more reasonable to speak of "percolation" rather than of flow. Evidently, these events are rather rare. Transpositions of genes between bacterial and animal cells, and between bacteria and yeast, have been carried out experimentally. The natural transfer of Cu-Zn genes (the superoxide dismutase of fish and the bacteria *Photobacter leiognathi*) has been thoroughly studied.

The possibility of gene transposition points to the low species-specificity of the corresponding proteins and may thus be regarded as an argument in favour of the neutral theory.

Khesin writes: "There is no doubt that the transposition of genes between organisms occurs at all times. But alien genes are seldom fixed since they are beneficial to an organism only under special conditions and are not fixed by selection.... Therefore, the idea of the gene pool being common to all organisms implies only their potential ability to borrow the necessary genes from any most remote species. This potential ability is realized rather seldom."

These words are based on a comprehensive, thorough and detailed study of all problems associated with the instability of the genome.

Meanwhile, attempts are being made to explain the entire process of biological evolution by a "horizontal" transposition of genes from organism to organism. This superficial idea is discussed by Kordium [7.28]. His book has been cogently criticized by Tatarinov [7.29] (see also [7.30]). It should be noted that by ascribing the determining role in evolution to a horizontal transposition we approach Lamarckism: "An evolutionist who supports the Lamarck illusion protects evolution by means of viral infection" [7.31].

As we see (Sect. 1.2), there are firm grounds for comparing biological evolution with the evolution of languages. In contrast to biology, a horizontal transfer really plays an important role in the origin and development of languages. The Norman conquest of England, as a result of which numerous French words were introduced into a language of Germanic origin, is a striking example. The appearance of the Russian words "sputnik", "perestroika", "glasnost" in a number of foreign languages is the result of a horizontal transfer. Of course, this is possible as far as languages are concerned; a language is not developed at the genetic level. However, in these cases too, a foundation is retained which arises by way of evolution.

7.3 Gene Duplication

In the preceding sections we have not touched upon the problems of speciation. We have only become convinced that point mutations do not, as a rule, serve as material for evolution under invariable conditions. As was pointed out in Chap. 6, there are four important propositions of the neutral theory. Let us repeat the fourth one: The appearance of a gene with a new function must always be preceded by the duplication of the gene.

The doubling (duplication) or multiplication (amplification) of genes make possible changes in genes without upsetting the functioning of the original gene. As Frazzetta put it in his *Complex Adaptations in Evolving Populations* (see [7.32]), "The evolutionary problem is, in a real sense, the gradual improvement of a machine while it is running."

The allelic mutations of existing genes evidently cannot be responsible for great evolutionary changes, i.e. speciation or, especially, macroevolution. Ohno [7.31] has given detailed and convincing arguments in favour of the determining role of duplication. The presence of a redundancy, i.e. copies of the original structural gene, makes possible the synthesis of an altered protein with a simultaneous undisturbed synthesis of protein by the original gene. The corresponding changes of the original gene would have been impossible, as these mutations are lethal in the long run. Ohno writes: "Only by the accumulation of forbidden mutations at the active sites can the gene locus change its basic character and become a new gene locus. An escape from the ruthless pressure of natural selection is provided by the mechanism of gene duplication." Natural selection rejects forbidden mutations only if the genome contains structural genes in one copy. It is duplication that leads directly to the polymorphism of enzymes, to isozymes. We may say that at the molecular level, speciation begins with polymorphism. In the course of further evolution, because of the duplication being followed by the development of various genetic regulatory mechanisms, the redundant genes may undergo change. Molecular evolution begins with natural selection ignoring a redundant gene, which is thereby free for the accumulation of forbidden mutations, these being determined, in particular, by an intragenic recombination (see Sect. 7.4). The presence of duplicated, redundant genes causes their accelerated evolution, which of course does not contradict in any way the neutral theory.

The general scheme that explains the evolution of a gene as a result of its duplication is shown in Fig. 7.2 (borrowed from Dickerson [7.33]).

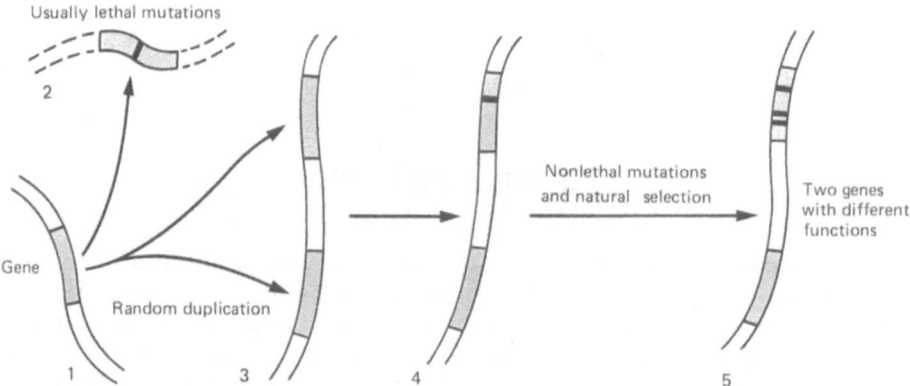

Fig. 7.2. Duplication and consequent changes of gene [7.33]

The important question is: What is the cause of such an acceleration –
beneficial mutations or a reduction in constraints on selection? A study of accelerated evolution associated with gene duplication has been carried out by Li
and Gojobori [7.34] (see also [7.35]), who examined the duplicated genes of the
β-globin of goat and sheep. Let us consider their work.

In goats, just as in humans, there occurs a switch of the synthesis of haemoglobin (Hb). First, an embryonic haemoglobin, $\alpha_2\varepsilon_2$, is synthesized, this being
followed by normal fetal Hb, $\alpha_2\gamma_2$. In humans, the synthesis of $\alpha_2\gamma_2$ is followed
by the synthesis of normal adult Hb, $\alpha_2\beta_2$, and in goats pre-adult Hb, $\alpha_2\beta_2^c$, is
formed first, followed by the synthesis of adult Hb ($\alpha_2\beta_2^A$). Presumably, 8 Hb
genes of goats arise by way of the block duplication of four genes ε-ε-$\psi\beta$-β,
which is shown in Fig. 7.3 [7.36]. The $\psi\beta$ gene is a pseudogene, i.e. a repeatedly
mutated gene, which has lost its function. It is assumed that the gene γ is located
farther along the chromosome and is a member of the third group of genes,
including ε^V and produced by an earlier duplication. An estimate of the rates of
non-synonymous substitutions in β^A and β^C, made by Li and Gojobori [7.34],
shows, in fact, that acceleration did occur. Based on nucleotide sequences, these
authors estimated the time of divergence T_1 between the goat and sheep as
7×10^6 years, the time of gene duplication T_2 separating β^A and β^C as 15×10^6
years and the time of duplication separating γ and the ancestor β^A and $\beta^C(T_3)$
as 3×10^7 years. This is shown schematically in Fig. 7.4 [7.35]. From these data
and also from the number of substitutions at non-synonymous sites we have
results given in Table 7.1.

In order to establish the main cause of acceleration, Li and Gojobori calculated the rates of non-synonymous substitutions in various functional groups of

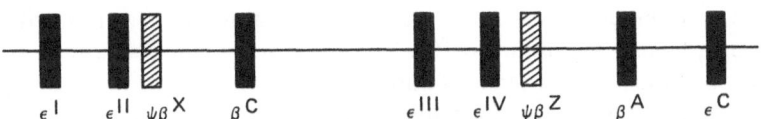

Fig. 7.3. Location of 8 genes of goat haemoglobin [7.36]

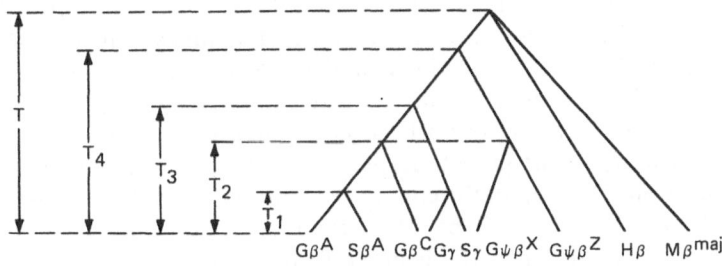

Fig. 7.4. Schematic representation of the evolution of haemoglobin genes. S = sheep; G = goat;
H = man; M = mouse [7.35]

Table 7.1. Annual rates of non-synonymous substitutions

Genes compared	Time of divergence for 10⁶ years	Rate (× 10⁹)	Acceleration factor
Hβ and Mβ^{maj}	80	0.8 ± 0.1	(1.0)
Gβ^A and Sβ^A	7	0.9 ± 0.5	1.1
Gγ and Sγ	7	1.4 ± 0.6	1.8
Gβ^C and Gβ^A, Sβ^A	15	2.3 ± 0.5	2.9
Gγ, Sγ and Gβ^A, Sβ^A, Gβ^C	30	2.3 ± 0.3	2.9

the β-chain of Hb. Here, of prime importance are the amino-acid residues that come into contact with the haem group and the residues responsible for inter-chain co-operation. It has been established that acceleration occurred only in the less important functional groups of the β-chain, primarily in the amino-acid residues located at the surface. We may conclude that the acceleration is first of all accounted for by relaxation, i.e. a decrease in constraints on selection, since a certain degree of relaxation allows for an acceleration of substitutions in less important but not in more important regions of the gene.

A similar treatment carried out for the genes of somatostatin, cytochrome c, growth hormone, somatomammotropin and insulins I and II of the rat, has also shown that the acceleration of evolution occurs in a single gene in a dupli-cate. Thus, accelerated evolution often follows gene duplication, which can be accounted for by the relaxation of selection constraints rather than by the ap-pearance of beneficial mutations.

The phenomena described above are intimately connected with gene con-version, which will be considered in the next section.

Flavell has studied the amplification of genes using a number of cereals [7.37]. The number of repeats in the corresponding genes varies within very wide limits, from a few units to 10^6. The repeats are often arranged into long tandems, but most of the repeat families are organized in a more complicated way. Some members of these families alternate with short, non-repeating por-tions of DNA or with dissimilar, repeated sequences. Either is possible. Figure 7.5 shows a scheme which illustrates the origin and evolution of repetitions [7.37]. New groups of repeats arise as a result of the amplification of the non-repeating DNA (NR) or of the repeated amplification of the repeat (R). It is obvious that in such a system the appearance of complex sequences is possible due to substitutions or conversions of the old sequences to new ones. Re-arrangement of the sequences may take place in portions of the genome in which R and NR alternate, in portions containing only R groups, or between these two portions. At the same time, deletions (D) occur in portions with differ-ent organization. Thereby, in the course of evolution a certain exchange of DNA occurs. The Flavell scheme is based on the assumption that sequences of one family may be organized in the genome in different ways. Amplification and deletions lead to the accumulation in chromosomes of significant changes, which may be called macromutations. Of course, part of these macromutations may be

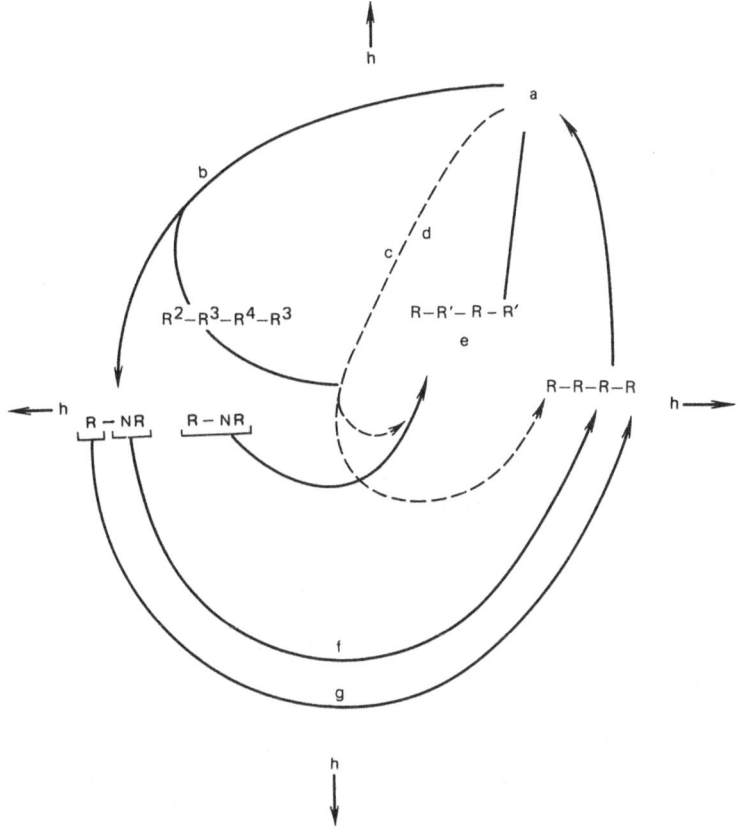

Fig. 7.5. Diagram illustrating the origin and evolution of repeating sequences; (*a*) divergent groups of sequences; (*b*) rearrangements of sequences; (*c*) repeated amplification; (*d*) substitution; (*e*) complex groups; (*f*) amplification; (*g*) repeated amplification; (*h*) deletions [7.37]

neutral, but there is every ground for believing that speciation and macro-evolution are determined at the molecular level exactly by macromutations. Flavell maintains that ncDNA plays a significant role in the origination of macromutations and has put forward the hypothesis of a more accelerated molecular evolution of organisms with a high content of the total DNA.

7.4 Unequal Crossing-Over and Gene Conversion

A change in the number of repetitions in tandems occurs by way of unequal crossing-over.

Crossing-over is a phenomenon which has been thoroughly studied. It is a process by which parts of homologous chromosomes are interchanged during

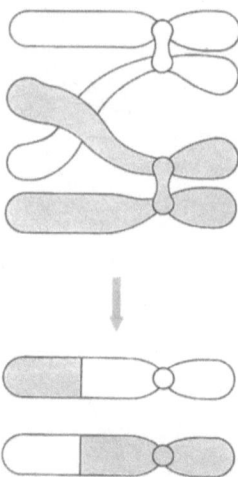

Fig. 7.6. Schematic representation of normal crossing-over after doubling of chromosomes

meiosis. Meiosis involves the conjunction of homologous parental chromosomes, which is known as synapsis. The greater the distance between the two parts, the more likely it is that they will be exchanged. The scheme of normal, i.e. symmetric or equal, crossing-over is shown in Fig. 7.6 [7.1]. As a result of crossing-over, the progeny genome is a recombinant of paternal and maternal genomes. Recombinations constitute the main factor in the variability of organisms. Recombinations are not deterministic; their occurrence and distribution are random. The recombinant combinations of parental features in normal crossing-over do not lead to any departures from Mendel's laws.

However, unequal crossing-over, in which the chromosomes that have undergone such an exchange of parts appear to be non-homologous, also takes place. Thus, a pair of chromosomes may be formed, one of which contains one recombinant copy of the gene, while the other contains three gene copies: two from the parents, and one recombinant copy. The relevant scheme is shown in Fig. 7.7 [7.1].

Unequal crossing-over results from an incorrect conjunction of the parental chromosomes. The behaviour of the resultant non-homologous chromosomes does not obey Mendel's laws.

From what has just been said it follows that "the amplification of a sequence, however rarely it occurs in evolution, implies that the sequence departs from strict Mendelian inheritance over evolutionary time periods" [7.37]. Speciation does not obey these laws; it does not boil down to the reshuffle of the alleles of stable genes. The genes themselves are altered due to duplication and associated macromutations.

In fact, the mobility or instability of genomes manifests itself in both duplications and unequal crossing-over. Let us consider one more important phenomenon which is intimately associated with those just mentioned. This is gene

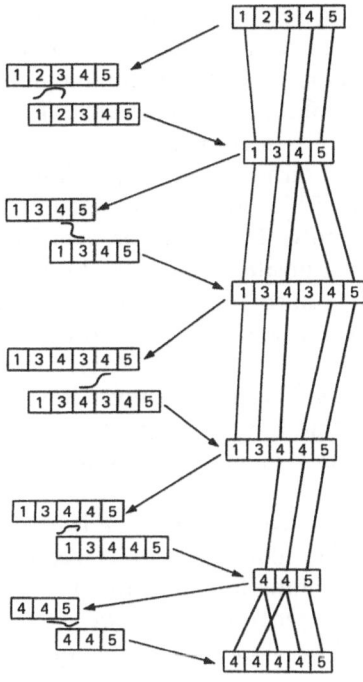

Fig. 7.7. Schematic representation of unequal crossing-over [7.1]

conversion. The definition of gene conversion given by Ayala and Kiger [7.19] is as follows: "A process as a result of which an allele on the intact chromosome is lost and replaced by another allele from the same or a homologous chromosome".

The instability of the genome was first discovered in maize, and gene conversion was first observed in mushrooms characterized by a specific process of meiosis. In a number of mushrooms, meiosis occurs in a cell called an ascus, in which spores are formed. Their regular disposition allows one to reveal the specific features and sequence of crossovers (Chap. 5 in [7.19]). In the asci of the mushrooms *Sordaria fimicola* variants are observed. If the parental forms carrying two different alleles are crossed, the eight spores present in the offspring asci consist, as a rule, of four wild-type spores (+) and four mutant-type spores (m). However, an aberrant segregation is observed in a small fraction of asci. For 127,000 cases of 4+:4m there are 98 of 6+:2m, 108 of 5+:3m and 15 of 4+:4m, with a changed sequence of spores, and 20 of 3+:5m and 13 of 2+:6m (Chap. 14 of [7.19]). These rare aberrations do not result from the mutational conversions of some spores to others. A detailed analysis has shown that gene conversion occurs, a phenomenon that has been studied in many other organisms, in particular *Drosophila*.

Ohta [7.38–41] has explored in detail the mechanisms by which genetic variability is maintained in multigene families which are widespread in the genomes of higher organisms. Such families undergo joint evolution, also called concerted evolution. The family of genes of immunoglobulins is a good example. The above mechanisms for multigene families are different from the mechanisms by which genetic variability is maintained for ordinary models of some loci. Gene conversion should be considered in conjunction with duplicative transposition. Ohta is concerned with a randomly crossed population. There is a dispersed repetitive DNA family. The family is evolving under mutation, gene conversion, duplicative transposition, and interchromosomal recombination.

Gene conversion is assumed to occur both within and between the two genomes of a diploid cell. Let w be the proportion of conversions that occur within a genome, and let $1 - w$ be the proportion between genomes in a cell. If λ_c is the rate per generation at which gene conversion takes place, then $w\lambda_c$ is the rate of gene conversion within the genome and $(1 - w)\lambda_c$ is the rate at which it occurs between the genomes. The model is given in Fig. 7.8. The conversion is not symmetric.

It is assumed that duplicative transposition is accompanied by the deletion of another gene, so that the number of copies per genome does not change. Assuming that the corresponding fractions are also equal to w and $1 - w$ and the rate of transposition per generation is λ_t, we see that the duplicative transposition inside the genome occurs at a rate $w\lambda_t$ and that between the genomes at a rate $(1 - w)\lambda_t$.

Ohta has carried out a population-genetics analysis of concerted evolution, i.e. joint effects of mutations, conversions and transpositions of genes. A method has been developed for calculating "allelism", i.e. the probability that if two genomes are randomly chosen in a population, and a certain unit of one of the genomes is also randomly chosen, then there is a homologous copy of this unit at the same site of the chromosome of another genome. Allelism is designated as F. Its change in one generation is equal to

$$\Delta F = -2\lambda_t F + \frac{1}{2N_e}(1 - F) \tag{7.1}$$

where N_e is the effective population size. At equilibrium $\Delta F = 0$ and

$$F_{eq} = \frac{1}{1 - 4N_e\lambda_t} . \tag{7.2}$$

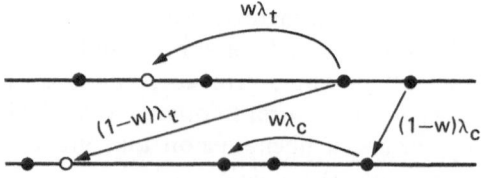

Fig. 7.8. Gene conversion

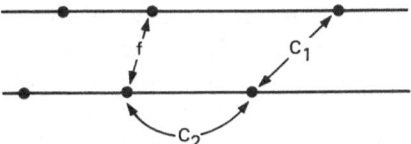

Fig. 7.9. Identity coefficients

Three identity coefficients are calculated: the allelic identity f; the identity coefficient for the units within a genome, C_1, and the identity coefficient for units belonging to different genomes, C_2 (see Fig. 7.9). The method of calculation is described by Ohta [7.41]. The theory of identity coefficients allows one not only to predict the average identity but also to estimate the time elapsed up to the fixation of a mutant that belongs to the multigene family. For example, for the variable part of immunoglobulins this time is 3×10^8 years and for ribosomal RNA (rRNA) genes it is a few millions of years.

Ohta writes that the propagation of genetic information not only in a population but also within each genome plays an important role in the evolution of the complexity of higher organisms. In relevant calculations the multigene family should be regarded as a unit.

We have briefly considered the phenomena that lead to non-Mendelian inheritance. It is these phenomena that appear in the molecular drive concept.

7.5 Molecular Drive

What is biological evolution – speciation and macroevolution – at the molecular level? The determining role of natural selection is obvious. Darwinian selection deals with phenotypes and thus indirectly with proteins. What are the phenomena that underlie the variability of proteins?

Proteins undergo change as a result of the accumulation of point mutations, i.e. as a result of molecular mutations. This is a fact rather than an evolutionary factor, as was shown in Chap. 6.

A large set of phenomena determined by the duplication and mobility of genes (transposition, unequal crossing-over, conversion, which in turn are determined by the presence of ncDNA and by the exon-intron structure of eukaryotic genomes) give rise to the diversity of proteins and, hence, of material for natural selection at the molecular level.

In 1982 Dover summarized these phenomena for the first time in his concept of molecular drive [7.42–44]. In subsequent years he continued to develop this concept [7.45–49].

Today it is common to say that molecular evolution is governed by three main factors: molecular (or genetic) drift, molecular drive, and natural selection. This is correct but not quite logical; in fact, the drift and drive form material for natural selection.

Let us cite the definition of molecular drive as given by Dover [7.46]: "Molecular drive is a process, operationally distinct from natural selection and genetic drift, for changing the average genotype of a population. It is a consequence of a variety of genomic mechanisms of turnover which have the effect of simultaneously spreading new variants through a multigene family (homogenization) and through a sexual population (fixation). The cohesive dynamics of the process, and the opportunity for molecular co-evolution, permit new ways for the evolution of biological novelties."

The mechanisms of gene turnover considered by Dover may be responsible for both the preservation and divergence of sequences. This follows from the facts which were discussed in the preceding sections.

Mendelian genetics, which is one of the foundations of the synthetic theory of evolution, is based on the identity of the dynamics of genes and chromosomes; genes occupy permanent loci on chromosomes and can only mutate or undergo (together with chromosomes) equal crossing over. As we have seen, the actual situation here is different, as the shift of genes may not coincide with the movement of chromosomes. Deviations from Mendel's laws, which are inconspicuous in small populations, are enhanced during short periods of time in a number of generations in large populations.

Figure 7.10 shows a scheme of molecular drive phenomena that arise within homologous and non-homologous chromosomes and also between them. A mutant of one of the genes of the family can replace wild-type genes either by chance or as a result of a directed transposition. The chromosomes then enter new individuals and a fixation of mutants in the population occurs, which is closely associated with the homogenization of the family. Molecular drive involves processes of gene conversion, unequal crossing over, transposition, slippage replication and exchanges by means of RNA. A certain fraction of accessible variability in nuclear genomes is generated as a result of the action of the enzyme reverse transcriptase, which transcribes RNA into the complementary DNA (cDNA), the latter being incorporated into various sites of the genome. In this way large families of DNA arose, for example, the Alu-family containing up

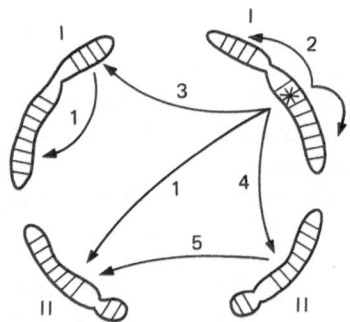

Fig. 7.10. Schematic representation of the phenomena of molecular drive: (*1*) gene conversion; (*2*) slippage replication; (*3*) unequal exchange; (*4*) transposition; (*5*) transfer by means of RNA [7.48]

to 500,000 copies per human being. Such a mechanism can lead to rapid accumulation of repetitions and pseudogenes.

An elementary scheme, according to which genes and corresponding proteins function independently, is insufficient. The most important processes in a cell, involving the organization and doubling of chromosomes, cell division, etc., are governed by the concerted action of numerous proteins synthesized in multigene families. This also refers to all functionally important RNAs. The mechanisms of molecular drive lead to the appearance of new genes and to the loss of the old ones. This makes possible a random propagation of altered genes in both a family and in a population. Dover gives the following example. Let us assume that a family of 100 genes contains 99 wild-type copies and one mutant in a given organism. The drive mechanisms can change this ratio with the formation of either 98:2 or 100:0. In the first case two mutant copies may find their way into two new individuals of the next generation and the processes may be repeated. Of course, the probability of the loss of a new gene is high, but a random replacement of a wild-type gene by a mutant gene in the family and population is not impossible. A new amplification of a new family will lead to a random accumulation of new genes. Of course, these probabilities depend on the size of the family and population and on the rate of an irreversible transposition.

The segregation of chromosomes at meiosis and their redistribution in a new generation is the main mechanism of propagation of altered genes. Since processes of various combinations of chromosomes always occur, the homogenization and fixation of genes in a family and population are intimately associated with each other.

The rate of generation of new copies of a mutant gene as a result of a drive (gene turnover) is of the order of 10^{-2} to 10^{-5} per generation. The propagation of genes due to sexual reproduction occurs much faster. The average number of copies of a mutant gene per individual changes gradually in a population from zero to complete homogenization. Figure 7.11 shows the corre-

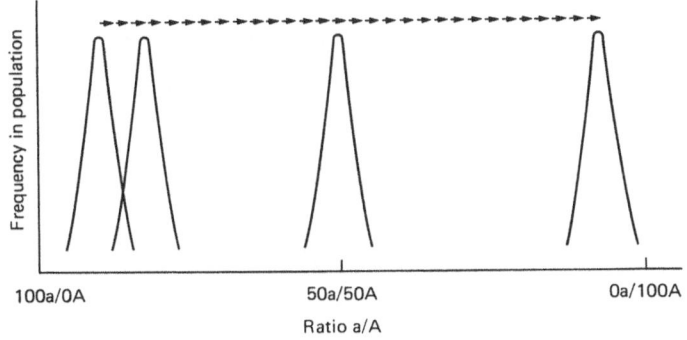

Fig. 7.11. Diagram showing the gradual change of the content of wild-type and mutant in a population due to molecular drive [7.46]

sponding scheme of gradual change of the content of two genes $a:A$ (wild-type and mutant) in the course of population development.

It must be emphasized that the graduality of such fixation is relative. The appearance of new genes and their propagation in a population presumably requires much less time than the time of stasis, the apparent persistence of species.

The drive, the interconnected significant changes of the genome, provides material for natural selection. Natural selection depends on two factors: the possibility of adaptation to a new ecological niche, and the internal mutual fitness of interacting molecules during ontogeny. As a result of selection only those species survive which are characterized by a wide homogeneity of variants within a given species. Concerted evolution takes place, owing to which the similarity of the sequences in a family of genes within a species is higher than between the members of different species. A relevant scheme is given in Fig. 7.12. Molecular drive explains the observed picture of concerted evolution. What has been said in no way contradicts the polymorphism determined by neutral mutations. Such polymorphism does not imply the appearance of new genes with appreciably altered properties.

In contrast to natural selection and drift, molecular drive propagates new information in population that multiplies by sexual reproduction, even in the absence of any advantages of given variant. It is only molecular drive that can account for the recurrence of homogenization cycles, which spread independently in gene families.

Molecular drive is a process which organizes the separate evolution of each gene family, which is independent of the distribution of a gene family in chromosomes and in organisms.

Molecular drive does not boil down merely to a directed gene conversion similar to that observed in mushrooms, but it involves this process. There are gene and non-gene families in which directed and undirected turnover mechanisms operate simultaneously.

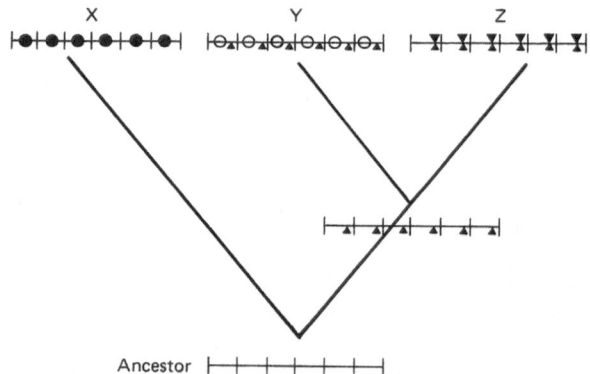

Fig. 7.12. Concerted evolution. A family of tandem genes in the original state evolves, leading to final species X, Y, Z [7.46]

Of course, both coding and non-coding DNAs participate in molecular drive. The relevant problems have been explored insufficiently, but the ability of any DNA to play a significant role in speciation seems to be beyond doubt.

It is clear that genetic changes leading to the generation of new functions and, hence, to biological evolution, can develop and spread due to a successful interaction with the existing molecular components. This is valid for both natural selection and molecular drive. Such an interaction implies co-evolution, the joint evolution of various molecular factors.

Thus, molecular drive expresses the intrinsic ability of organisms to influence the mutual fitness of molecular changes. Both a cell and an organism are a sort of ecological niches for molecules (called molecular cenoses). From this it follows that we cannot say organisms appear as a result of exact fitting, i.e. adaptation to a biotic and physical environment. Apart from adaptation, an internal process also occurs, which Dover has termed "adoptation". In the case of adaptation, "the environment technically sets the 'problem', and the 'solution' is provided by selection from a subset of existing variant individuals." In the case of adoptation, "the 'problem' is posed by the internally driven changes in a population of phenotypes, with the environment providing a 'solution'" [7.46,47].

Roughly speaking, in the first case the environment chooses organisms, and in the second case organisms choose the environment. The problems associated with the fitness are discussed in Sect. 2.4.

Let us now consider, following Dover [7.47], some concrete examples of the action of molecular drive. The first example is the evolution of the membrane of silkworm eggs. Such a membrane (chorion) serves to protect the content of the egg, prevents drying and provides gas exchange. It contains more than 100 proteins secreted at certain stages of development. The proteins form fibrils which are specifically packed and oriented. The chorion structure is species-specific (Fig. 7.13, cf. [7.50]). These proteins are the products of a multigene family organized into a large superfamily, which is differentiated into families and subfamilies with a decreasing degree of similarity in the sequences.

The difference between the wild silkworm *Antheraea polyphemus* and the common silkworm *Bombyx mori* manifests itself in the structure of the egg surface. In *A. polyphemus* the outer layer of the chorion contains gas-exchange tubes and in *B. mori* the surface is covered with a continuous layer of cystein-rich proteins Hc-A and Hc-B, coded for by two subfamilies. Each of these proteins consists of about 15 almost identical genes located in pairs (see Fig. 7.14). The sequence of the central region of each gene shows that 15 genes of each of the subfamilies originated from the predecessor subfamilies, either anew during the amplification process, or by way of gradual substitution.

How did the genes coding for Hc-A and Hc-B evolve and spread in a population? This could have occurred as a result of natural selection if the first individual gene with the first motif Cys/Gly in the left or right domain had affected the phenotype. Dover writes that this is difficult to assume because a single gene cannot exert an appreciable effect on the surface structure. And

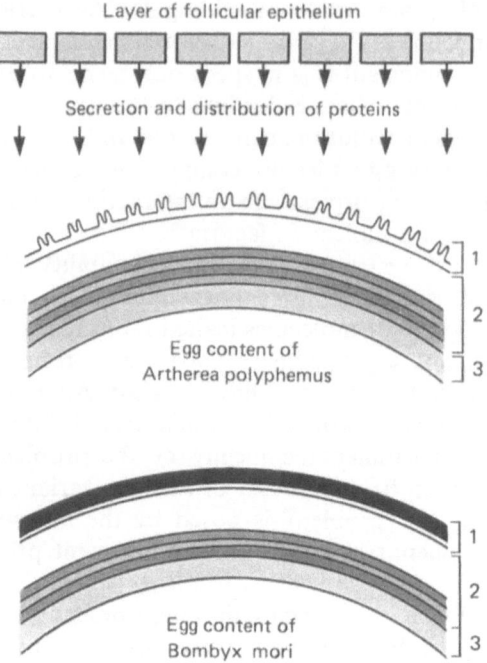

Fig. 7.13. Structures of chorion of two species of silkworm: (*1*) later layers; (*2*) intermediate layers; (*3*) early layers [7.47]

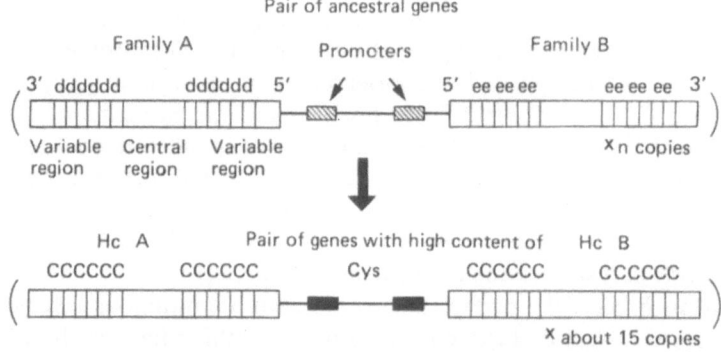

Fig. 7.14. Schematic representation of a pair of genes of two Cys-rich subfamilies responsible for the external layer of chorion of *Bombyx mori* [7.47]

what mechanism could it be, whose result is that natural selection might contribute to a horizontal distribution of the Cys/Gly motif within the gene and in all non-allelic loci of the subfamilies A and B? It is much more probable that an appreciable number of such motifs have been accumulated in other ways. Investigations have shown that the system experiences fluctuations in the number of

copies produced by slippage replication [7.51]. Presumably, a random accumulation of Cys/Gly occurred as a result of molecular drive. Gene conversion
took part in this process; the genes Hc-A and Hc-B underwent parallel evolution [7.50]. Such pairs could have accumulated due to unequal crossing-over.

This example refers directly to a species-specific feature and even to a feature of the genus since *Antheraea* and *Bombyx* are different genera. We are thus
speaking here of macroevolution. In this case the internal factors appear to be
more important than the external ones.

The second example considered by Dover is the molecular co-evolution of
genes of ribosomal RNA (rRNA). In the family of genes that code for rRNA
molecular drive operates and gives rise to a driving force for natural selection.

Figure 7.15 shows two major genes of rRNA, 18S and 28S, which form a
single system together with a DNA portion located between them. Such a system often recurs hundreds of times. Concerted evolution is evidenced by a high
level of species-specific homogeneity in the DNA family. Unequal crossing-over

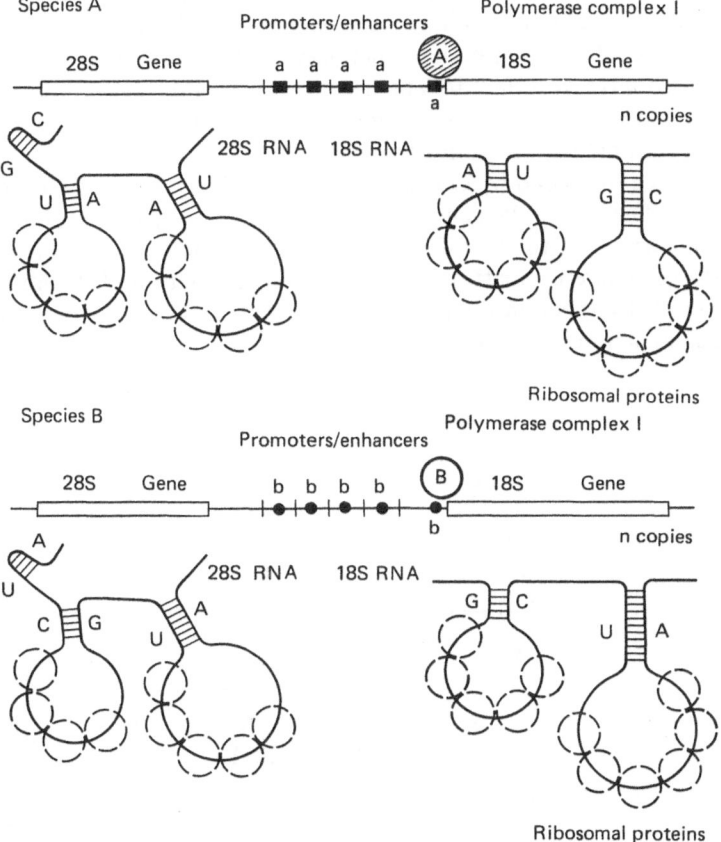

Fig. 7.15. Two genes of 18S and 28S rRNA separated by a spacer [7.47]

is the mechanism of genome turnover, which is responsible for molecular drive [7.42].

The majority of homogenizable mutations of rDNA are located in an inter-genic spacer (Fig. 7.15). There are data indicating that this region of the genome contains signals required for the enhancement, promotion and termination of transcription, for the initiation of replication, for the differential amplification of genes, etc. [7.52]. These events take place with the participation of proteins and, possibly, RNA coded for by other parts of the genome.

The RNA polymerase I and its co-factors exhibit an interspecies divergence. As regards rRNA, it is characterized by the strong conservatism of the second-ary structure with considerable differences in the primary structure. The second-ary structures of the rRNA of bacteria, slime moulds, mushrooms, plants, in-sects, amphibians, and mammals appear to be very similar. In this sense rRNA resembles proteins.

The third example discussed by Dover consists in the suppression of a so-called P element ('paternal' factor) in the genome of *Drosophila*. The presence of this element is responsible for dysgenesis, i.e. the complex syndrome of non-adaptive effects, including rudimentary, sterile gonads. Dysgenesis is usually observed in the offspring of males which have a P factor and in the offspring of females which have no P factor. The irreversibility of dysgenesis (its absence in succeeding generations) and the stability of wild populations are accounted for by the cytoplasmic suppression of the mobility of P elements. A hypothetical model has been proposed by O'Hare and Rubin [7.53]. Intact P factors consist of two genes. The first of the genes is responsible for the synthesis of transposase, which is necessary for the transfer of the gene, and the second for the synthesis of a repressor. There is positive feedback in the production of the latter protein and a negative feedback in the production of transposase. This two-component model is sketched in Fig. 7.16.

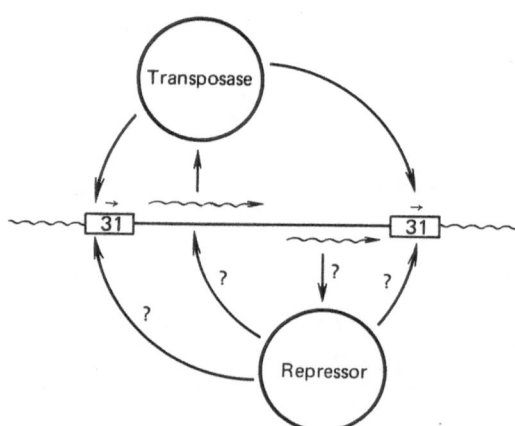

Fig. 7.16. Two-component model of suppression of the mobility of the P-element in *D. melano-gaster* [7.47]

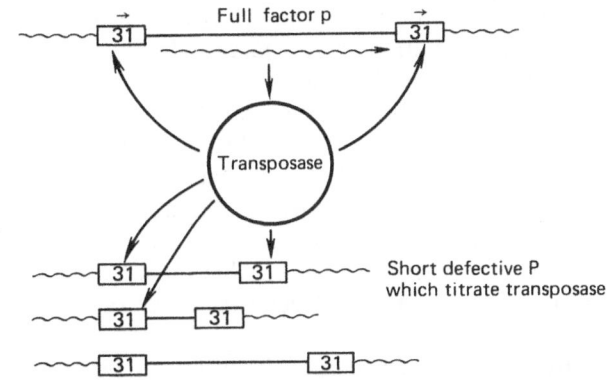

Fig. 7.17. One-component model of suppression of the P-element [7.47]

The one-component model in which the intact P factor produces only transposase is perhaps more substantiated (Fig. 7.17, [7.54]). In the first model the suppression of dysgenesis occurs as a result of the accumulation of defective P elements. In the second model, a progressive "titration" of transposase occurs by the ends of the elements which underwent internal deletions due to translocations and which are therefore incapable of producing transposase. As a matter of fact, both dysgenesis and its inevitable suppression are consequences of molecular drive.

The three examples given above show that important genetic and evolutionary events occur due to the mobility of genes and their joint action; coevolution takes place. Of course, molecular drive is not independent of the other two evolutionary factors, molecular drift and natural selection.

We have seen that point mutations, which are responsible for molecular drift, are mostly neutral and have no direct relation to speciation or macroevolution. Neutralism disappears upon interaction between the drift and the drive. Imagine a mutation which is quite neutral in a stable genome. If this mutation finds its way into the region of initiation upon duplication or transposition of the gene, it is no longer neutral. The same refers to mutations adjacent to the sites of turning-on of transposons.

Let us now consider the role of natural selection [7.47]. We begin with the third example reported by Dover. Hybrid dysgenesis occurs between males with intact P elements and females with no means for their suppression. How does dysgenesis spread in a population in which the P elements are accumulated originally? If the fitness undergoes change, natural selection starts operating. The extent of dysgenesis in a given population depends on the differences between males and females. As a result of duplicative transposition, in the following generations a shift in the number of copies with intact and partial P factors occurs. Natural selection can eliminate a progeny with dysgenesis, which differs significantly in properties from the average. Molecular drive provides material for natural selection.

In the case of the egg membrane of the silkworm, negative selection rejects populations with an accumulation of too large a layer of alternate proteins, which may prevent gas exchange. A certain threshold must exist, with which the accumulation of a variant gene due to molecular drive is harmful to adaptation. Dover writes [7.47] that natural selection always operates at the individual level, but the manner in which a population is developing introduces an element of intrapopulation selection, the dynamics of which requires further analysis. Despite such possibilities, the common silkworm *Bombyx mori* provides evidence that the accumulation of genes responsible for a high content of cystein in proteins is the result of the internal gene turnover rather than of natural selection.

The general situation that determines the interplay between the drive and natural selection may be represented in the form of the scheme proposed by Dover [7.47]; it is shown in Fig. 7.18. To what extent does natural selection support the results of an internal molecular drive? The force that drives natural selection is the survival pressure created by ecological stresses. In the case of the silkworm egg, natural selection could have directly speeded up the gene turnover mechanism or could have chosen individuals with the largest number of copies of a new gene in any generation. The scheme gives pathways provided by both natural selection and molecular drive for the "solution" of the "tasks" encountered. The relative role of either of the solutions depends on environmental conditions. There is an interesting question associated with the winter diapause, the cessation of egg activity. If we confined ourselves to natural selection, we would be forced to introduce the concept of "pre-adaptation". Based on molecular drive, we should speak of the diversity of ecological niches accessible for utilization at each stage ("adoption"). The diapause is the final outcome of

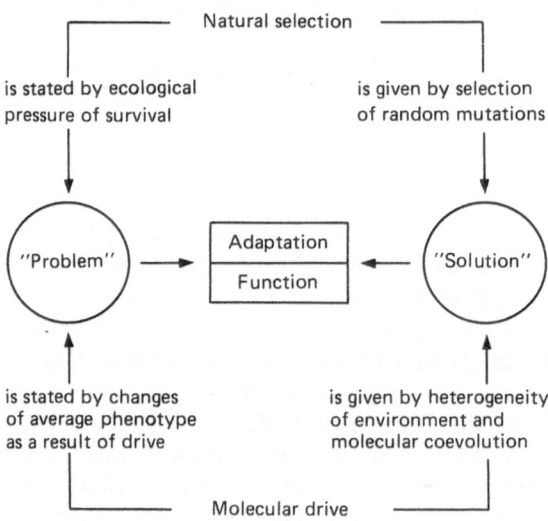

Fig. 7.18. The interactions of evolutionary events – different origins of functions and adaptations

the interaction between genes (such as Hc-A and Hc-B) controlled by molecular drive.

As has already been said, molecular drive is associated with molecular co-evolution. Thus, co-evolution occurs between promoter/enhancer complexes and polymerase I. Molecular drive can provide the switching of the efficiency of the given polymerase as a result of a change in the fraction of old and new promoters.

The interplay between the drive and selection may be changed at various stages due to changes in the sequences in multigene families.

Molecular drive implies non-Darwinian and non-Mendelian evolution at the molecular level. The study of these phenomena is at the initial stage but there is no doubt that molecular evolution and speciation at the molecular level cannot be visualized without recourse to the drive concepts.

The main features of molecular drive can be reduced to the facts that the fixation of new genetic variants is influenced by internal forces, and that the population changes "in unison" (homogenization) with (rather than via) the differential survival of individuals. The concept of molecular drive is directly applicable to families of coding and non-coding repeats. Homogenization occurs as a result of the constant turnover of repeated sequences, which boils down to unequal crossing over, transposition and conversion of genes (cf. [7.44]).

No mathematical models of molecular drive have been constructed so far, though their development cannot be halted. Theoretical population genetics of multigene families, dispersed in two or a larger number of chromosomes, have been developed, and a more general investigation has been carried out in this area [7.55]. However, there are no equations (supposedly, nonlinear) describing changes in hereditary traits with time due to molecular drive.

The concept of molecular drive is directly tied up with the general problems of evolutionary theory, namely the problems of gradual or punctuated evolution and with the problems of adaptation. The theory of molecular drive is consistent with the corresponding concepts of synergetics (see Chap. 9).

7.6 Molecular Evolution

Concluding this second part of the book, which is devoted to the molecular foundations of evolution, it is now necessary to consider the most important features of protein evolution, which have not yet been discussed. We have only considered the protein structures and the functions which they determine. However, of no less importance to molecular evolution are precise answers to three other questions: (a) in what amount, (b) where and (c) when is a protein synthesized? Convincing proof of the importance of these questions is furnished by King and Wilson [7.56], whose work called "Evolution at two levels in humans and chimpanzees" may be regarded as classical.

Table 7.2. Comparison of some proteins of Man and chimpanzee

Protein	Number of residues	Difference
Fibrinopeptides A and B	30	0
Cytochrome c	104	0
Lysozyme	130	0
Haemoglobin (Hb) α	141	0
Hbβ	146	0
Hb$^{A\gamma}$	146	0
Hb$^{G\gamma}$	146	0
Hbδ	146	1
Myoglobin	153	1
Carboanhydrase	264	3
Plasma albumin	580	6
Transferrin	647	9

The two species *Homo sapiens* (modern man) and *Pan troglodytes* (the common chimpanzee) belong to the same superfamily (Hominoidea); chimpanzees are members of the *Pongidae* family, while humans are in the *Hominidae* family. The chimp differs very strongly from man in both morphology and behaviour; however, it has been shown that the genetic distance between the two species is small, and that of all the anthropods, the chimpanzee is the most closely related to man. The genetic distance was determined by King and Wilson [7.56] from the amino-acid composition of 44 proteins. Some data are given in Table 7.2.

Accordingly, these proteins have a similar primary structure, which means that structural genes are also similar. The mitochondrial DNAs of the two species are identical, the average difference between the proteins is 7.2 amino-acid residues per 1000, and the average human polypeptide coincides with that of the chimpanzee by 99 percent. The differences do not exceed this figure in many sibling species, i.e. species morphologically indistinguishable but with different genotypes. At the same time, appreciable differences have been revealed in the allele frequencies of the polymorphs by means of electrophoresis.

These results (and also many others) show that phenotypic evolution is significantly different from molecular evolution. It may be presumed that molecular changes were accumulated, at approximately the same rate, in two separate lines which led to the *Pan* and *Homo* genera. At the same time, phenotypic changes occurred at markedly different rates, the rate of change in *Homo* being considerably higher than that in *Pan*, according to palaeontological findings. This is schematically shown in Fig. 7.19. These data thus serve as a direct and independent confirmation of the neutral theory, the basic proposition of which is based on the difference between molecular and phenotypic evolution.

Incidentally, clear-cut differences exist even at the supramolecular level of chromosomes. The *Homo* genus has 46 chromosomes and the *Pan* genus has 48 chromosomes.

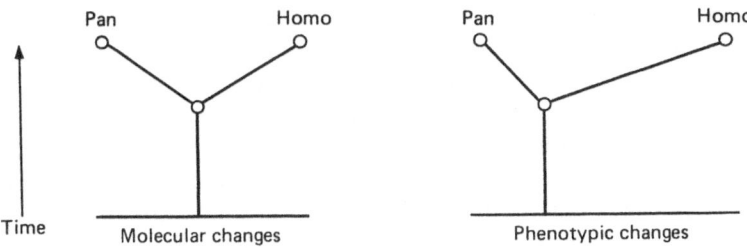

Fig. 7.19. The divergence of man and chimpanzee

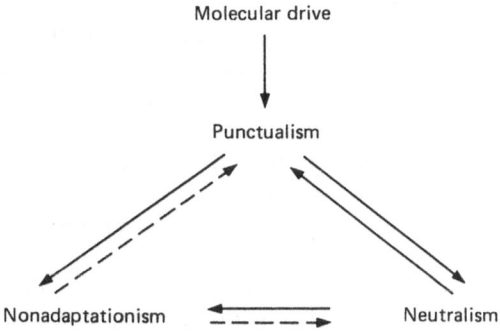

Fig. 7.20. The evolutionary triad

A question arises at once concerning the causes of such differences. They are accounted for as follows: as has already been discussed, the genome is not a fixed "blueprint" or a plan for constructing an organism. In the course of individual development, processes of self-organization of the system occur; these are directly associated with gene regulation and with the mobility of the genome, i.e. with molecular drive. In these processes, proteins are synthesized in different amounts in different cells and tissues, and at different stages of development.

In the long run, these processes are also molecular events. However, they cannot be reduced to the determination of the structure of proteins. We are speaking here of the interactions of DNA and RNA with proteins, in particular regulatory proteins, and of the interactions of genes. The problem of the relationship between genotype and phenotype revolves around the relationship between phylogeny and ontogeny, the understanding of which has only begun. The problems of the nature and tempo of evolution are closely connected with the events that take place in the genome. We may state that molecular drive occurs appreciably faster than molecular drift. Neutral point mutations are gradual; they are accumulated slowly, one after another, usually without exerting any substantial influence on phenotypes. On the other hand, the phenomena that determine molecular drive have no gradual character in the genome. As a result of the drive, homogenization and fixation in a population can be realized at a greater or lesser rate, depending on both internal and external factors. The external factors are what we call the pressure of natural selection. Molecular drive does not contradict the punctualism concept, whereas molecular drift is quite gradual. So, the neutral theory correlates with punctualism, according to which speciation implies not gradual but discontinuous evolution, thus resembling a phase transition (see Chaps. 8 and 9).

From the concept of punctualism, and also from a consideration of molecular drive, it follows that numerous traits are non-adaptive (see Chap. 4). If there are internal factors responsible for changes in the genome, then these changes cannot be adaptive from the beginning. Thus, the concept of non-adaptationism, which is taken to mean a permanently present multitude of non-adaptive characters, agrees with the molecular theory.

Non-adaptationism logically follows from punctualism, but not vice versa. The presence of non-adaptive features in itself does not yet mean that evolution is discontinuous.

At the same time, the neutralism of point mutations, which logically stems from punctualism, is also consistent with the non-adaptationism of a multitude of characters determined by non-gradual events occurring in the genome. Again the converse is not true; here the facts of non-adaptationism do not yet give rise to neutral theory. In this case, too, a one-way logical relationship is realized.

What has just been said may be represented by a scheme shown in Fig. 7.20. The word "punctualism" in this scheme implies only the discontinuity of speciation and macroevolution, which points to the similarity between these processes and phase transitions.

The punctualism-neutralism-non-adaptationism triad characterizes, to a certain extent, the present-day state of the theory of evolution. Important features of the evolutionary process have been marked out, a step has been made towards the union of the synthetic theory of evolution with molecular biology [7.57,58].

"Molecular drive" has now been added to the triad. Without the molecular drive concept we would not have been equipped with molecular approaches to discontinuous, punctuated evolution.

Irrespective of the concept of molecular drive being developed at supramolecular and organismic levels, substantiated conceptions have appeared regarding the role of chromosome rearrangements and macromutations in speciation and macroevolution. Even within a single species, variants of karyotypes are possible. These conceptions have been expounded with special clarity in a number of works by Vorontsov, who has also analyzed the character of evolution in different mammalian lines, and shown that different degrees of gradualism and punctualism are observed in them.

The future tasks of the theory will be the treatment of evolution on the basis of the physics of dissipative systems or synergetics, and the union of phenomenological and molecular theories. These problems have already been solved with respect to pre-biotic evolution (see the next chapter).

Part III

Physics of Biological Evolution

8. Physical Theory of Molecular Evolution

8.1 The Eigen Theory

Everything that has been expounded in the preceding chapters is based on Darwin's natural selection of eukaryotic phenotypes. The existence of natural selection is documented by an enormous number of facts established in biology, but not substantiated by more general and detailed propositions of physics. Moreover, one may encounter statements to the effect that natural selection is not amenable to physical interpretation and that Darwin's theory contradicts physics.

We have considered the relationship between physics and biology (Chap. 1). There is no doubt that physics has every right and duty to encroach upon the field of the fundamental problems of theoretical biology, into evolutionary theory.

The Kimura neutral theory and the concepts of molecular drive (Chaps. 6 and 7) are of prime importance, but they are predominantly concerned with genetic aspects. Here physics is confined to investigations into the structure and dynamics of biopolymers, i.e. proteins and nucleic acids. Physical approaches to the solution of the fundamental problem of the relationship between phenotype and genotype are only in first stages of development.

In the long run, the development of the physics of living nature is associated with the solution of the problem of the origin of life on earth. We are convinced of the abiogenic origin of life, but before the publication of Manfred Eigen's work there had been no convincing model of the transition from inanimate to animate nature.

Eigen's "Self-organization of matter and the evolution of biological macromolecules" was published in 1971 [8.1], 112 years after the appearance of *The Origin of Life*. In the preface to Eigen's book published in Russian [8.2], I wrote that his work may be compared to Schrödinger's book "What is Life from the Standpoint of Physics?"; these two books stimulate the development of molecular biology and biophysics. Later it was found, however, that Eigen's work is of much greater importance, going beyond mere stimulation. Eigen has worked out a well-grounded physical theory of self-organization and evolution of informational macromolecules, which has been fundamentally developed and supported experimentally. Eigen has created a new stage in the construction of evolutionary theory; he has for the first time truly united evolutionary theory with physics and chemistry.

The Eigen theory is well known; it has been expounded in a number of his work [8.3–9] and in the works of other authors (see [8.10–12]). It is nevertheless necessary to outline it briefly in the present book.

The Eigen theory is based on the concepts of information theory (see Chap. 10). It is clear that the beginning of molecular events that led to the appearance of life must have been quite random. Two arguments can be given in favour of this statement. Firstly, the original system, i.e. the archaic earth with a reducing atmosphere, contained a chaotic mixture of small molecules, from which monomers of nucleic acids and proteins were then built. Secondly, the converse statement to the effect that the original events were non-random would have meant a certain determinateness had existed even earlier, possibly before the earth was formed. There is no ground whatsoever for such a statement.

Randomness may lead to the fixation of information, to the self-organization of informational macromolecules, polymeric chains, which are "texts", if the emerging information self-reproduces and multiplies. Jumping ahead, it may be said that information is contained in an ordered collection of dissimilar elements, which can be fixed and memorized. As is known, the unit of information, one bit, is obtained upon choosing either one of two equally probable possibilities, say in coin-tossing. In this case, the dissimilar elements are the heads and tails and a set is formed by a series of sequential throws. New information is always generated as a result of the "accidental choice remembered" [8.13]. A chaotic mixture of monomers gives rise to various informational macromolecules, that is, various types of "texts" capable of self-reproduction. Template synthesis of polymeric chains occurs. This most important chemical phenomenon was ignored by Wigner, who asserted that biology is not consistent with quantum mechanics [8.14].

These rather simple propositions exhaust the possibilities of canonical information theory, which definitely ignores semantics, i.e. the content, meaning or value of information. The Eigen theory calls for a consideration of just the value of information, which is called selective value. The point is that informational macromolecules self-reproduce and break down at different rates. Therefore, they compete with one another and Darwinian natural selection arises at the prebiotic, molecular level. It cannot be said *a priori* which of the chains will be chosen and will "survive", i.e. will possess the greatest selective value. This means that natural selection itself creates new information.

This kind of molecular Darwinian system can be realized only under special conditions. First of all, a certain chemical structure of macromolecules is required if they are to be capable of self-reproduction. This structure is exhibited by nucleic acids. Special conditions exist, in that the small molecules that combine into macromolecular chains must have a reserve of energy; nucleic acids are synthesized from nucleoside triphosphates by way of their polycondensation with release of free energy of the order of 10 kcal/mole per unit. Thus, a source of free energy is required.

In order for a set of macromolecules to evolve, creating new information, the self-reproduction of macromolecules must be realized with random errors, i.e. with mutations. This is what actually occurs with nucleic acids.

However, evolution which implies the creation of new order, new information from chaos, is possible only in an open far-from-equilibrium system that is, a dissipative system. Near equilibrium the entropy increases with time, and the order is destroyed. Periodic processes and instability, which could have brought about substantial changes, are impossible near equilibrium. The nearness to equilibrium is the nearness to death.

The open nature of the system invariably implies metabolism – the exchange of matter and energy with the surroundings.

Thus, the physical conditions that must serve as a basis for a model of self-organization and evolution of macromolecules boil down to self-reproduction, mutability, to the presence of a dissipative system (which provides metabolism) and to the presence of a source of energy.

The theoretical model proposed by Eigen is a flow reactor, whose scheme is shown in Fig. 8.1 (taken from a book by Küppers [8.12]). In the reaction vessel the energy-rich monomers are united into macromolecular chains. For precision, we will speak of nucleic acids. In turn the macromolecules decompose into monomers which are now devoid of redundant energy. The conditions are imposed upon the reaction by means of the regulation of the influx and efflux of monomers and by the presence of a global dilution flux. The system can operate at a constant flow or at constant concentrations. Thus, the reactor is open and far from equilibrium. It is supplied with free energy in the form of energy-rich

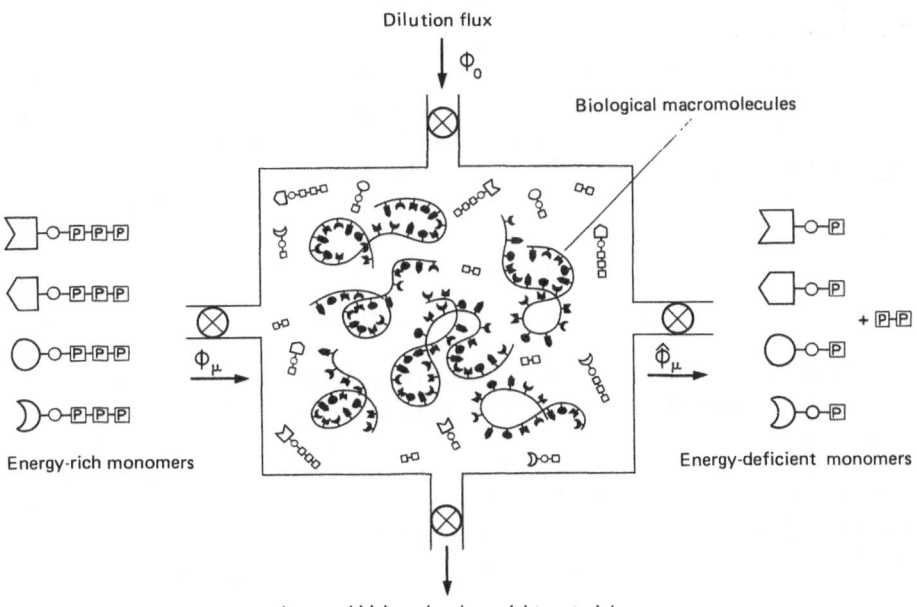

Fig. 8.1. The Eigen reactor [8.12]

monomers (nucleoside triphosphates). Macromolecules are informational since their chains are built up of k types of monomers ($k = 4$ for nucleic acids), which are arranged in this or that sequence. The number of chains per unit volume containing v units, with a certain sequence characterized by the index j, is equal to x_{v_j}. The total number of chains of length v per unit volume is

$$n_v = \sum_{j=1}^{N_v} x_{v_j} \ , \tag{8.1}$$

where N_v is the number of possible sequences consisting of k types of units with the number of units being equal to v

$$N_v = k^v \quad (=4^v) \ . \tag{8.2}$$

The number N_v is very large; at $v = 100$ and $k = 4$, $N_v = 4^{100} = 10^{60}$. Hence, the total number of chains of given length v in the reactor volume V is very small as compared to N_v:

$$n_v V \ll N_v \ . \tag{8.3}$$

Let us write the simplest kinetic equation for each information carrier, taking into account the possibility of inaccurate copying, i.e. mutability. We have

$$\dot{x}_i = F_i x_i - R_i x_i + \sum_{j \neq i} W_{ij} x_j \ . \tag{8.4}$$

The first term describes the rate of template autocatalytic self-reproduction of chains, and the second one the rate of decrease of the number of chains as a result of their breakdown and of the dilution of the solution. It is obvious that the total increase in the number of chains is possible only with $F_i > R_i$. The third term describes the appearance of chains of type i due to mutations which convert chains of type j to those of type i. The rate F_i depends on the concentration of monomers $m_1, m_2, \ldots, m_k$ ($k = 4$). Equations (8.4) are non-linear since F_i and R_i depend on all x_i.

We introduce a rate constant k_0 measured in s^{-1}:

$$F_i = k_0 A_i Q_i \ , \quad R_i = k_0 D_i + W_{0i} \ . \tag{8.5}$$

The dimensionless quantity $A_i Q_i$ expresses the rate of formation of the ith chain; A_i is the enhancement factor; Q_i is the quality factor, a number which shows that part of template synthesis provides accurate copies of the ith information carrier. Accordingly, $1 - Q_i$ characterizes a fraction of resultant mutants which are different from the ith sequence. The term D_i expresses the degradation of chains and W_{0i} the decrease in their concentration due to dilution. In the simplest case W_{0i} is related to the overall dilution flux Φ_0 as follows:

$$W_0 = \Phi_0 \Big/ \sum_{j=1}^{N} x_j \ . \tag{8.6}$$

The subscript i may be omitted here. Now Eqs. (8.4) assume the following form:

$$\dot{x}_i = k_0(A_iQ_i - D_i)x_i + \sum_{j \neq i} W_{ij}x_j - W_{0i}x_i \; . \tag{8.7}$$

Informational macromolecules for which $A_iQ_i > D_i$ increase in number, i.e. they "survive"; those for which $A_iQ_i < D_i$ "die out". Thus, segregation, the separation of the system into two parts, takes place, but this is not yet selection. For selection to occur selective constraints must exist. In the reactor, in the chemical system, such constraints may be imposed by the stationarity condition. Let us say, the concentrations of monomers and polymers remain constant:

$$m_1, m_2, \ldots, m_k = \text{const} \tag{8.8}$$

$$\sum_{j=1}^{N} x_j = n = \text{const} \; . \tag{8.9}$$

Simultaneously, the dilution flux Φ_0 is such that it balances the total excess production of macromolecules:

$$\Phi_0 = k_0 \sum_{j=1}^{N} (A_j - D_j)x_j \; . \tag{8.10}$$

Another constraint may consist of the constancy of fluxes rather than of concentrations (generalized forces).

We introduce the definitions:

$$E_i = A_i - D_i \quad \text{(productivity)} \tag{8.11}$$

$$\bar{E} = \sum_j E_j x_j \Big/ \sum_j x_j \quad \text{(mean productivity)} \tag{8.12}$$

$$W_{ii} = A_iQ_i - D_i \quad \text{(selective value)} \; . \tag{8.13}$$

With the concentrations being constant, comparing (8.9), (8.10) and (8.11), we get

$$\bar{E} = \Phi_0/k_0 n \; . \tag{8.14}$$

The selective value has the meaning of information value, and determines the selection of informational macromolecules. In the notation of (8.12) and (8.13), Eqs. (8.7) take the form

$$\dot{x}_i = k_0(W_{ii} - \bar{E})x_i + \sum_{j \neq i} W_{ij}x_j \; . \tag{8.15}$$

This is a system of non-linear equations since $\bar{E}$ contains all x_j. The behaviour of dissipative systems is described by strictly non-linear equations. The presence of $\bar{E}$ in (8.15) determines the slippage-like self-regulating threshold of self-organization. The number of information carriers, whose selective value W_i is above the threshold $\bar{E}$, increases. An increase in their number shifts the threshold $\bar{E}$ towards ever greater values until an optimum of $\bar{E}$ is reached, which is equal to the maximum selective value W_{max} of all existing species:

$$\bar{E} \rightarrow W_{max} \; .$$

The system will tend to assume a state of "selective equilibrium", which is unstable. It will be upset when, as a result of mutations, a new copy appears with a higher selective value, $W_{m+1} > W_m$. A new state of equilibrium arises, and so on.

The information value is thus expressed in terms of real physical quantities; it characterizes the rate of multiplication of macromolecules. We are speaking here of the value as applied to the evolution of macromolecules by way of natural selection. As will be seen at a later time (see Sect. 10.3), the phrase "as applied" is essential here; in contrast to the amount of information, its value or meaning or semantics is of no importance, irrespective of the act of reception of information, or its memorizing and utilization.

A mutation leading to an increase of selective value implies a decrease in entropy and points to the instability of the preceding stationary state.

The mean productivity may be represented in the form

$$\bar{E} = \frac{E_m x_m + \sum\limits_{j \neq m} E_j x_j}{\sum x_j} = \bar{E}_{j \neq m} + \frac{x_m}{n}(E_m - \bar{E}_{j \neq m}) \; , \tag{8.16}$$

where

$$\bar{E}_{j \neq m} = \sum\limits_{j \neq m} E_j x_j \bigg/ \sum\limits_{j \neq m} x_j \tag{8.17}$$

expresses the mean residual productivity. It is obvious that

$$\sum\limits_{j \neq m} x_j = n - x_m \; .$$

The equilibrium condition $\bar{E} = W_m$ gives the "equilibrium fraction" of the selected species, i.e. its relative survival. From (8.16) it follows that

$$\frac{x_m}{n} = \frac{W_m - \bar{E}_{j \neq m}}{E_m - \bar{E}_{j \neq m}} \; . \tag{8.18}$$

The stationary error fraction $1 - x_m/n$ is proportional to $1 - Q_m$:

$$1 - \frac{x_m}{n} = \frac{A_m(1 - Q_m)}{E_m - \bar{E}_{j \neq m}} \; . \tag{8.19}$$

If $Q_m = 1$, then $x_m = 1$, so that a compete selection of "main copies" i_m would have taken place, but further evolution would have stopped. For evolution to progress, a value of the quality factor Q_m of less than unity is necessary, but it should be greater than a minimum value $Q_{\min}$ at which survival is still possible. The value $Q_{\min}$ is found from the condition $x_m = 0$, i.e. $W_m = \bar{E}_{j \neq m}$. We have

$$1 > Q_m > Q_{\min} = \frac{\bar{A}_{j \neq m} + D_m - \bar{D}_{j \neq m}}{A_m} = \gamma \ . \tag{8.20}$$

Due to the presence of the factor Q_m which is different from unity, a large variety of mutants is present in the system. This makes it possible to attain a greater optimal value of W_m and to increase the rate of evolution. According to (8.18), the general criterion of selection is given by

$$W_m > \bar{E}_{j \neq m} \ , \tag{8.21}$$

which can be realized with appropriate values of the parameters, depending on the conditions imposed.

The quality factor Q is determined by the accuracy of recognition of a unit of a given type during template reduplication. Timofeeff-Ressovsky called the DNA reduplication covariant, meaning the fixation in the replica of random errors (mutations). If the reduplication is not co-operative, i.e. the fate of each unit is independent of the other units, and the probability of accurate reproduction of the unit is q, then the probability of formation of a correct copy containing v units is given by

$$Q = Q_{v0} = q^v \ . \tag{8.22}$$

The distribution of the probabilities of appearance of a chain with j errors is binomial:

$$Q_{vj} = q^{v-j} (1 - q)^j \binom{v}{j} \ , \tag{8.23}$$

since the probability of an error is $1 - q$. With $(1 - q) \ll 1$ the distribution (8.23) is close to a Poisson distribution:

$$Q_{vj} \cong \frac{[v(1 - q)^j] \exp[-v(1 - q)]}{j!} \ . \tag{8.24}$$

Hence, the probability of appearance of a correct copy at $j = 1$ is

$$Q_{v0} \cong \exp[-v(1 - q)] \ . \tag{8.25}$$

According to (8.22) and (8.25), Q_{v0} falls off with increasing chain length v. Taking the logarithm of (8.25), we find an approximate value of the minimum quantity Q_{v0} at a given value of q:

$$|\ln Q_{\min}| > v(1 - q) \ .$$

The same result is obtained directly from (8.22). So, we have

$$v_{max} = \frac{\ln Q_{min}}{\ln q} = \frac{\ln \gamma}{\ln q}$$

and with q close to unity,

$$v_{max} \cong \frac{\ln \gamma}{1 - q} . \tag{8.26}$$

The number of monomeric units in a self-replicating chain, i.e. the amount of information contained in it, is limited. The limiting value of v is inversely proportional to the mean probability of an error per monomeric unit $1 - q$.

　　Now let us leave mutations, assuming that $Q = 1$, and consider the competition of chains on the basis of the simplified equations (8.15):

$$\dot{x}_i = k_0(W_{ii} - \bar{E}) . \tag{8.27}$$

The solution of (8.27) has the form (cf. [8.1,2,10,12])

$$x_i(t) = \frac{x_i(0)n \exp(k_0 W_{ii} t)}{\sum\limits_{j=1}^{N} x_j(0) \exp(k_0 W_{jj} t)}$$

$$\equiv \frac{nx_i(0)}{\sum\limits_{j=1}^{N} x_j(0) \exp[k_0(W_{jj} - W_{ii})t]} . \tag{8.28}$$

At $t = 0$ each $x_i(t) = x_i(0)$. With $t \to \infty$ the sum of the exponentials tends towards a value corresponding to a species with the highest selective value W_m, i.e. only one term is retained in the sum:

$$x_i(t) \cong n \frac{x_i(0)}{x_m(0)} \exp[(W_{ii} - W_m)k_0 t] \tag{8.29}$$

and with $t \to \infty$, only $x_m(t)$ remains to be different from zero and is equal to n; the selection of one main copy takes place. Macromolecules with $W_{ii} < W_m$ will eventually disappear and the main copies will increase in number at their expense. Figure 8.2 shows the results of simple calculations in accordance with formula (8.29) for four chains with $W_1 = 1$, $W_2 = 4$, $W_3 = 9$ and $W_4 = 10$. The maximum selective value corresponds to W_4. Chains with W_3 will disappear non-monotonically; their number passes through a maximum.

　　In what follows we will continue to expound the Eigen theory, which will be presented in this section in the form of a zeroth approximation. Further development of the theory considered in Sects. 8.2 to 8.5 has made it possible to reach very important assumptions; but even at this point significant results have been obtained.

　　It has been shown for the first time that a simple physical model, based on the template replication of macromolecules, allows one to describe and explain

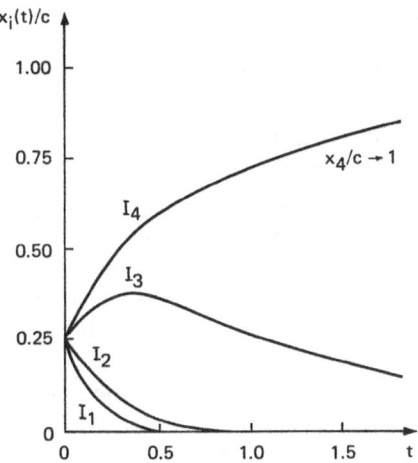

Fig. 8.2. Selection in the system of four competing species. $W_1 = 1$; $W_2 = 4$; $W_3 = 9$; $W_4 = 10$ [8.1]

the process of natural selection (Darwinian evolution) at the molecular, prebiotic level. The seeming contradiction between natural selection and physics has been removed. Natural selection operates with respect to information value which has the meaning of selective value. No new physics has been required for the solution of this problem.

As has already been pointed out, Darwinian evolution was treated by some scientists as a tautology, "the survival of the survivor." In fact, there is no tautology in the theory of natural selection. The Eigen theory is also free from it, since selection occurs with respect to selective value determined by adaptation to the imposed constraints, that is to the conditions of permanent organization in the case under consideration. Best fitness implies the maximum selective value under these constraints.

The Eigen theory has for the first time united evolutionary theory not only with molecular biology, but also with the physics of dissipative systems (synergetics) and with information theory. The problem to be discussed in this book has been solved in application to molecular evolution.

8.2 The Quasispecies and the Hypercycle

When Lifshitz acquainted himself with this first version of the Eigen theory in 1971, he noted that the selective value W_{ii} does not correlate unambiguously with the primary structure of the chain. In a macromolecule mutations may arise, and template synthesis may occur with errors, without any appreciable changes in the parameter W_{ii}, the selective value. Template reduplication is based on the recognition of separate units and their nearest neighbours. Thus, a double mutational substitution of the type

$$\cdots \text{CATTG} \cdots \text{CACTG} \cdots \rightarrow \cdots \text{CACTG} \cdots \text{CATTG} \cdots$$

(for example) does not at all affect the value of W_{ii}, i.e. the rate of synthesis of a macromolecule as a whole in the long run.

The probability of double mutation is, of course, very low. A single replacement of a unit changes W_{ii}, but with a large number of units this change is insignificant. A relative change of the rate of template synthesis with one unit being replaced has an order of magnitude equal to v^{-1}. Suppose that, as a result of a substitution, W_{ii} increased by a correspondingly small amount. Selection takes time. If, during the time of predominant survival of main copies, a second mutation occurs which brings W_{ii} to its former value, then a new, changed main copy will not appear.

However small the probabilities of mutations are, after a sufficient period of time chains with close values of W_m but different primary structures will accumulate in the system. In other words, the correlation between selective value and primary structure will be degenerate. If the chains are long enough, the number of such degenerate macromolecules will become large. In this sense, the situation is similar to evolution established in the neutral theory. The relationship between Kimura's theory and Eigen's theory will be considered in Sect. 8.5. In the final analysis the maximum diversity of primary structures of macromolecules which have equal or almost equal values of W_m will represent the most probable state of the system. Thus, the selection equilibrium proves unstable. The selection described in the preceding section may lead to the transition to a strongly degenerate state, in which the relative content of main copies will be small (see [8.15]).

While working out his theory, Eigen introduced the concept of the quasi-species, whose content is to a certain extent close to what has been described, but leads to important quantitative results.

According to Eigen, the main copy "drags along the cometic tail of mutants" [8.1]. A single species is not an independent, separately selected unit. With the overall population of molecules being retained, all species compete with one another, and the presence of mutations determines the co-operation between closely related species, i.e. between species i and j, for which W_{ij} and W_{ji} are different from zero. A quasispecies is selected that is an organized assembly of differing, though related, species with a certain distribution of probabilities. Mathematically this implies the replacement of the subdivision of the whole population into N species by a system of N quasispecies, whose population variables y_i are linear combinations of the original variables x_i, their sum remaining constant:

$$\sum_{j=1}^{N} x_j = \sum_{j=1}^{N} y_i \; . \tag{8.30}$$

New variables are introduced in such a manner that the basic equations (8.15), which contain the mutational factors W_{ij} different from zero, are transformed by way of an affine transformation to a form similar to (8.27):

$$\dot{y}_i = (\lambda_i - \bar{E}(t))y_i \; . \tag{8.31}$$

The quantity $\bar{E}(t)$, which determines the non-linearity of the system, remains invariant upon transformation:

$$\bar{E}(t) = \sum \lambda_j y_j / \sum y_j \; . \tag{8.32}$$

It is obvious that the quantities W_{ii} and W_{ij} in Eqs. (8.15) are the elements of a matrix. They are all positive by definition. Equations (8.31) are obtained as a result of the transformation of the matrix into a diagonal form; λ_i represents the eigenvalues of the matrix, and the variables y_i, which represent quasispecies, are normal modes.

If the matrix W is symmetric, $W_{ij} = W_{ji}$, the eigenvalues λ_i are material; if it is asymmetric, λ_i are complexed with the positive material part. In this case, periodic vibrations arise.

We once again emphasize that all these calculations are performed under conditions of constant concentrations. An alternative approach, which does not alter the basic physical inferences, is determined by the condition of constant fluxes. This approach has been described, in particular, by Eigen and Küppers [8.1,12].

Finding the eigenvalues y_i for a large number of competing species is complicated. We can, however, make use of an approximate method for the realistic case of relatively small off-diagonal terms $W_{ij} \ll W_{ii}$.

Second-order perturbation theory gives

$$\lambda_i \cong W_{ii} + \sum_{j \neq i} \frac{W_{ij} W_{ji}}{W_{jj} - W_{ii}} \; . \tag{8.33}$$

We see that with the off-diagonal terms being equal to zero, i.e. in the absence of mutations, unambiguous selection is realized, and only one species survives (cf. Fig. 8.2). In the presence of mutations, a quasispecies rather than a species is selected. Figure 8.3 presents the results of calculations for four species with W_{ii} equal to 1, 2, 3, 4, with all the non-diagonal terms of the matrix being small, with the exception of W_{14}, W_{24}, W_{34}. The matrix has the form

$$W = \begin{bmatrix} 1 & 0.001 & 0.001 & 1 \\ 0.1 & 2 & 0.01 & 1 \\ 0.001 & 0.1 & 3 & 1 \\ 0.001 & 0.001 & 0.001 & 4 \end{bmatrix} \; .$$

The fourth species I_4 has the largest selective value, but a quasispecies is selected which consists predominantly of species I_3 and I_4, in which case the steady-state concentration of I_3 with $t \to \infty$ appears to be even greater than the concentration of I_4 [8.12,16].

Thus, the competition of quasispecies is completed by the selection of one of them, for which $\lambda = \lambda_{max}$. The average productivity $\bar{E}(t)$ will increase until

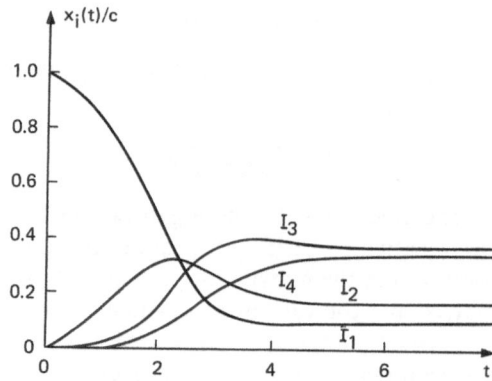

Fig. 8.3. Competition among four species [8.12]

it becomes equal to this value:

$$\bar{E}(t) \to \lambda_{max} \, . \tag{8.34}$$

Any biological species is a quasispecies, which is a set of mutants responsible for intraspecies variability. A large set of mutants is present in any population which is large enough. A quasispecies therefore possesses enormous possibilities for evolution and adaptation. We may ask the following question: "Is not the difference between species determined by specialists and generalists, different scales of quasispecies, and by different variabilities?" No relevant investigations of this subject have been carried out.

Within a quasispecies there is no selection of equivalent reversible mutants that are neutral in the sense of the Kimura theory (Chap. 6). The competition of such mutants is, of course, a neutral drift with "the survival of the survivors".

A quasispecies is stable up to a certain minimal value of the quality factor, i.e. up to a maximum value of the probability of mutations. If this threshold is passed, the quasispecies and the information it carries will gradually disappear in the successive generations of macromolecules. The threshold is specified by relation (8.26). For real polynucleotide chains, the probabilities of errors $1 - q$ are estimated at 0.01–0.10. The lengths of reliably replicated chains are of the order of $v = 100$. In this case, $Q = q^v = (0.99)^{100} = 0.36$. Eigen and Schuster [8.4] have described the corresponding computer models in detail.

Polynucleotide macromolecules do not autocopy but they synthesize complementary chains. The system is described by the alternation of "plus" and "minus" copies ($+I, -I$). Selection occurs among $\pm I$ assemblages represented by the cycles

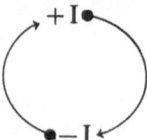

According to Eigen and Schuster [8.4], the oldest replicative macromolecules that reproduce in the absence of special catalysts might have been relatively short RNA chains similar to modern tRNA or their precursors. Types of RNA have recently been discovered which are capable of catalyzing various biochemical processes, particularly their template synthesis [8.17].

Subsequent development of the Eigen theory has led to the concept of the hypercycle [8.4]. In the first stages of molecular evolution, the amount of reproducible information was limited to v_{max} of the order of 100. This is much less than the amount of information contained in present-day genomes, which are reproduced with the invariable participation of the enzymes RNA- and DNA-polymerases. The transition to systems of this kind called for the integration of several units of the cycle type $+I$ into a co-operative system. From a set of cycles hypercycles were formed which were capable of competing with one another. Reproduction occurs with the participation of enzymes synthesized with the aid of polynucleotides present in the hypercycle. The scheme of the hypercycle is shown in Fig. 8.4. The information carriers are cycles I, which are responsible for the synthesis of the polymerases that catalyze the formation of the subsequent cycles. The closing of the hypercycle provides the co-existence of all cycles and, hence, the stability of the information contained in it. Hypercycles exhibit a much higher accuracy of reproduction. Table 8.1, which was borrowed from Küppers [8.12], presents data that characterize the accuracy of reproduction in polynucleotide systems operating with and without enzymes. Due to enzymes a radical increase in the quality factor (and hence in the value of v_{max}) occurs. The quantities listed in Table 8.1 express values of $1 - q$ and v_{max} calculated on the basis of the threshold condition (8.26).

It should be stressed here that the replication of RNA in the absence of enzymes under normal laboratory conditions has not been practically observed.

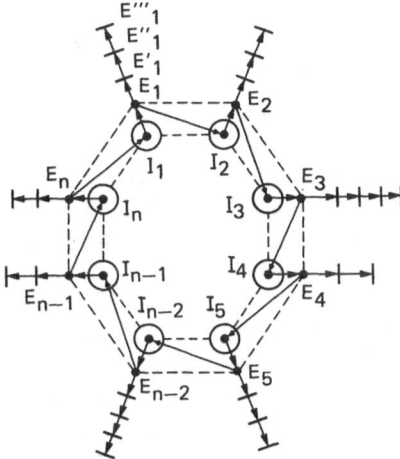

Fig. 8.4. Self-instructed catalytic hypercycle: $I_j \pm j$ = RNA cycles; E_j are enzymes encoded by I_j and catalyzing the formation of I_{j+1}; E_j', E_j'' ... are enzymes of parasitic processes [8.3]

Table 8.1. Error rate for individual units and values of Q_{min} and v_{max}

Biological entity	Molecular mechanism	Error rate $1 - q$	Superiority Q_{min}	Greatest chain length v_{max}	Examples in nature
Prebiotic RNA molecules	Replication of RNA in absence of enzymes	5×10^{-2}	2 20 200	14 60 106	Precursors of tRNA ($v \approx 80$)
Viruses	RNA-dependent synthesis by means of specific replicases	5×10^{-4}	2 20 200	1386 5991 10597	$Q\beta$ phage RNA ($v \approx 4500$)
Bacteria	Replication of DNA with the aid of polymerases and correction enzymes	1×10^{-6}	2 20	0.7×10^6 3.0×10^6 5.3×10^6	E. coli bacterium ($v \approx 4 \times 10^6$)
Vertebrates	Replication of DNA and recombination in eukaryotic cells	1×10^{-9}	2 20 200	0.7×10^9 3.0×10^9 5.3×10^9	Homo sapiens ($v \approx 3 \times 10^9$)

However, the role of catalysts may be played not only by enzymes but also by ions of some metals and catalytically active surfaces (minerals such as silicates and phosphates).

The human genome contains enormous quantities of non-coding DNA (see Chap. 7). Evidently, the accuracy of ncDNA replication must play a much lesser role than the replication of DNA coding for proteins.

The hypercycles make use of a large part, or all, of the information stored in polynucleotide chains. The hypercycle has the following properties:

1. It provides stable and controlled co-existence of all species connected via the cyclic linkage.
2. It allows for coherent growth of all its members.
3. The hypercycle competes successfully with any single replicative unit that does not belong to the cycle.
4. The hypercycle is capable of increasing or reducing its size if this modification offers any selective advantage.
5. Hypercycles do not easily link up in networks of higher orders.
6. Hypercycles compete with one another. The internal linkages and co-operative properties of the hypercycle, which are responsible for its integral behaviour, can evolve to optimal function.

For hypercycles the selective advantages depend on the size of the population due to non-linearity. Once established, the hypercycle is therefore not easy to replace by a new one, since a new species always appears in a small number of copies. The information stored in the hypercycle cannot be easily lost.

These properties of hypercycles are demonstrated by model calculations. Hypercycles that contain a sufficient number of units are characterized by limit cycles on phase portraits. This means that all the components of the hypercycle experience correlated oscillations. This provides a mutual stabilization of the components [8.4].

Ratner has proposed an alternative model called a "syser" (the abbreviation for a system of self-replication) [8.18]. This model is more complicated than the hypercycle since it incorporates structures similar to ones which actually exist, such as translation, transcription and replication systems. The intricate problems pertaining to the origin of such systems are far from being solved. According to Eigen and Schuster, the "hen" preceded the "egg", the "hen" being molecules of the tRNA type with v of the order of 100. These molecules of "pre-tRNA" were selected as a quasispecies. In their three-dimensional structure (see, for example, [8.11]) there was fixed chirality, i.e. a certain stereo-isomer was chosen. We may remark in passing that such selection implies the appearance of little information, only one bit per monomeric unit. Presumably, chirality has a chance origin. The three-dimensional structure of pre-tRNA provided the stability of these molecules and particularly, their resistance to hydrolysis. It may be presumed that several such molecules exhibiting different specificity were required to construct a translation system. This system may have arisen from a single quasispecies but the partners involved in it must have developed simultaneously. Such development called for the establishment of functional linkages between all the units, i.e. their integration. The union of these chains into a single large macromolecule has no future because of the threshold for errors. If prebiotic evolution really occurred in such a way, then the system providing the co-operation of informational macromolecules, a competition with erroneous copies and with any other less effective systems, could have been the hypercycle.

We will not analyze here other possible models, such as sysers. In the first place it is necessary, on the basis of the quasispecies concept, to examine the behaviour of self-reproducing and mutable macromolecules. As we have seen, the theory has been confirmed experimentally. A cornerstone has thus been laid for constructing a physical theory of molecular evolution.

8.3 Quasispecies in Sequence Space

Further investigation of the evolutionary behaviour of a quasispecies called for the development of a new concept: the sequence space. Along with the quasispecies concept, this concept was an important novelty in evolutionary theory. This concept is described in detail in the fundamental work by Eigen et al. [8.9], called "Molecular Quasi-Species".

Darwin's principle of the survival of the fittest calls for independent definition of the two words "survival" and "fitness". Otherwise, as the authors cited above write, "we inevitably get caught in the tautological loop of survival of the survivor."

Survival is defined in terms of the population or concentration. Fitness is characterized by a certain parameter which expresses the type that can survive, say efficient reproduction. In a system consisting of evolving macromolecules, in accordance with the Eigen theory, survival is expressed by the numbers x_i and fitness by the selective value W_i. As we have seen, the relationship between these quantities is far from trivial. The "zeroth approximation", according to which only one species with a maximum selective value W_m survives, is insufficient, as has already been shown. If the "zeroth approximation" were valid, evolution would have stopped whenever it did not correspond to the maximum value W_m, since beneficial mutations leading to this value occur very seldom. This difficulty is overcome by the quasispecies concept. It turns out that the evolution of a quasispecies has a directed, deterministic character. The evolution of a quasi-species is a higher approximation of reality.

The sequence space contains an assembly of chains of identical length v built up of different monomers. The mutants, whose total number is equal to k, may be arranged in accordance with Hamming distances $d(i, j)$ [8.19], which represent the number of differences in the sequences i and j.

Suppose the system is binary, $k = 2$, and that the nucleotides contain purines A, G, or pyrimidines C, U. The corresponding symbols are 0 and 1. First, let us consider one position in the chain, to which two points in sequence space, the points 0 and 1, correspond. The straight line connecting these two points is the unit Hamming distance. If we consider two positions, the straight line is doubled; there are four possibilities: 00, 01, 10, 11. At $v = 3$ and $k = 2$ the sequence space contains eight (2^3) points located at the vertices of a cube. With

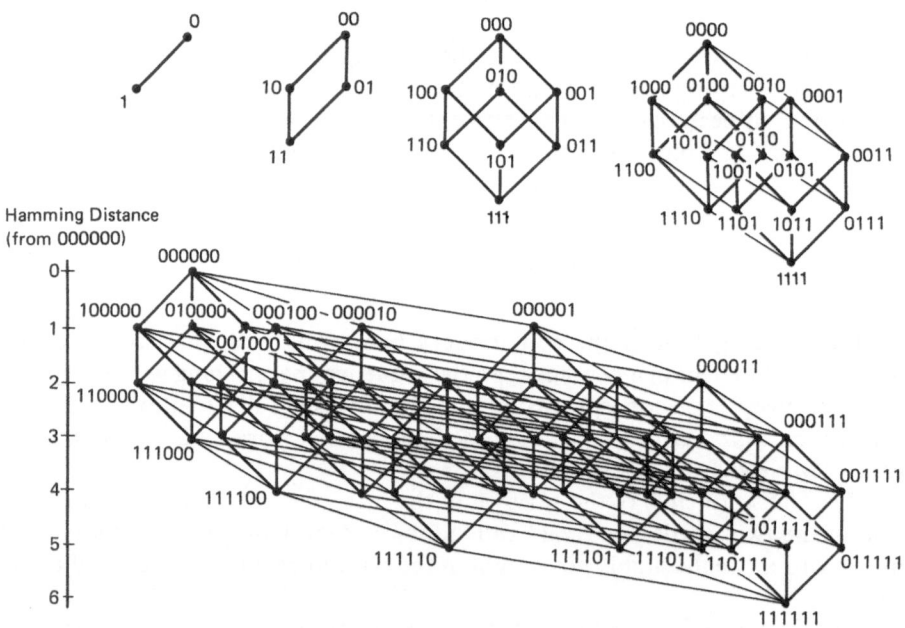

Fig. 8.5. Sequence space at $v = 1, 2, 3, 4, 6$ [8.6]

$v = 3$ we deal with hypercubes. Figure 8.5 shows the cases of $v = 1, 2, 3, 4, 6$ [8.6,9]. The number of vertices in the hypercube (hypercube at $v = 1, 2$) is 2^v. The lines connecting these vertices represent mutational pathways. The number of the shortest mutational paths connecting two vertices, i.e. two sequences, increases rapidly with increasing Hamming distance. There are $d!$ paths connecting 2^d states. Each state has numerous nearest neighbours. Most of the distances between the vertices of the hypercube are very small, so that any of them can be reached from the original one as a result of a limited number of steps of unit length. All the vertices, i.e. the states, are connected by numerous alternative paths.

The sequence space is, as it were, superimposed upon the evolutionary landscape; for each vertex of the hypercube there is a corresponding selective value. A landscape of this kind was first considered by Wright [8.20]. It is a mountainous landscape; larger selective or fitness values correspond to higher mountains. The alternative paths may pass through saddle points, i.e. the selective value may increase in μ directions and fall off in v-μ directions.

Suppose that the probability of mutations $1 - q$ is the same for all positions in all sequences. The probabilities of mutations are related to the Hamming distances.

The number of sequences in a certain class of mutants, i.e. the number of sequences with Hamming distances $d(0, i)$ from a certain original species I_0, is equal to

$$x_d = \binom{v}{d}(k - 1)^d \ . \tag{8.35}$$

The summation over all possible values of d from 0 to v naturally gives

$$n = \sum_{d=0}^{v} \binom{v}{d}(k - 1)^d = k^v \ . \tag{8.36}$$

The frequency with which a set of sequences with the Hamming distance d arises as erroneous copies of the matrix I_0 is given by (cf. (8.23))

$$Q_d = \binom{v}{d}q^{v-d}(1 - q)^d \equiv \binom{v}{d}q^v(q^{-1} - 1)^d \ . \tag{8.37}$$

The frequency of a correct reproduction (the quality factor) is given by

$$Q_0 = q^v \ . \tag{8.38}$$

The frequency with which a given polypeptide is produced as an erroneous copy of the original chain I_0 is

$$Q_{i0} = \frac{Q_{d(0,i)}}{x_{d(0,i)}} = Q_0 \left(\frac{q^{-1} - 1}{k - 1}\right)^{d(0,i)} \tag{8.39}$$

and, in general, the frequency of appearance of I_i from any sequence I_j is given by

$$Q_{ij} = \frac{Q_{d(j,i)}}{x_{d(j,i)}} = Q_0 \left(\frac{q^{-1} - 1}{k - 1} \right)^{d(j,i)} . \tag{8.40}$$

The values of Q_{ij} form a mutational matrix. A matrix of values W_{ij} simultaneously exists as has been said above. The off-diagonal elements of W_{ij} depend directly on the matrix Q and the Hamming distance $d(i, j)$. We have

$$W_{ij} = \left(\frac{q^{-1} - 1}{k - 1} \right)^{d(j,i)} A_j Q_j . \tag{8.41}$$

Let us consider the construction of the matrix Q.

The matrix $Q(v + 1)$ for binary sequences of length $v + 1$ is constructed recursively from the matrix $Q(v)$. We make $v = 1$ and introduce the notations

$$\varepsilon = (q^{-1} - 1) , \quad B = q^{-v} Q(v) .$$

The two smallest matrices $B(1)$ and $B(2)$ are as follows:

$$B(1) = \begin{matrix} & 0 & 1 & \\ \begin{pmatrix} 1 & \varepsilon \\ \varepsilon & 1 \end{pmatrix} & & \begin{matrix} 0 \\ 1 \end{matrix} \end{matrix}$$

$$B(2) = \begin{matrix} 00 & 01 & 10 & 11 & \\ \begin{bmatrix} 1 & \varepsilon & \varepsilon & \varepsilon^2 \\ \varepsilon & 1 & \varepsilon^2 & \varepsilon \\ \varepsilon & \varepsilon^2 & 1 & \varepsilon \\ \varepsilon^2 & \varepsilon & \varepsilon & 1 \end{bmatrix} & \begin{matrix} 00 \\ 01 \\ 10 \\ 11 \end{matrix} \end{matrix}$$

and, in general,

$$B(v + 1) = \begin{pmatrix} B(v) & \varepsilon B(v) \\ \varepsilon B(v) & B(v) \end{pmatrix} . \tag{8.42}$$

The eigenvalues $\lambda_i(v + 1)$ of the matrix $B(v + 1)$ are equal to

$$\lambda_i(v + 1) = \lambda_i(v)(1 + \varepsilon) .$$

The model under consideration, which unites the sequence space with the fitness landscape, and the quality matrix Q with the selective value matrix W, leads to concepts that are basically novel in evolutionary theory [8.6,9]. The size of populations and the concentrations of mutants specify the direction in the production of new mutants. The appearance of new mutants is not random, since these arise from the preceding ones. The Hamming distances gradually increase upon advancement in the sequence space; the process occurs step by step. The fitness of a given sequence is determined not only by its own selective value but also by the presence of neighbouring sequences capable of mutation.

As has been said above, the fitness landscape represents a mountainous country. Due to the multidimensionality of the population space there are numerous ways of reaching the highest peak along mountain ridges, without losing height, and using alpinist routes. Eigen and co-workers have drawn an analogy with the downward flow of water from mountains. Let us turn the landscape inside out, and thus replace the peaks with valleys. This analogy may be more spectacular; the best adaptation implies the minimum of free energy. The water accumulated in a groove finds its way into deeper valleys and eventually flows downhill. The flow direction is specified by the amount of water and by the terrain relief.

The concepts of the quasispecies and selection space show that deleterious mutations located along high ridges in the landscape play an important role. They lead the populations to peaks of high selective values.

The model considered is devoid of truly neutral mutations; the Kimura theory is not valid here. Even slightly harmful mutations lead to peaks. The main result of the theory is the deterministic directedness of the evolution of macromolecules.

Due to the selection of quasispecies the number of main copies may be small compared to the overall number of all mutants. A quasispecies may reach the maximum W_m, starting from an arbitrary initial position. The larger the size of the population with respect to v, and the more time selection takes, the more deterministic, not random, is the behaviour of the system. A mutation represented by a large number of copies serves by itself as a basis for new mutations. It is this factor that specifies the deterministic progress to the highest peaks of the landscape of selective values. The concept of the sequence space implies the construction of a kind of statistical geometry. The mutational behaviour of macromolecular populations is, of course, statistical, but it leads to deterministic evolution as a result of the superposition of numerous unit processes of motion from some vertices of the hypercube to others. This means that mutants arise not randomly but predominantly from precursors that are represented more often. This is explained in Fig. 8.6, which shows three fragments of the RNA chain. Fragment 3 (which differs from fragment 1 by two replacements) is produced from fragment 1 with a very low probability of a simultaneous double mutation, but the path $1 \rightarrow 2 \rightarrow 3$ is much more probable. The frequency of appearance of the precursors mentioned naturally depends on their selective value. The selective values are arranged in the form of a mountainous landscape, with the peaks connected by ridges. Practically the only peaks which appear to be occupied are those which are interconnected by multidimensional ridges.

The behaviour of the system thus depends directly on the probability of erroneous reproduction of macromolecules $1 - q$. The longer the chain, the more accurate must be the reproduction; otherwise, the errors will be accumulated in succeeding generations and the original information will disappear.

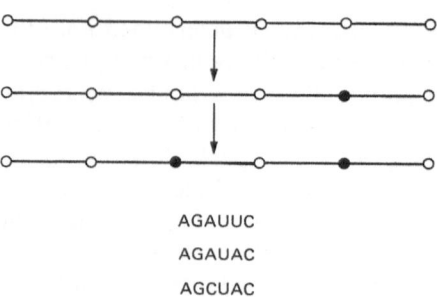

AGAUUC

AGAUAC

AGCUAC

Fig. 8.6. Schematic representation of successive mutations

What happens when the probability of errors increases? This problem has been explored on a computer model by Schuster and Swetina [8.21]. The model consists of binary chains ($k = 2$) and has a length $v = 50$. At first there is one best-adapted species, reproducing at a rate 10 times faster than any other species. The authors calculated the dependence on the wild type of the relative size of populations of wild type and mutants x_d with the Hamming distance d. The values of x_d with $d = 1$ to $d = 50$ represent the sums of all mutants belonging to these classes. The relative values of x_d vary from zero to unity. Figure 8.7 shows the curves obtained. Only at $1 - q = 0$ is the survival of the fittest a reality. As $1 - q$ increases, the main copy I_0 dies out, and the mutant copies increase in number, but at the critical value $1 - q = 0.046$ these mutant copies are also represented by a very small number. This is shown more clearly in Fig. 8.8. At the threshold value $q = 0.954$, all the information is lost and a sharp transition occurs to a state in the which the system is quite disordered. The number of mutants with $d = 0$ and 50 is the least ($x_0 = x_{50} \cong 10^{-15}$), and that of mutants with $d = 25$ is the largest ($x_{25} = 0.1$).

The practically complete destruction of the master copy and the wild type, and the distribution of all mutants in small relative amounts, thus occur at a threshold, like a phase transition. Let us consider this important analogy in more detail, following the fundamental work of Eigen et al. [8.9].

Demetrius, and also Leuthäusser [8.22] have analyzed the analogy between replication dynamics and the Ising two-dimensional model: the spin lattice (see, for example, [8.23,24]). Replication dynamics may be represented by a system of difference equations for discrete time:

$$x_i(t + \Delta t) - x_i(t) = \sum_{j=1}^{N} W_{ij}x_j(t) - [D_i + \phi(t)]x_i(t) . \qquad (8.43)$$

These equations coincide with differential equations (8.15) and (8.31) if $t \to \infty$.

The discrete dynamic model (8.43) may be pictured as a scheme of genealogies, that is time-ordered series of sequences, each of which arises as a result of correct or erroneous copying of the original one. The replication of polynucleotides is treated as a process of multiple branching. This scheme is shown in Fig.

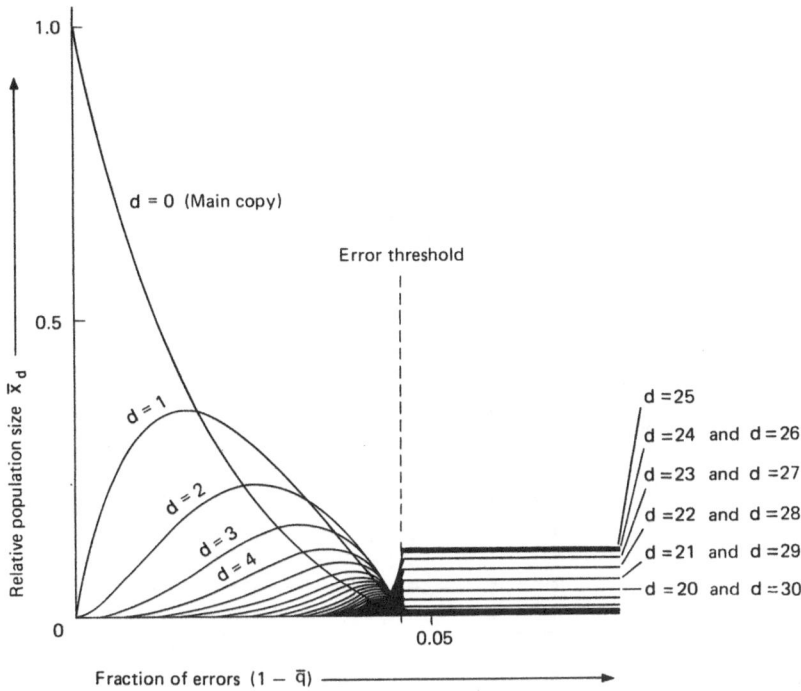

Fig. 8.7. The behaviour of a system of macromolecules at $v = 50$ and $k = 2$. Dependence of the population size on replication errors [8.6]

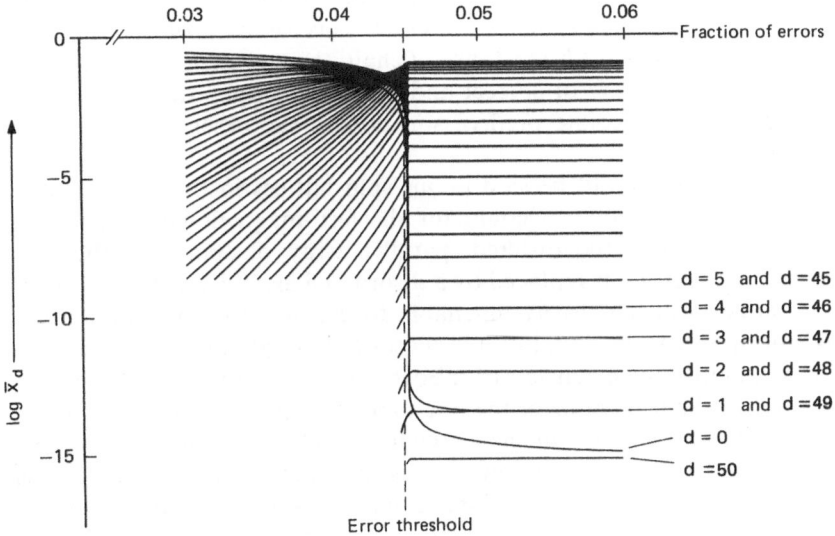

Fig. 8.8. The behaviour of the system of Fig. 8.7. at large magnification [8.6]

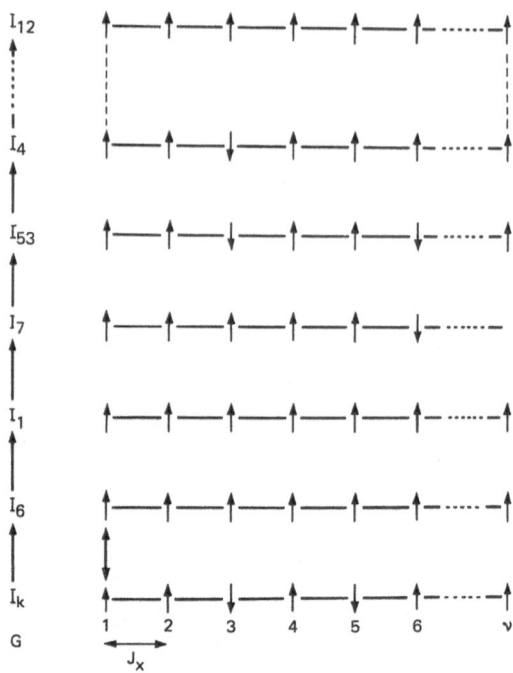

Fig. 8.9. A two-dimensional spin lattice. Each line consists of v spins and is a polynucleotide chain at $k = 2$. J_x is the coupling constant in the chain; J_y is the measure of mutation frequency [8.9]

8.9. The scheme is similar to a one-dimensional lattice of spins i.e. arrows which may be in N positions, $N = k^v$. In the two-dimensional Ising model, only pairs of directly neighbouring spins can interact. This follows from the properties of the system associated with the Markov chain theory, namely that the distribution of the probabilities of correct and erroneous copies depends only on the template of the immediate predecessor, not on the other predecessors in the genealogy.

A few words should be said about spin lattices in the Ising model. This model was introduced to account for the ferromagnetic-paramagnetic phase transition, in which the ordered, parallel array of spins, i.e. the magnetic moments of electrons, is replaced by a disordered arrangement. The one-dimensional model, which is easily amenable to calculation, does not provide the sought-for phase transition, but it is successfully used in macromolecular physics. For example, the extension of rubber with its entropy elasticity is found to be similar to the behaviour of the one-dimensional model; this is also the case in the theory of such phenomena as the helix-coil transitions [8.11]. The model takes account of the interaction of neighbouring spins, i.e. the co-operative nature of the system. The two-dimensional Ising model makes it possible to interpret the ferromagnetic-paramagnetic phase transition [8.23,24].

The principal eigenvalue of the matrix $W(\lambda_m)$ is formally identical to the minimum of free energy in statistical mechanics, $F \sim \log \lambda_m$. The threshold value $1 - q$ corresponds to the phase transition temperature.

The analogy between the Eigen evolutionary model and the two-dimensional Ising model was disclosed by Leuthäusser [8.22] (see [8.9]). Let us describe it.

We rewrite Eq. (8.7) in the form

$$\dot{x} = A_i x_i - \sum_{j \neq 1} W_{ji} x_i + \sum_{j \neq i} W_{ij} x_j - x_i \sum_j A_j x_j \ . \tag{8.45}$$

Here A_i is the rate of replication, which includes both correct and erroneous copies; the last term on the right-hand side is associated with the dilution flux and is responsible for keeping the number of molecules constant. We have

$$W_{ji} = A_i q^{\nu - d_{ij}} (1 - q)^{d_{ij}} \ , \tag{8.46}$$

where d_{ij} is the Hamming distance between chains i and j.

We introduce new variables:

$$z_i(t) = x_i(t) \exp \left[\int_0^t dt' \sum_j A_j x_j(t') \right] \ . \tag{8.47}$$

We obtain the transformed linear differential equations

$$\dot{z}_i = A_i z_i - \sum_{j \neq i} W_{ji} z_i - \sum_{j \neq i} W_{ij} z_j \tag{8.48}$$

or in vectorial form

$$\dot{z}(t) = W z(t) \tag{8.49}$$

with the solution

$$z(t) = e^{Wt} z(0) \ . \tag{8.50}$$

For a binary chain, (8.50) is a matrix of order 2^ν ($2^\nu \times 2^\nu$). Purines and pyrimidines may be likened to the two directions of spins ± 1. If

$$I_i = (s_1^{(i)}, s_2^{(i)}, \ldots, s_\nu^{(i)})$$

is the sequence of nucleotides, the Hamming distance will be written in the following form:

$$d_{ij} = \frac{1}{2} \left(\nu - \sum_{l=1}^\nu s_l^{(i)} s_l^{(j)} \right) \ . \tag{8.51}$$

Formula (8.51) may be explained by a simple example. Figure 8.10 shows the chains I_1, I_2, I_3, I_4 at $\nu = 4$. According to (8.51),

$$d_{12} = d_{23} = d_{34} = I, \quad d_{13} = d_{24} = 2, \quad d_{14} = 3 \ .$$

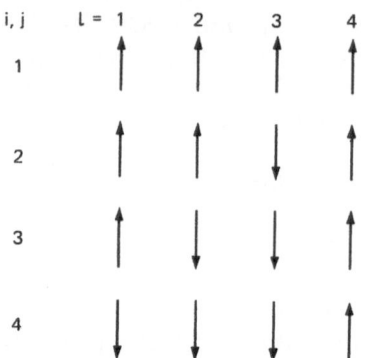

Fig. 8.10. The spin lattice for $v = 4$ and $k = 2$

Substituting (8.51) into (8.46), we obtain

$$W_{ji} = A_i[q(1-q)]^{v/2}\left(\frac{1-q}{q}\right)^{-1/2} \sum_{l=1}^{v} s_l^{(i)} s_l^{(j)} \qquad (8.52)$$

or

$$W_{ji} = A_i[q(1-q)]^{v/2} \exp\left(-K \sum_{l=1}^{v} s_l^{(i)} s_l^{(j)}\right) \qquad (8.53)$$

where

$$K = \frac{1}{2}\ln\frac{1-q}{q} . \qquad (8.54)$$

Expression (8.53) is identical to the transfer matrix for the two-dimensional Ising model (8.23,25]:

$$Z(\sigma',\sigma) = \exp(-\beta E(\sigma,\sigma))\exp\left(-\beta J \sum_{l=1}^{v} \sigma_l \sigma'_l\right). \qquad (8.55)$$

$\sigma = (\sigma_1{}^,\sigma_2,\ldots,\sigma_v)$, $\sigma' = (\sigma'_1,\sigma'_2,\ldots,\sigma'_v)$ represent the spin states of two neighbouring horizontal rows in a two-dimensional lattice. $E(\sigma,\sigma)$ represents the interactions between spins of one row; J represents the interaction between neighbouring spins in the vertical row, $\beta = 1/kT$. Comparison of (8.55) with (8.54) shows that K also includes the ferromagnetic and antiferromagnetic interactions of spins. Indeed, $K < 0$ with $1/2 \leqslant q \leqslant 1$, this being equivalent to $J < 0$, i.e. we are speaking here of the ferromagnetic interaction. On the other hand, with $0 \leqslant q \leqslant 1/2$ we have $K > 0$ and, hence, $J > 0$, the interaction being antiferromagnetic. Since the system of macromolecules under consideration q is always close to unity, we are dealing with the first case.

In the Ising model, the matrix $Z(\sigma',\sigma)$ connects the neighbouring rows of spins. In the theory considered, the matrix $\exp(W\Delta t)$ links macromolecules to the next generation, which arises in a time interval Δt.

The matrix W may be generalized with a view to taking into account the theory of different mutability at different points of the chain, i.e. the dependence of q on the index $I \leqslant i \leqslant v$. In terms of the Ising model this means the dependence of the interactions in the vertical direction on the co-ordinate in the horizontal direction. The transition from a binary to a quaternary system (A, C, G, U, $k = 4$) also leads to an exact solution, although this does not alter the basic conclusions.

Thus, at the critical value of $1 - q$, a second-order phase transition occurs in the macromolecular system. The analogy with the transition temperature is expressed by the equivalency

$$\frac{1}{T} \sim \left| \ln \frac{1 - q}{q} \right| .$$

The picture shown in Figs. 8.7 and 8.8 corresponds to the symmetrical distribution of mutants with respect to the wild type, which replicates 10 times faster. Other situations are also possible. Figure 8.11 (see [8.6]) shows the effect of the fitness landscape on evolution. On the left-hand side of the upper part of Fig. 8.11 the fitness distribution for the case shown in Figs. 8.7 and 8.8 is

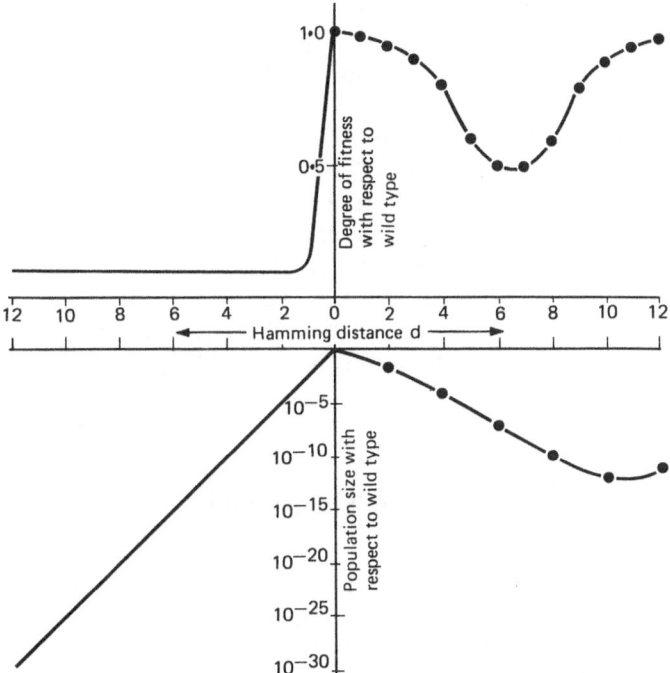

Fig. 8.11. See text for explanation

reproduced. The left-hand side of the lower part of Fig. 8.11 gives the relative numbers of the corresponding population (the wild type population is taken to be equal to unity). In the right-hand part of the upper figure a rather typical mountainous fitness landscape is shown. When the Hamming distance increases, the reproduction rate first decreases by 50% and then again increases to a value close to that for the wild-type. In the bottom part the corresponding relative population sizes are given. These curves were calculated for $v = 316$, $k = 2$. The population sizes were obtained for $1 - q = 3.16 \times 10^{-3}$.

The qualitative conclusions following from the curves of Fig. 8.11 are very important. The left-hand part shows that the probabilities of appearance of mutants with rather high values of d are negligibly low if the only sharp fitness maximum corresponds to the wild-type. For example, a mutant that differs by 12 units out of 316 from the wild-type appears with a population size less by a factor of 10^{30} than the wild-type. The populations of viruses with which we have to deal in laboratory experiments are less by many orders of magnitude. Hence, a 12-fold mutant, not to mention macromolecules, containing a larger number of errors, has no chance whatsoever of being present in the system.

On the other hand, with a mountainous fitness landscape, for a 12-fold mutant there is a corresponding population which is 10^{20} times larger than that in the system under consideration. This situation is quite different. It is determined by a multiplicative amplification: 12-fold mutants arise from 11-fold ones, 11-fold mutants arise from 10-fold ones, etc. But this means that evolution towards peaks occurs by way of motion up a mountain ridge, without descending to valleys, as has already been pointed out.

During a phase transition the information present is completely lost in the sequence space. The wild-type practically disappears and the diverse mutants are found to be represented in small but appreciable numbers. This gives rise to favourable conditions for accelerated evolution. The information that has disappeared is replaced by new information, in another region of the sequence space. Changes in the environment, i.e. changes in the fitness landscape, have an immediate effect on speciation (see also [8.25]).

The Eigen theory outlined here has been supported experimentally. The theory may be applied directly to viruses. Whether viruses (and bacteria) are alive or not is a matter of definition. In any case, they become viable in living cells.

8.4 The Eigen Theory and Viruses

It may seem that the Eigen theory is an abstract model of Darwinian prebiotic evolution, and self-organization of macromolecules. Even if this were so, the Eigen theory would still have been of very great importance, since it shows for the first time the possibility of constructing reasonable models of these phenomena on the basis of normal physics and chemistry. As a matter of fact, this theory actually goes far beyond this; it is confirmed by experiment and is directly applicable to the study of viruses, their variability and evolution.

Even in his first fundamental publication [8.1,2], Eigen described the "evolution *in vitro*". Spiegelman has conducted test-tube experiments on phage $Q\beta$, which multiplies in *Escherichia coli* bacteria [8.26–30]. This phage contains RNA and makes use of a special replication enzyme, $Q\beta$ replicase, which was isolated and purified. The experiments were performed as follows [8.27]. A reaction mixture was used, 0.25 ml of which contained four nucleoside triphosphates of RNA, 200 mµ-moles of each, 40 µg of $Q\beta$ replicase and 2×10^{-2} M $MgCl_2$. UTP was labeled with ^{32}P. The mixture was incubated with the phage RNA. The synthesis of the enzyme was initiated, following which a series of dilutions were carried out. Each time, after a certain period of incubation time, 0.02 ml of the reaction mixture was transferred to 0.25 ml of fresh standard solution. The first reaction was initiated by 0.2 µg of phage RNA, which was incubated for 20 min. The incubation time was then reduced from 20 min (transfers 1–13) to 15 min (transfers 14–29), to 10 min (transfers 30–38), to 7 min (transfers 39–52), and, finally, to 5 min (transfers 53–74). The molecules of phage RNA produced as a result of these procedures, their infectivity and length (v) were studied. The infectiousness was lost after the fourth transfer. The molecular mass of RNA decreased as the transfers were carried out; after the 74th transfer v decreased to 0.17 of the original value, which was 3600. Simultaneously, the rate of incorporation of tagged phosphor increased, i.e. the rate of replication increased. In other words, the system constantly selects a species with the highest selective value (shorter chains in the case given). Figure 8.12 shows the

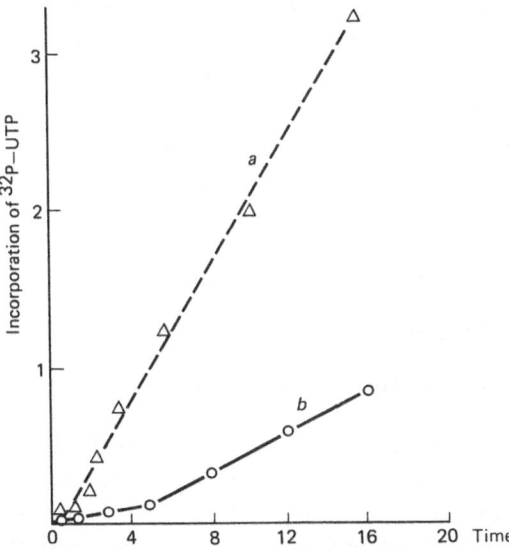

Fig. 8.12. The kinetics of synthesis of RNA $Q\beta$-phage after the 74th transfer (*a*) and the kinetics of replication of $Q\beta$-RNA (*b*) [8.3]

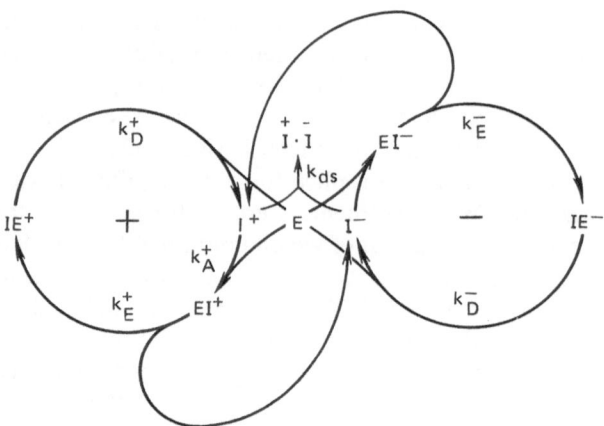

Fig. 8.13. Simplified representation of the three-step replication mechanism [8.32]

kinetics of synthesis after the 74th transfer and the kinetics of replication of the original Qβ-RNA.

A detailed study of this kinetics has been carried out by Biebricher in the Eigen laboratory [8.31,32]. The process may be represented by a three-step mechanism of RNA replication, i.e. an irreversible cross-catalytic cycle (Fig. 8.13). We denote the RNA species with a certain sequence by ^iI. The complementary chain belongs to the same species. RNA first binds to the replicase E with a rate constant ^ik_A, forming an active replication complex ^iEI, which synthesizes and liberates the replica RNA strand with a rate constant ^ik_E. The remaining inactive complex ^iIE dissociates into its components, ^iI and E, at a rate constant ^ik_D. The kinetic equations are as follows (square brackets represent concentrations):

$$d[^iI]/dt = {^ik_E}[^iEI] + {^ik_D}[^iIE] - {^ik_A}[E][^iI]$$

$$d[^iEI]/dt = {^ik_A}[E][^iI] - {^ik_E}[^iEI]$$

$$d[^iIE]/dt = {^ik_E}[^iEI] - {^ik_D}[^iIE]$$

$$\overline{d[^iI_0]/dt = {^ik_E}[^iEI]}$$

The overall equation expresses the kinetics of variation of the concentration of species i, $[^iI_0]$.

The kinetic relationships like those given in Fig. 8.12 characterize the overall replication rates. Figure 8.14a shows experimental rates for 10^{10}, 10^7 and 10^4 chains and Fig. 8.14b gives the rates calculated on the basis of the model. Two growth phases are clearly distinguished in accordance with theory: an exponential one where the enzyme is in excess, and a linear one where the enzyme is saturated with template chains. In the exponential growth phase a coherent

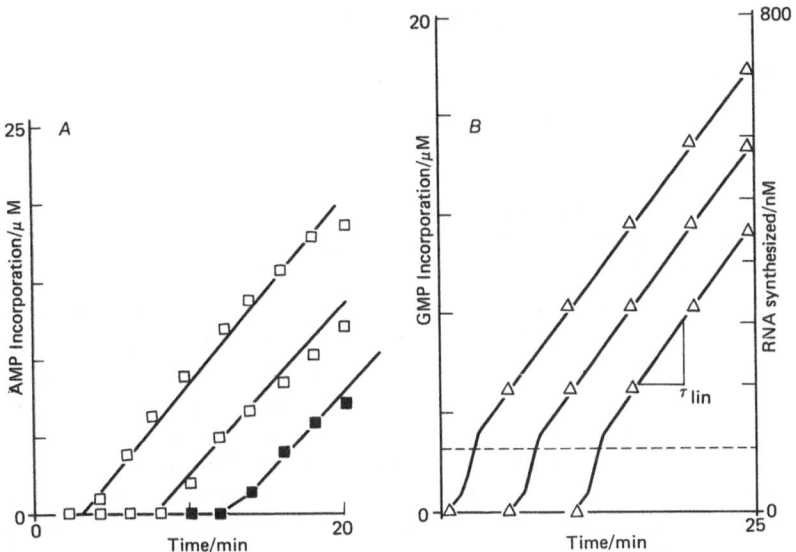

Fig. 8.14. Kinetic dependences for replication of RNA $Q\beta$-phage. (*A*) The rate for chains 10^{10}, 10^{7}, 10^{4}. (*B*) Theoretical dependences [8.32]

growth of the total population and of all intermediate complexes occurs:

$$d[^iI]/[^iI]\,dt = d[^iEI]/[^iEI]\,dt = d[^iIE]/[^iIE]\,dt = d[^iI_0]/[^iI_0]\,dt = {}^iK \ .$$

We denote the relative growth of the population due to replication by $A_i = {}^ik_E = [^iEI]/[^iI_0]$, and find that in the exponential growth phase $A_i = {}^iK$. In the linear growth phase a steady state is rapidly established:

$$v = A_i[^iI_0] = {}^ik_E[^iEI] = {}^ik_A[^iI][E] = {}^ik_D[^iIE] = {}^i\varrho[E_0] \ ,$$

where v and $^i\varrho$ are constants, while A_i varies with time.

Between different $Q\beta$ RNA species a competition takes place. In the exponential growth phase each species multiplies, independently of the other ones, but the relative populations of the species appear to be different at different rates of multiplication. In complete agreement with the Eigen theory, the relative populations $x_i = [^iI_0]/\sum_i [^iI_0]$ depend on time according to the law

$$\zeta_i \equiv dx_i/x_i\,dt = E_i - \bar{E} \ , \tag{8.56}$$

where ζ_i is the selection rate value. We have

$$x_i(t) = x_i(0)e^{(^iK - \bar{E})t} \ . \tag{8.57}$$

The absolute fitness, the selective value, is expressed by the product of the probability of survival to reproduce and the number of offspring produced [8.33]. If τ is the average generation time, then we obtain for the absolute fitness

$$e^{A_i \tau} e^{-D_i \tau} = e^{W_{ii} \tau} \; , \tag{8.58}$$

while for the relative fitness we have

$$e^{A_i \tau} e^{-D_i \tau} e^{-\bar{E} \tau} = e^{\zeta_i \tau} \; . \tag{8.59}$$

In the exponential growth phase, in the long run, only the fittest species with the highest selective value remains. However, in real systems the conditions for the exponential growth are rarely fulfilled; the growth rates usually fall off with increasing population. Accordingly, there is no selection of the most rapidly multiplying macromolecules, the rate of growth in the linear phase is limited by reserves of free enzyme and the selection is determined by the competition for binding with this enzyme. In this case the selective value does not correlate with the rate of replication and the more slowly-growing species often appears to be the winner in the competition. Experimental investigations have shown that selection in the linear growth phase is rather specific. In the first place, in the steady state of one species the other one may be present in a small number of copies and multiply exponentially. The relative concentrations of the species rapidly vary in this case. In the second place, a temporary appearance of negative selective values is possible if the decreasing number of templates that should have been bound to the replicase molecules no longer compensates for the losses determined by double-strand formation. In the third place, a balanced co-existence of distant RNA species can be set up if the species attain steady-state concentrations, which depend only on the rates of enzyme binding and double-strand formation [8.34].

It has thus been demonstrated experimentally that the evolution of RNA is a direct result of the interplay of mutation and selection in full quantitative agreement with the Eigen theory. All those details of the behaviour of $Q\beta$ RNA have been studied [8.31,32,34] which appear to be nontrivial. A kinetic analysis of the replication of mutant populations allows us to determine selective values and construct an evolutionary landscape in the sequence space.

A biological reactor has been built in the Eigen laboratory at the Institute of Biophysical Chemistry of the Max Planck Society in Göttingen, to carry out such experimental investigations (see [8.6]). The scheme of this bioreactor is shown in Fig. 8.15. A bacterium culture (*E. coli*) is maintained in the turbidostat at a constant population density, which is determined by studying the turbidity of the medium. The turbidostat supplies a flow reactor in which a virus infection is realized at a constant influx of fresh bacterial cells and efflux of the products of their degradation. The *E. coli* cells are infected by $Q\beta$ phage at time $t = 0$. In about 40 minutes up to 10,000 phage particles form in each cell. These events are repeated regularly since the number of phage particles decreases between two consecutive multiplications as a result of the dilution of the phage by the efflux of the reaction solution and influx of fresh *E. coli* cells. In about 3 hours a steady state is attained, in which the formation of a virus is compensated for in a precise way by dilution.

The set-up makes it possible to monitor the process by introducing various

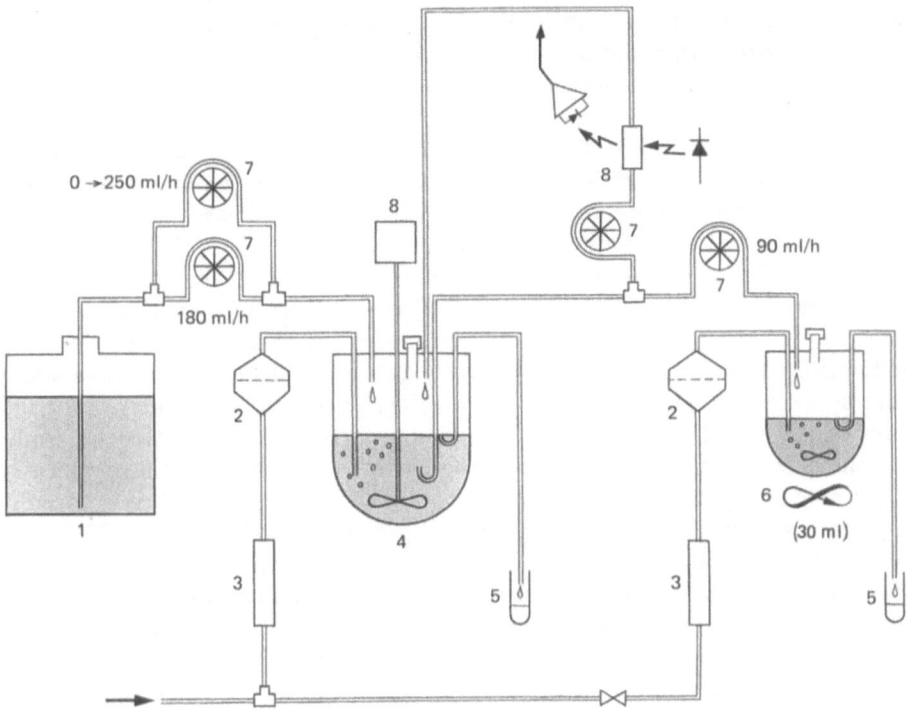

Fig. 8.15. Biological reactor: (*1*) nutrient solution; (*2*) filter; (*3*) flowmeter; (*4*) turbistat (300 ml), bacterial culture; (*5*) drain of excess; (*6*) viral culture (30 ml); (*7*) pump; (*8*) stirrer; (*9*) nephelometer, measurement of turbidity [8.6]

substances that affect its steps. The reactor is convenient for work with large quantities. But experiments with microliter volumes are also conducted in the Eigen Laboratory. In this case, serial transfer is used, just as in Spiegelman experiments. By changing the environmental conditions, it is possible to alter the course of the evolutionary process, in particular to approach the error threshold and to cross it for a short period of time. In accordance with the theory considered, it is in precisely this region that the evolutionary process appears to be especially efficient. The rate of evolution can be changed by many orders of magnitude. This is accomplished with the aid of a computer-controlled automatic device which provides serial transfer. The program developed by Eigen involves the following steps:

1. The choice of samples and setting the rate of mutations, which depends on the ratio of nucloside triphosphates.
2. The transfer of the RNA template by means of a programmed automatic pipette to the cooled reaction mixture.
3. A temperature jump from 0 to 37°C and from 37 to 0°C during a second to initiate and terminate the replication.

4. Control and measurement of the rate of growth by means of a laser fluori-
    meter with fibrous optics.
5. Exact establishment of physico-chemical conditions.

The machine was constructed for evolutionary biotechnology. Instead of
individual samples, work is done on a whole plate for cloning, which has a
capacity of 1000 samples with a volume of $1\mu l$. This allows one to study the
special properties of RNA that are not directly associated with the rate of reproduction. Variation of mutation rates produces a mutational spectrum with hier-
archically distributed distances. The multichannel laser fluorimeter records the
efficiency of reaction products, and the system of pipettes moving in three
dimensions allows artificial selection to be realized. This procedure, which has a
number of advantages as compared to serial dilution, leads to the reconstruction of the evolutionary landscape. A computer is used to construct a topo-
graphical map of the distribution of the selective values of cloned assemblies.

These experiments, which are so far confined to $Q\beta$ phage, are very promising. Both artificial selection and evolution are simulated experimentally at the
molecular level. The simplicity of the $Q\beta$-RNA system studied opened up enormous possibilities. In this connection, it should be pointed out that there is a
characteristic feature of all the branches of biology. It is always possible to
choose, among an enormous variety of living things, organisms that make it
possible to solve the fundamental problems of science most efficiently, and serve
as models. The garden pea proved an ideal model for establishing Mendel's
laws; the subsequent development of the genetics of multi-cellular organisms is
first of all associated with the fruitfly *Drosophila melanogaster*. Developmental
biology was originally studied on newts, the nerve conduction on the squid
axon, and muscle contraction on the frog sartorius muscle, which had long
before been used in Galvani experiments. Molecular biology arose from experiments with *E. coli*, and for molecular evolution $Q\beta$ phage appeared to be an
ideal model.

Eigen and his coworkers [8.35] have given a detailed interpretation of the
theory and have conducted experiments which directly support the hypercycle
model. The hypercycle implies a feedback loop, which connects the genotype
with the phenotype. The $Q\beta$ phage system studied in the laboratory is the
prototype of the hypercycle. The viral genome is a dual source of information.
Being a template, it instructs its own reproduction and as a messenger it determines protein information by means of translation. Its products create a specific
feedback loop which regulates the growth of the virus.

In an experiment, the hypercycle manifests itself in an explosive growth of
both RNA and protein. These "explosions" are synchronous and co-operative.
Here the protein being synthesized exerts an inhibiting action and, hence, serves
as a regulator.

The RNA replicated is selected in accordance with the theory. The most
efficiently replicated mutant is selected in a competition. In this case, the mutant
genotype also plays the role of phenotype, being the target of selection. How-

ever, coding simultaneously for proteins, the genotype is selected in the composition of virus particles. The role of the phenotype is eventually played by the hypercycle as a whole.

It had been earlier assumed that to eliminate parasites involved in infection, the hypercycle must be compartmentalized. However, in the host cell (*E. coli*) all the genotypes that are suitable as targets for replication are enhanced, irrespective of their functional properties.

Originally the hypercycle was introduced as the principle of organization, which enables the constraints on the size of genotypes determined by the error threshold to be overcome.

In order to incorporate a higher information content, many of the genes should have been involved in co-operation rather than competition. Such co-operation, which is evidently being realized in eukaryotic viruses, may arise as a result of the combination of hypercyclic linkages with compartmentation.

The $Q\beta$-phage system studied shows directly that the hypercyclic organization functions optimally if it contains both enhancing and inhibiting feedback loops. The models that have been investigated demonstrate that the presence of a large excess of membrane proteins in the earliest stage of infection completely hinders the hyperbolic growth of the phage. This phenomenon can be used in a novel anti-virus strategy (see the next section).

Thus, the detailed experimental investigations of $Q\beta$-phage in the Eigen laboratory have confirmed the theory of the evolution of macromolecules, including the hypercycle, proposed by Eigen.

Viruses that are dealt with by man, such as polio viruses or rabies virus, influenza or AIDS viruses, function near the error threshold value $1 - q$, which is of the order of 10^{-3} to 10^{-4}. Therefore, they are represented by a wide set of mutants. Whereas efficient vaccines were obtained for poliomyelitis or rabies, not to mention smallpox, the production of vaccines for influenza and AIDS appears to be very difficult because of the extreme variability of these viruses. As has been pointed out by Eigen [8.36], in the 50 years after 1933, the influenza virus changed under the influence of immune pressure as strongly as a portion of the *E. coli* genome of the same length did in many millions of years. Viruses that can be isolated from the organism of one patient and cloned appear to be so different that at first the presence of multiple infection was presumed. It is exactly this similarity to a phase transition that accounts for a considerable proportion of the difficulties in the combat against AIDS.

In his lecture "Quasispecies of Viruses, the Pandora's Box" read on July 4, 1990 at the Symposium of Nobel Prize Winners (in Lindau on Lake Constance), Eigen expounded his theory and dwelt in detail on the problem of the high variability of viruses, which is especially significant in the AIDS virus. In this case, the variability is revealed in all three nucleotides of codons responsible for the incorporation of amino-acid residues into protein chains. Smallpox was successfully eliminated by vaccination, and the incomparably more difficult combat against AIDS has only recently begun (see also [8.37]).

The Eigen theory and the corresponding experimental investigations open up new paths for the development of molecular biotechnology. Modern bio-technology reduces mainly to genetic engineering. By introducing foreign genes, it becomes possible to teach the *E. coli* culture to synthesize insulin or growth hormone. This requires neither great ingenuity nor a physico-chemical theory. The situation is different in the field of protein engineering (see Sect. 8.5). The creation of new proteins, i.e. proteins that do not exist in nature, is possible only on the basis of genuine protein physics, an area where numerous problems are to be solved.

A new stage in molecular biotechnology is evolutionary biotechnology – the application of the selection principle to populations of replicable molecules, viruses, and micro-organisms. We have seen how this principle is implemented in application to $Q\beta$ phage. The selection pressure in such systems can be con-trolled from without. Selection can be realized by external constraints that act on populations. Direct intervention is also possible – the isolation of specific mutants, specific phenotypes and their clones. This is artificial selection at the molecular level. The experimental technique worked out in the Eigen laboratory can be directly used for these purposes.

The experiments carried out with RNA indicate new ways of combating viruses since they make it possible to apply specific RNAs for directed interac-tions with RNA viruses and with polymerases. The point is that deterministic selection of a population becomes possible. The mutational process is random, but the population structure of RNA mutants is deterministic. This makes it possible to avoid the basic problem of natural selection, in which feedback occurs between the phenotypic function and the selection of the genotype.

The combination of the methods developed with protein engineering is very promising for the solution of numerous fundamental and applied problems. The 1st International Conference devoted to natural and artifical selection in biotechnology was held in April, 1991. A brief summary has been published ([8.38], see also [8.39]).

8.5 The Importance and Prospects of the Eigen Theory

The great importance of Eigen's theory for natural sciences is that it provides for the first time a quantitative physical interpretation of the evolution of macro-molecules and, hence, of viruses. The theory also opens up new possibilities for the understanding of the evolution of multicellular organisms.

First of all, attention should be focused on stochastic approaches to the evolution of macromolecules. The mathematical apparatus of the theory in question is one of differential equations based on a continuous change of the variables x_i. As a matter of fact, the appearance or disappearance of a fractional number of molecules of a given species is, of course, impossible. Macromole-cules are synthesized or break down as entities. The treatment of such processes rests on the probabilities of appearance or disappearance and the corresponding equations are stochastic. The deterministic model described, which ignores randomness, may, in principle, give rise to errors. As we have seen, the model predicts a nearly uniform stationary distribution of all sequences beyond the threshold, which leads to a phase transition. However, this is impossible, because the system can contain at least one molecule of each type. The stationary distribution proves unstable; the assembly of macromolecules varies and moves randomly in sequence space. We also encounter non-stationarity in phenomena associated with the finite size of populations, especially if they are small. In a stochastic treatment of replication and mutation processes, evolution may be looked upon as a sequence of discrete steps in the direction of increasing fitness [8.37]. Prolonged "quasistationary" phases are interrupted by short periods of active selection with increasing fitness. The situation is similar to punctuational evolution of multicellular organisms. During these short periods a quasispecies is localized in a restricted part of the sequence space, in which the error threshold is crossed.

As pointed out above, the Markov theory of chains can be efficiently applied to evolutionary processes. It is possible to write the corresponding master equations, whose solutions are approximate. Some attempts have been made to write and solve master equations for evolution, but they have failed to provide a stochastic version of the phase transition described, except for some special cases (see [8.9]). The results obtained in these cases are equivalent to those obtained by means of the deterministic model.

The doctoral thesis of Weinberger is devoted to a stochastic generalization of Eigen's model of natural selection [8.38] (see also [8.39]). The author has supplemented the Eigen theory by taking account of spatial effects, which play no part in the flow reactor shown in Fig. 8.1, but are essential to biological evolution. In other words, the vector x should be treated as being dependent on both the time and position in real space, $x(r, t)$. The two-dimensional real space was considered, in which macromolecules move and undergo evolution. We may speak of the behaviour of these molecules on the surface of clay or stone particles. It has been shown that taking account of spatial effects does not change the basic inferences arrived at on the basis of the Eigen model. Consideration of stochastic effects requires a study of the joint distribution of the probabilities of a set of sub-populations. Random processes of the type of birth and death occur in these sub-populations. The condition of the constancy of the total population is imposed. Equations have been derived for the statistical characteristics of the system. The equation for the average distribution is identical to the basic deterministic equation in the Eigen theory. Weinberger points out that it is very tempting to apply an analysis of the same kind to populations of organisms rather than of molecules, with the Eigen species being identified with real biological species or with genotypes within a species. Naturally, only a species that was originally present in the system, even if only in a very small amount, can win in competition. The thermodynamic conclusions from the above treatment are that the average fitness of the system increases during evolution and the entropy decreases. The behaviour is similar to that of the system being cooled. In this sense, the differential replication of macromolecules operates like Maxwell's demon. Of course, here the second law is not violated; the process is paid for by the energy supplied to the system in the form of energized monomers, i.e. nucleoside triphosphates.

Weinberger [8.40] has analyzed the fitness landscapes, taking into account their autocorrelation which arises during a random walk in this mountainous country. Corresponding models have been proposed. We will return to a consideration of the landscapes in Sect. 9.5. The notion of a rugged fitness landscape has been suggested by Kauffman and Levin [8.41]. These two authors, together with Weinberger [8.39,41], have dealt with various forms of the model of spin glasses, which has recently gained wide application in the various areas of physics and biophysics: the theory of protein structure, the theory of neuron networks, etc. An overview of experimental facts and theoretical conceptions associated with spin glasses has been given by Binder and Young [8.42].

Spin glass is a disordered system of binary variables which assume two values ± 1; the system is described by a function of the type

$$H(s) = \sum_{i,j=1}^{N} A_{ij}s_i s_j \tag{8.60}$$

where A_{ij} are random variables. We have already encountered the spin model but not with "glass". In spin glass there is a frozen disorder and long-range order is impossible. Nevertheless, in such a system transitions similar to phase transitions are also possible. A direct similarity to spin glasses is observed in some magnetic materials. The theoretical apparatus of the corresponding models are applied, not without success, to a number of biological problems, in particular neuron networks and protein molecules. The application of the spin glass model to the problem of the origin of life has been described by Anderson [8.43] and to the problem of biological evolution by Rokshar et al. [8.44] (see also [8.45]). Population dynamics in the evolution of macromolecules has been studied by Peliti and co-workers on the basis of the spin glass model [8.46,47]. This is a stochastic theory which deals with the asymptotic dependence of the variability of populations on the environmental parameters, on the correlational properties of the fitness landscape. Exact calculations have been performed by Derrida and Peliti for a flat fitness landscape [8.47]. These authors emphasize that the Eigen theory is good if optimal fitnesses are strongly pronounced and rare. A stochastic analysis is required for an understanding of certain special features of finite populations in dealing with neutral evolution. In this case the deterministic equations for quasispecies are solved in a trivial way; in a steady state all species are represented identically. However, the finite populations sizes imply that at a given moment only a few species are present in the system.

It should be stressed that neither of these interesting works points to the necessity of a revision of the Eigen theory, which is noted for its beauty and integrity, properties that are inherent in a true physical theory. As was shown in Sect. 8.4, the theory is supported by direct experiments.

It should be noted that the Eigen theory allows one to put forward reasonable arguments concerning the origin of life. Evidently, the question of whether the egg or the hen (i.e. nucleic acids or proteins) appeared first, is solved in favour of RNA, which performed both the template and the catalytic function. Then hypercycles appeared. The subsequent development took place with an invariable participation of compartmentation. Compartments could have appeared in pores of clay particles; later they were built by proteins synthesized in hypercycles. No theoretical model of the origin of life, using the compartmentation concept, has yet been worked out.

Eigen and his coworkers have published an article [8.48] under a remarkable title: "How Old is the Genetic Code? Statistical Geometry of tRNA Provides an Answer." A comparative analysis of transfer RNAs carried out by means of the methods of statistical geometry in the sequence space permits us to show that approximately one third of differences in the present-day sequences of tRNA existed at the "prokingdom" level. This occurred during the divergence of archaebacteria and eubacteria. The age of the genetic code is close to the age of our planet and does not exceed it. In other words, life was not brought to the Earth from outside. The age of the Earth is about 4.5×10^9 years and the age of the genetic code is estimated at $3.8-0.6 \times 10^9$ years [8.49].

In this book we have repeatedly encountered the basic problem of theoretical biology: the genotype-phenotype relationship. Let us consider this problem within the framework of Eigen's theory. As has already been pointed out, in the original theory, the genotype (i.e. the nucleotide sequence in RNA) plays the role of the phenotype. In experiments, however, we deal not with RNA but with a virus, i.e. the biological process also involves proteins. The construction of a theory taking account of translation is a formidable and still unrealized task; no attempts have even been made to solve it. We can, however, go farther along the path from genotype to phenotype, by taking account of the secondary structure of RNA, which is determined by the formation of Watson-Crick pairs of nucleotides, A-U, G-C. A large variety of such RNA structures are known from experiments and for a number of tRNAs three-dimensional, spatial structures (Rich) have also been determined. Relevant theoretical investigations and computations have been carried out in a number of works by Schuster and his colleagues [8.49-53].

The presence of the secondary, conformational structure in RNA is directly reflected in its properties: the ability to undergo replication, and stability toward hydrolysis. Base pairing lowers the replicating ability since this requires free chains, and also increases the stability to hydrolysis.

Taking account of the tertiary structure of RNA is a more complicated task than in the case of proteins (Chap. 5). Schuster confines himself to a consideration of planar, two-dimensional structures of RNA. These are graphs devoid of nodes (the presence of nodes would have implied a tertiary structure). The elements of the secondary structure are portions of double-stranded RNA, in which one finds stacking, loops and also free ends and bonds formed by external nucleotides.

There is every ground for believing that the principal features of the conformational behaviour of RNA are reflected no less by the secondary structure than by the tertiary structure. Calculation of secondary structures from primary ones is performed with the aid of an algorithm that makes use of the criterion of the minimum free energy. In other words, account is taken of base pairing in double-stranded stretches and also of the corresponding stacking. Thus, the free energies and rate constants for replication and degradation are evaluated for secondary structures. The algorithm describes the co-operative nature of the "melting" of double-stranded portions. Each double-stranded portion, i.e. a stacking portion, is weighed "sigmoidally", depending on the length, and gives an additive contribution to the overall slowing-down of replication:

$$A_k = K_R - K_1 \sum_j \frac{n_j^{(k)}(1 + n_j^{(k)})^3}{(1 + n_j^{(k)})^4 + L} , \qquad (8.61)$$

where $n_j^{(k)}$ is the number of double-stranded bases in the jth double-stranded portion, k being the number of the chain; L is a constant. Of course, the third and fourth powers of $n_j^{(k)}$ in the formula are rather arbitrary, but it is just in this way that co-operation is taken into account. On the other hand, each unpaired loop or the outer region increases the rate of degradation of the chain in a non-cooperative way, thus:

$$D_k = K_D + K_2 \sum_j \frac{u_j^{(k)}}{u_m} \exp\left[\frac{(u_j^{(k)} - u_m)}{u_m} \right] + K_3 \frac{1}{v} \sum_l v_l^{(k)} , \qquad (8.62)$$

where $u_j^{(k)}$ is the number of unpaired units in the jth loop; u_m is the maximum number of units in the loop; $v_l^{(k)}$ is the length of the lth external element.

Finding the minimum of free energy may be treated as a kind of annealing i.e. the cooling of a system similar to spin glass until the least-energy configuration is attained.

The fitness landscape for RNA macromolecules characterized by both the primary and the secondary structure is very inhomogeneous, rugged, since these fitnesses expressed in the long run by replication kinetics and the kinetics of chain consumption, depends directly on the phenotype, the secondary structure. It is evident that to one and the same primary structure numerous secondary structures may correspond, including those which greatly differ from one another. The relationship between the fitness landscape and the sequence space appears to be specific; closely related sequences may possess strongly different secondary structures. At the same time, a large number of secondary structures of different chains may be very similar to one another. We see that a consideration of the difference between genotype and phenotype even in application to the simplest molecular model is demonstrated by the degeneracy of neutralism. A large number of RNA chains are neutral in the sense that, while differing in primary structure, they have similar secondary structures and, hence, similar physical and biological properties. In the Schuster model the degeneracy of the genotype-phenotype relationship, which has been considered in a discussion of

Kimura's theory (Chap. 6), is characterized by an entire hierarchy of levels. Firstly, many of the primary structures lead to the same two-dimensional structure. Secondly, many structures have identical sets of u_j and/or n_j. Thirdly, different sets of these quantities may lead to identical values of A_k and D_k and with their different values identical values of $E_k = A_k - D_k$ are possible.

Thus, in a system characterized by both primary and secondary structures degeneracy and, hence, neutral mutations exist. With the error probability being higher than the critical probability, the quasispecies is no longer localized in some definite region of the sequence space but undergoes a random drift. Evolution is no longer directional. Schuster has established two ways for evolutionary optimization which depend on the population size, the accuracy of replication and the structure of the fitness landscape, i.e. selective values. The population may reside for a long time in some region of the sequence space and then rapidly, jumpwise, move into another region. The optimum solution is achieved by intermittent improvements. The second way consists in the gradual achievement of the optimum. The first way is inherent in small populations and low error probabilities. The second one is realized in the opposite situation. There is a steady-state transition from one scenario to the other; say, the gradual transition is transformed to a sequence of small jumps with decreasing q. Large jumps occur with further lowering of the error probability and eventually the population is localized in a certain local peak of the landscape.

We see that punctuational and gradual evolution is realized even at the level of macromolecules. The threshold values of q lead to punctualism. It seems probable that punctuational evolution is realized in limited populations.

At first sight there seems to be a contradiction between Eigen's theory and Kiumra's theory (the latter having been discussed in Chap. 6). Indeed, the evolution of molecular chains in the sequence space has a directed rather than a random character. This contradiction, however, is easily removed. The Eigen theory was originally built for macromolecules, i.e. of viral genomes. These genomes are relatively small and the populations of the objects under consideration are very large. On the other hand, the genomes of multicellular organisms dealt with in the Kimura theory are very large (in particular, they contain many non-coding DNAs), and the populations are limited. As we have seen, the selectionally neutral mutations multiply independently of the population size, as a result of which a random drift occurs. This appears to be possible as a result of two circumstances: the degenerate correspondence between genotype and phenotype, and limited population size. Certain positions of the neutral theory prove to be applicable to bacteria as well. In Sect. 9.2 of his monograph [8.54], Kimura considers the results pertaining to haploid organisms, and in particular to *Escherichia coli* bacteria, whose genomes are small and whose populations are enormous. The level of genetic variability of *E. coli* does not differ too greatly from that of diploid organisms. There is no sexual reproduction; *E. coli* cannot be heterozygous. Kimura points out, however, that for bacteria we can introduce the notion of heterozygosity: the fraction of heterozygotes among imaginary zygotes, each of which arises from a random combination of two haploid cells. The virtual heterozygosity of *E. coli* attains 24%, which even exceeds the heterozygosity of diploid organisms. The neutral theory appears to be applicable to *E. coli* too.

Gojobori, Moriyama and Kimura have recently demonstrated the agreement between the neutral theory and the evolution of viruses [8.55]. Viruses evolve millions of times faster than organisms do. It has therefore been found to be possible to directly trace their evolution. It appeared that in accordance with the neutral theory the synonymous point substitutions in viral RNA occur several times faster than substitutions that change the amino-acid residues in proteins.

The systems that have been studied both theoretically and experimentally in the works of Eigen and his associates are characterized by relatively small sizes of the genome, i.e. low values of v, and by large populations. In these cases too, the populations are much less than the number of vertices in the hypercube of the sequence space:

$$N \ll k^v .$$

Indeed, $k = 4^{1000} = 10^{600}$ at $k = 4$ and $v = 1000$. The number of vertices in the hypercube is enormous and any mutant may be represented in it. Neutral mutants are close to the main copy and form a quasispecies together with it. Neutral mutants that have large Hamming distances from the main copy, i.e. are remote from it, seldom appear. In typical populations with N of the order of 10^{10}–10^{15}, mutants that are separated from the wild type by more than 4–5 substitutions are practically non-existent. It is for this reason that during prebiotic evolution in RNA or in $Q\beta$-phage directional evolution occurs. The larger the population as compared to v, and the longer the time period allotted to the system, the more limited is the random behaviour: the system behaves in a deterministic manner. The region of the sequence space occupied by many mutants itself produces a large number of mutants. Deterministic, directed evolution is the more strongly pronounced, the larger is the population.

The basic problem that we encounter here is the problem of the genotype-phenotype relationship, which has been repeatedly mentioned in this book. It is the genotype that is inherited, genetically fixed, while phenotypes are subject to natural selection. Thus, the evolutionary landscape is a mountainous country of phenotypes connected with the mountainous country of genotypes in a non-trivial way. At the molecular level, the genotype is determined by the nucleic acids DNA and RNA, and the phenotype by proteins. As has been pointed out a number of times, neutralism is based on the degenerate correlation between the primary structure of protein specified by the genotype and the spatial structure of protein, and on the degenerate correlation between the spatial structure of protein and its biological function. Viewed in a different way, proteins with different primary structures may have the same spatial structure. The correlation between the three-dimensional structure and the biological function is also ambiguous; certain differences in spatial structure may be of practically no importance in biology.

The molecular basis of the neutral theory reduces to the degeneracy of the correlation between the primary and the spatial structure of protein, i.e. to the degeneracy of the genotype-phenotype correlation.

As we have seen, such degeneracy is described in works by Schuster, who introduces the secondary structures of RNA macromolecules. There is no contradiction whatsoever of the neutral theory; moreover, Eigen's theory and the Schuster theory offer, in principle, efficient approaches to the evolution of multicellular eukaryotes.

The neutral theory may and must be described by a stochastic rather than a deterministic theory. We have already mentioned the stochastic modification of Eigen's theory. Stochastic approaches have also been developed by McCaskill [8.37] and Schuster [8.49]. The stochastic reaction network corresponding to RNA replication with mutation, in the presence of chain degradation and a dilution flux, is calculated by a computer. Here account is taken of the secondary structures. The computer calculations performed by Fontana, Schuster and others [8.50,53] involved 2000 macromolecules with a chain length of $v = 70$. Despite the relatively small size of the system, features typical of real populations were observed; there are error thresholds for replication, evolutionary steps and quasistationary distributions of the sequences. The authors cited have examined quantitatively the role of selectively neutral or nearly neutral variants. They calculated four characteristics of the ensemble: the distribution entropy, the correlations in the ensemble, the average Hamming distance and the population diversity. The sensitivity of these characteristics to a change in a number of parameters has also been studied.

The Hamming distance between selectionally neutral sequences and the selective value of neighbouring sequences determine their content in the quasispecies. A random drift occurs if two distributions similar to quasispecies with the same fitness are separated by large Hamming distances [8.21,53].

Fontana et al. [8.53] have continued the statistical analysis of the fitness landscape. They established the statistical isotropy of the fitness landscapes and compared the results for chains containing four different nucleotides (GACU) to ones with only two (GC). In the first case the landscape is found to be somewhat smoother than in the second case.

In his recent work Wong [8.56] applied the method of spin glasses, which had been developed for neuron networks, to the pre-biotic evolution of macromolecules. In this work Wong has made an attempt to take into account the feedback between the population and the environment, i.e. the effect of the population on the fitness landscape. Perhaps this will also have to be done in the further development of the Eigen theory.

We see that the Eigen theory not only does not contradict the neutral theory, but also offers a rigorous interpretation of it at the level of molecular models. As we have seen, these models are fully supported by experiment and allow one to go farther along the important paths of virology. There are solid grounds for believing that this theory will make it possible to pass over to the physical treatment of evolution at the levels of cells and organisms.

The Eigen theory has disclosed the physical meaning of molecular biological evolution; it has united evolutionary theory with molecular biology, information theory and with synergetics. The most important philosophical value of Eigen's theory is that it has destroyed "the Berlin wall" separating non-living from living nature. It may be said that while molecules and even macromolecules remain inanimate, the $Q\beta$ phage is already alive in *Escherichia coli* cells.

9. Chaos and Order in Evolution

9.1 Chaos

Let us recall the words of Schrödinger: "An organism feeds on negative entropy". The outflow of entropy maintains the existence and development of the biosphere. Earlier we calculated the corresponding balance. This proposition serves as evidence of the validity of the second law of thermodynamics in living nature, and stems from this law. However, this statement is insufficient for an understanding of biological evolution.

As has been pointed out, in the course of individual development and biological evolution new information arises, i.e. a random choice is memorized. It is necessary to understand how and why this choice is realized.

The basic physical and philosophical problems of evolutionary theory, i.e. the process of development and formation of all living things, are far from being solved. Nonetheless, due to great advances in information theory and in the theory of dissipative systems or synergetics, the situation is starting to be clarified. It appears that the instructive source of thought is provided by the concepts of dynamic chaos. We are speaking here of rather complicated ideas, which in fact are rooted in ancient history, in the pre-scientific period of human thought.

The word "chaos" is a Greek word "khaos", meaning empty space. The deep thinker and expert on ancient philosophy, A.F. Losev, speaks of the dual meaning of this word [9.1]. For ancient philosophers chaos is, on the one hand, a physical space, either empty or filled with something. On the other hand, chaos is the basis of world life. This is one of the most original specimens of ancient mythological and philosophical thinking.

It is worth quoting a striking definition of this complex and many-sided notion in ancient ideology given by Losev:

"Chaos is the universal principle of continuous, infinite and limitless becoming. Ancient chaos is the maximum rarefaction and dispersion of matter and therefore it is eternal death for all living things. But it is also the maximal condensation of any matter. It is a continuum, devoid of any discontinuities, any empty gaps and even of any differences. Chaos therefore is the principle and source of any becoming, creating living things at all times; it is room for any living formations. Ancient chaos is omnipotent and faceless, it forms everything, while being itself formless".

Expressed in modern language, ancient chaos does not contain any information, either micro- or macroinformation. It serves as an unlimited source for creating new information, i.e. for memorizing a random choice. It is impossible, of course, to directly comprehend the content of these myphopoetic images physically. In what follows we will consider the modern conceptions of structural and dynamic chaos.

The myphological and philosophical ideas of chaos appeared also in other ancient cultures. The closely related conceptions of chaos as a source of all existing things were developed in Egypt and in Sumer, India and China [9.2]. The primitive, formless, unsplit chaos; an idea characteristic perhaps of any culture disposed to abstract thinking and pondering over the Universe.

In religions such as in Judaism, Christianity and Islam, the idea of primitive chaos is replaced by the idea of a God, a notion which is far from being formless. The Biblical chaos, the Flood, acts according to a program specified by God and is confined to it. Modern religions know nothing of chaos, but the new meaning of this notion is of great importance for the science of our century.

The notion of chaos as the source of the Universe again appeared in the cosmogony of Kant. The galaxy, the stars, the solar system all emerge from a completely disordered distribution of cosmic dust, the particles of which are capable of mutual attraction and repulsion. This theory was further developed in the works of Laplace. The modern concepts of the origin of cosmic bodies has gone well beyond the Kant-Laplace hypothesis, but this hypothesis is the first scientific model of the emergence of order from chaos under non-equilibrium conditions. In this sense Kant is the founder of synergetics or the physics of dissipative systems; he is the first thinker who had set out on the path which was a century later taken by Darwin.

In this connection it is appropriate to recall the words of Timofeev-Ressovsky, who said that Kant had arrived at agnosticism since in his time there was no Darwin's theory and the development of life was thought to be a mystery.

Nonetheless, it must be stressed that physical chaos, from which everything emerged, is not the original cause of the world according to Kant. The original source is God, just as it was for Newton (see [9.3]).

In our times the myphopoetic idea of chaos retains its meaning in art. In his speech "The Mission of a Poet" devoted to Pushkin, Aleksandr Blok said [9.4]:

"Harmony is the concordance of the world forces, the order of world life. Order is the cosmos as opposed to disorder, chaos. Chaos is the primitive, spontaneous infinity; the cosmos is organized harmony, culture; chaos gives rise to the cosmos; chaos bears the seeds of culture; harmony arises from the infinity."

"World life is the unceasing creation of new species, new breeds. They are lulled by chaos which has no beginning; they are brought up by culture which carries out a selection among them; harmony endows them with images and forms."

"A poet is the son of harmony; he is assigned a role in world culture. Three missions are delegated to him: firstly, to free sounds out of the beginningless element in which they reside; secondly, to bring these sounds into harmony, to give them a form; thirdly, to bring this harmony into the outer world."

These words of the great poet can be translated into the language of modern science. Order and harmony emerge from chaos, from utmost disorderliness. This generates information; the creation of a verse (or any other work of art) is the memorizing of a random choice, which is not logically deterministic [9.5]. This is the essence of creative work. For poetry there was the word at the beginning; for dramatic art there was the action at the beginning (cf. [9.6]). Each of these involves the emergence of order, information from chaos.

Today, "chaos" is one of the fundamental concepts of physics as opposed to structure and organization. Until recently, this notion was used only to characterize the positions and motion of molecules in a gas under equilibrium conditions. The notion of this molecular chaos was implicitly contained in Boltzmann's H-theorem. It was very important to the development of statistical physics.

Equilibrium molecular chaos corresponds to maximum entropy and, hence, to a minimum of microinformation. Molecular chaos cannot serve as a source of new information; it contains no differences, is homogeneous and structureless.

The transition from molecular chaos to order under equilibrium conditions (the union of the same gas molecules into a crystal lattice with a decrease of temperature) is not the creation of new information. Of course, the entropy of a crystal is lower than the entropy of a gas and, accordingly, there is more information. However, this information is not new; it is known in advance and there is no choice here. Equilibrium crystallization does not produce new information but merely brings out old information.

With reference to non-equilibrium open systems, the word "chaos" has various meanings. Following Klimantovich, we will make a distinction between physical chaos and dynamic chaos [9.7].

Physical chaos is a negative notion, as opposed to orderliness, self-organization. The system is characterized by a certain set of controlling parameters $a = (a_1, a_2, \ldots, a_n)$. They may be chosen in such a way that positive increments in the parameters Δa_i will correspond to self-organization. Physical chaos may be defined as a state in which there is no self-organization, $\Delta a = 0$. Hence, at $a = a_0 + \Delta a$ the values of a_0 correspond to physical chaos. This state may differ significantly from the state of equilibrium.

The notion of dynamic or deterministic chaos is much more informative and concrete. We are speaking here of the disordered, unpredictable behaviour of dynamical systems which can be described by deterministic non-linear equations. Such behaviour appears to be inherent even in very simple systems, simple in the sense that they are described with the aid of uncomplicated equations. Randomness is determined by high sensitivity to changes in initial conditions (the Butterfly Effect; see below) and is manifested in the observed randomness of behaviour. However, this is not noise but rather deterministic randomness, which is practically indistinguishable from noise. Chaos is a strange type of mathematical order, which seems to be random, but in reality follows precise rules. It is a sort of order that imitates disorder, "a wolf in sheep's clothing" [9.8].

As a matter of fact, we encounter deterministic chaos in all branches of the natural sciences and the humanities: meteorology, economics, physiology of the brain, astronomy, international politics, and traffic-control systems. All self-reproducing systems bear the seeds of chaos. Biological systems with positive feedback are capable of switching over to a random regime with a change in the parameters that characterize this feedback, which manifests itself in the non-linearity of the corresponding equations [9.9].

This kind of dynamic chaos reflects the fundamental dynamic instability of motion expressed by the exponential divergence of trajectories in phase space. The motion assumes stochastic character and is described by statistical rather than by dynamic regularities. It is this circumstance that has led to the development of statistical mechanics from ordinary mechanics, demonstrated by the simplest model of "Sinai billiards". There is a rectangular vessel with one convex wall. The only particle (the billiard ball) in such a vessel rebounds elastically from the walls. Two originally adjacent trajectories pull away from one another (Fig. 9.1). The transition from mechanics to statistics has been considered in the works of Sinai [9.10] and Kolmogorov [9.11] (see also [9.12] and a popular exposition [9.13]). Events of this kind can be observed directly upon motion of two barium ions in an electromagnetic trap [9.14,15]. Another striking example is the turbulent motion of a liquid [9.16,17].

In an open system, which is in the state of dynamic chaos, small fluctuations may increase to macroscopic values. This is a different expression of what is known as sensitive dependence on initial conditions, or half-jokingly as the Butterfly Effect [9.18]: the notion that a butterfly stirring the air today in Peking can transform storm systems next month in New York.

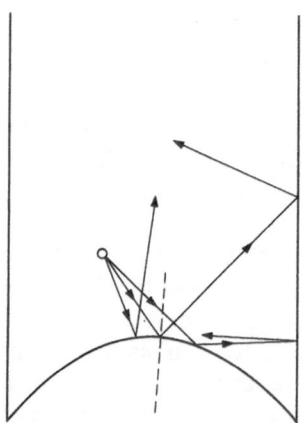

Fig. 9.1. The Sinai billiards: elastic rebounding of a particle from the concave wall of the vessel

As a matter of fact, the Butterfly Effect was not an altogether new notion: it was implied in English folklore:

"For want of a nail, the shoe was lost;
For want of a shoe, the horse was lost;
For want of a horse, the rider was lost;
For want of a rider, the battle was lost;
For want of a battle, the kingdom was lost!"

A small cause leads to catastrophic consequences under conditions of chaos.

Dynamic or deterministic chaos is realized in the famous Lorenz strange attractor. While solving certain hydrodynamic problems pertaining to meteorology, Lorenz studied the behaviour of the phase trajectories of a system of three differential equations with three variables. The equations are as follows [9.19]:

$$\dot{x} = -\sigma(x - y)$$

$$\dot{y} = -xz + rx - y \qquad (9.1)$$

$$\dot{z} = xy - bz$$

where the parameters σ, r and b are positive quantities. Two of these equations are non-linear. It appears that with $\sigma > b + 1$ and $r > \sigma(\sigma + b + 3)/(\sigma - b + 1)$, a rather peculiar phase trajectory corresponds to the Lorenz system in three-dimensional space x, y, z [9.20]. There are two stationary points; the zero values of the derivatives $\dot{x} = \dot{y} = \dot{z} = 0$ are obtained at $x_s = y_s = \pm(br - b)^{1/2}$, $z_s = r - 1$. The phase trajectory makes several circuits around one of these points, then does the same about the second point, and returns back, and so on (Fig. 9.2). The trajectory never intersects itself; it loops around and around forever, which implies the absence of periodicity. The trajectory is unstable; slight changes in the initial conditions or parameters alter it substantially. The system is a strange attractor; in phase space there is an infinite attracting line, which stays inside a finite space.

Similar properties are displayed in the Rössler system which contains only one second-order nonlinearity (see [9.21]):

$$\dot{x} = -y - z$$

$$\dot{y} = x + ay \qquad (9.2)$$

$$\dot{z} = bx - cz + xz$$

where a, b, c are positive constants. Here there are also two stationary points, $x_s = y_s = z_s = 0$ and $x_s = c - ab$, $y_s = b - c/a$, $z_s = c/a - b$. Figure 9.3 shows a phase trajectory at $a = 0.32$, $b = 0.30$, $c = 4.50$; a phase trajectory at $a = 0.38$, $b = 0.30$, $c = 4.50$ is shown in Fig. 9.4.

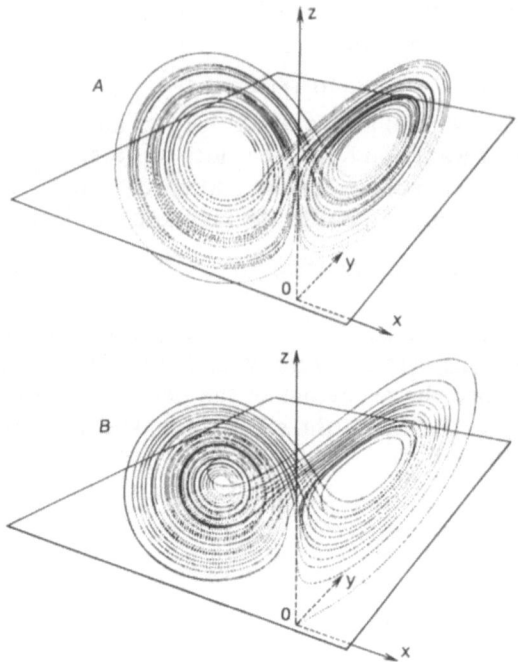

Fig. 9.2. The phase trajectory of the Lorenz strange attractor at $\sigma = 10$, $b = 8/3$ and values of $r = 28$ (*A*) and 40 (*B*)

Strange attractors depict dynamic chao; randomness is determined by the instability of the trajectory towards slight perturbations. A slight change in the parameter a in the Rössler system radically alters the behaviour of the strange attractor. According to Nicolis and Prigogine, Fig. 9.3 shows spiral chaos and Fig. 9.4 shows screwlike chaos [9.21].

It is necessary to emphasize that a strange attractor arises only in a three-dimensional rather than in a two-dimensional system. Such a chaotic attractor is characterized by non-periodic behaviour of the type of turbulence. This is dynamic chaos in the sense indicated above. As is shown in Figs. 9.2–9.4, the chaotic strange attractor cannot be considered to be completely devoid of order; spiral or screwlike behaviour is ordered in a certain sense. Accordingly, the transition from laminar to tubulent flow of a liquid also implies the appearance of dynamic chaos with a new specific order rather than the complete destruction of order.

The emergence of dynamic chaos in phase space occurs at certain critical points corresponding to special values of the parameters. The Lorentz attractor appears in the system (9.2) when $\sigma \geqslant b + 1$ and $r \geqslant \sigma (\sigma + b + 3)/(\sigma - b - 1)$. Instability arises.

Let us now consider an important example of a simple deterministic system, in which dynamic chaos can be realized [9.8,9]. The simplest population model,

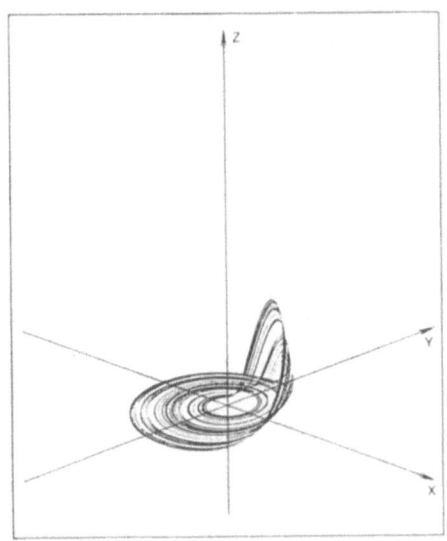 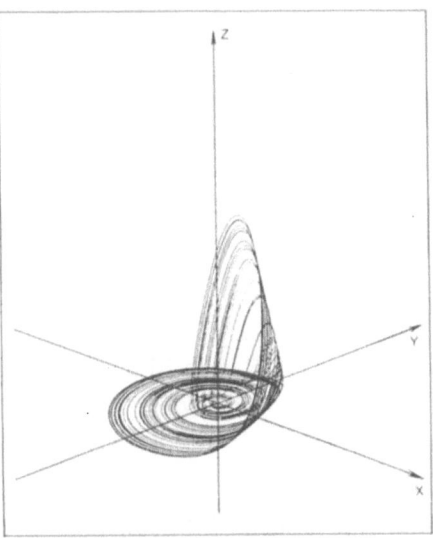

Fig. 9.3. The Rössler model. Spiral chaos [9.21]   **Fig. 9.4.** The Rössler model. Screw chaos [9.21]

which interrelates the population sizes of two successive generations, has the
form

$$x_{t+1} = ax_t \ .$$

This linear model describes an exponential growth of the population. The
population increases in proportion to its size. In reality, such a Malthusian
growth is far from being realized at all times. An increase in population size
lowers the rate of growth as a result of overpopulation. This can be described by
a logistic rather than a linear equation:

$$x_{t+1} = a(x_t - x_t^2) \ . \tag{9.3}$$

This is an iterative equation. The successive generations are iterations. The next
iteration gives

$$x_{t+2} = a(x_{t+1} - x_{t+1}^2) = a^2 x_t [1 - (1 + a)x_t + 2ax_t^2 - ax_t^3] \tag{9.4}$$

The power of the polynomical rapidly increases and the non-linearities become
complicated.

Let us rewrite Eq. (9.3) in the following form:

$$f(x) = ax(1 - x) \ . \tag{9.5}$$

We examine the properties of the function $f(x)$. In arbitrary units $0 < x < 1$
with $a > 0$ and $f(x) > 0$. We are interested in the behaviour of the system after
a large number of iterations. We find the so-called iteration numbers. Let us

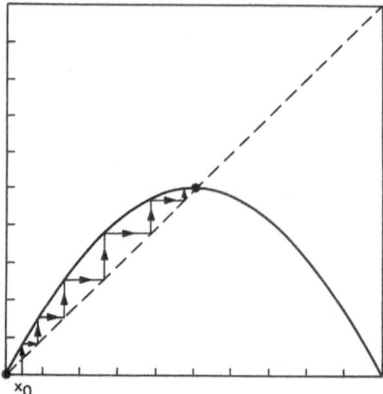

Fig. 9.5. Iterations of x_0 at $a = 2$ [9.9]

find fixed points of the function $f(x)$, the stable states of the system. In other words, we need to find values of x at which iterations do not alter the states, i.e. $x_{t+1} = x_t$. The function (9.5) has two such fixed points which satisfy the condition

$$x = ax(1 - x)$$

that is,

$$x = 0 \quad \text{and} \quad x = 1 - \tfrac{1}{2} \ .$$

The function (9.4) has a maximum at $x = 1/2$. If $a > 0$ and x lies in the interval $(0, 1)$, $f(x) > 0$, in this and in a more general case the fixed points are found as points of intersection of the straight line $y = x$ and the curve $y = f(x)$. Figure 9.5 shows the corresponding structure at $a = 2$. The successive iterations of a certain value x_0 are carried out as follows [9.10].

1. A vertical straight line is drawn from the point $y = 0$, $x = x_0$ until it intersects the $y = f(x)$ curve.
2. A horizontal straight line is drawn from that point until it intersects the $y = x$ straight line.
3. These operations are then repeated.

In the case shown in Fig. 9.5, the fixed point $x = 0$ is unstable. On the other hand, the point $x = 1 - 1/a$ is a stable attractor.

If $0 < a < 1$, only the point $x = 0$ is stable. If $a = 1$ the two fixed points vanish. With $1 < a < 3$ only the point $x = 1 - 1/a$ is stable. In this region of the value of a the population becomes stabilized at ever larger values of x. In the interval $3 < a < 4$ the attracting fixed points are absent. The effective value of x is periodically changed, first, in two successive generations and then the periods increase by 2, 4, 8 times, etc. Finally, at $a = 3.57$ chaotic behaviour sets in. This

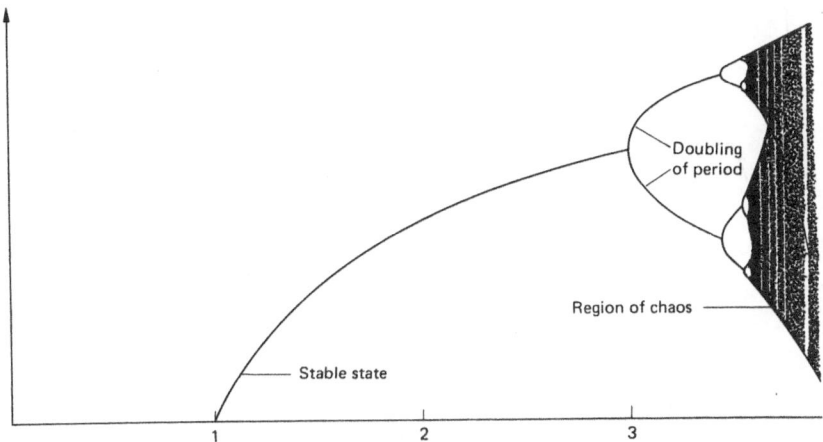

Fig. 9.6. The behaviour of the model (9.3). The abscissa represents a and the ordinate shows stable values of x_0

is shown in Fig. 9.6, which presents the dependence of the values of x that are obtained as a result of successive iterations, on the parameter a. Figure 9.7 shows the time dependence of x at $a = 2.7$, when the value of x is stable, at $a = 3.4$ (when periodic behaviour appears), and at $a = 3.99$ under the conditions of chaos. These figures were published in a special issue of the journal "La Recherche" devoted to chaos (No. 232, May, 1991, page 590).

Chaos and the associated instabilities can determine the emergence of new species; this has already been considered in discussing the evolution of macro-molecules in Eigen's theory.

The parameters, whose variations change the behaviour of the system, are bifurcational. Their role is played by the parameters σ and r in the Lorenz system and by the parameter a in the logistic system (9.5). Bifurcations are taken to mean quantitative changes in phase trajectories at which the topology of the phase diagram is changed. In particular, at bifurcations, branching of the solutions is possible, which is very important to the theory of evolution (see Sect. 9.4). Figure 9.8 shows a typical branching bifurcation, which implies a change of symmetry and topology [9.21]. The bifurcation point is a critical point at which dynamic chaos appears. Branching bifurcations arise, beginning with a cubic non-linearity.

The situations described by specific iterative equations introduce us into the world of fractals: systems that are characterized by self-similarity, invariance with a change of the scale. Let us consider so-called Julia sets, i.e. two-dimensional structures that arise when the following expression is iterated:

$$x_{t+1} = x_t^2 + c , \tag{9.6}$$

where c is a complex quantity. The behaviour of such a complex function was

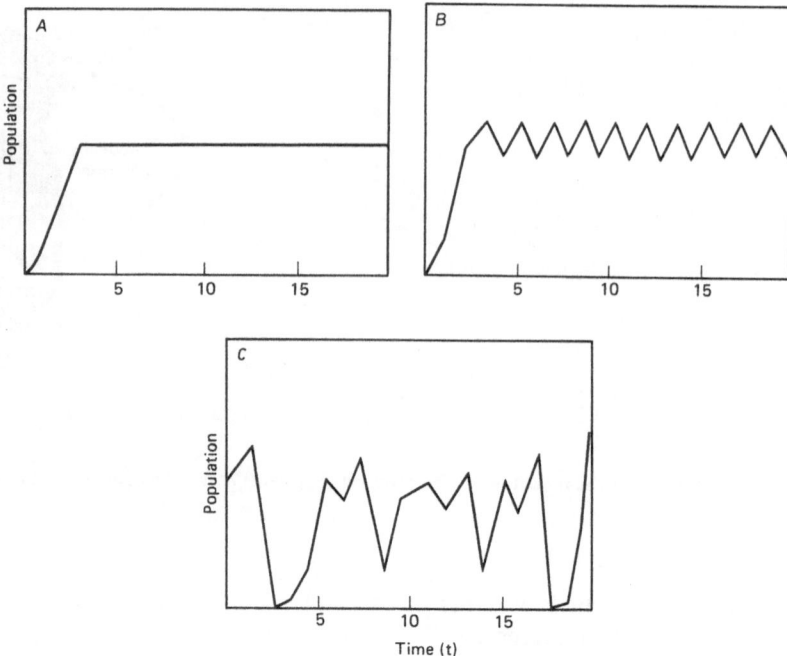

Fig. 9.7. The function $x(t)$ at $a = 2.7$ (A), $a = 3.4$ (B) and $a = 3.99$ (C)

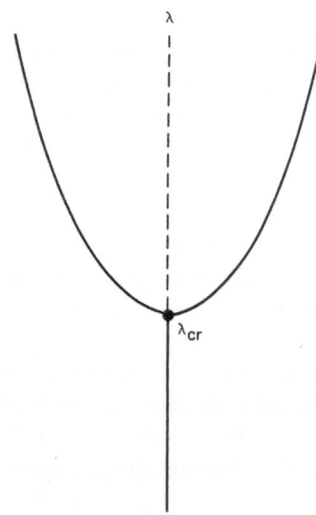

Fig. 9.8. Branching bifurcation. At the critical value of the controlling parameter λ the original solution becomes unstable and two new stable branches arise

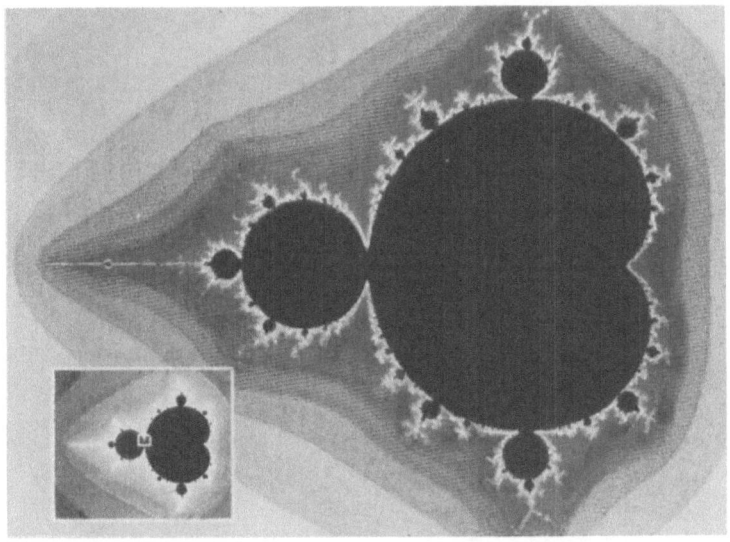

Fig. 9.9. The Julia fractal (Mandelbrot)

first studied by Julia as far back as 1918. At $c = 0$ we have the following series
of values:

$$x_0 \rightarrow x_0^2 \rightarrow x_0^4 \rightarrow x_0^8 \rightarrow \cdots .$$

With $x_0 < 1$ the attractor of such a sequence is zero, and with $x_0 > 1$ the attractor is infinity. The value of $x_0 = 1$ is the boundary between these two regions of
attraction, a bifurcation point. With $c \neq 0$ rather complex behaviour arises.
Mandelbrot examined this system at $c = -0.12375 + 0.56508i$. Figures 9.9, 9.10
and 9.11 show the corresponding structures in a complex plane at different
magnifications – upon affine linear transformations. Of course, these patterns
are obtained with the aid of computer graphics; fractal geometry could not have
been developed before the advent of computers. Fractal geometry is the geometry of deterministic dynamic chaos: the appearance of diverse highly complex
structures is determined by small changes in the parameters of the iterated
function, say the parameter c in the function (9.6). In a number of cases, fractals
are similar to structures observed in living nature, in the plant and animal
kingdoms; fractals have aesthetic value. The essence of deterministic chaos is
disclosed with utmost clarity by fractal geometry; extreme complexity and
unpredictability arise from very simple equations with the aid of very simple
algorithms. As Mandelbrot himself wrote: "Scientists will ... be surprised and
delighted to find that not a few shapes they had to call grainy, hydralike, in
between, pimply, pocky, ramified, seaweedy, strange, tangled, tortuous, wiggly,
wispy, wrinkled, and the like, can henceforth be approached in rigorous and
vigorous quantitative fashion" (quoted from [9.23]). The science of fractals has
been described elsewhere [9.18, 21–25].

Fig. 9.10. The Julia fractal at intermediate magnification

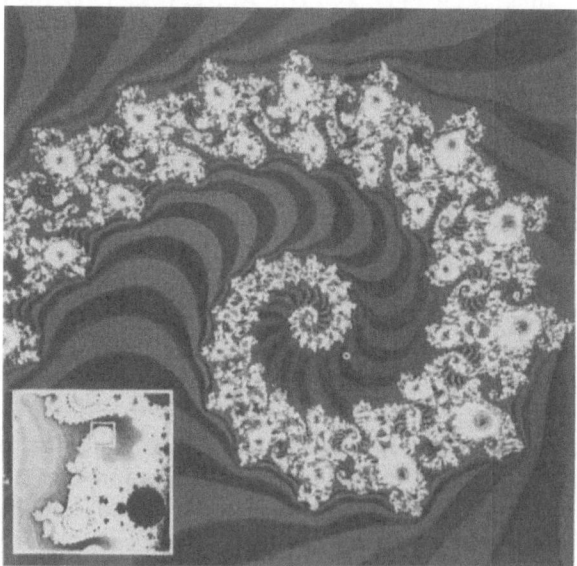

Fig. 9.11. The Julia fractal at high magnification

The quantitative criterion of dynamical chaos, i.e. the instability of phase trajectories towards small perturbations, may be represented by the average velocity of divergence of phase trajectories [9.20] (see Fig. 9.1). Suppose there are two trajectories, $x_i^{(1)}(t)$ and $x_i^{(2)}(t)$, which emerge at the initial moment of time $t = 0$ from the nearby points $x_i^{(1)}(0)$ and $x_i^{(2)}(0)$. The distance between the two trajectories at time t is given by

$$D(t) = \sqrt{\sum_i [x_i^{(1)}(t) - x_i^{(2)}(t)]^2} \ . \tag{9.7}$$

With an instability the trajectories diverge, usually exponentially. The quantity

$$K(t) = \frac{1}{t} \ln \frac{D(t)}{D(0)} \tag{9.8}$$

characterizes the behaviour of the trajectories. If with $t \to \infty$ auto-oscillations set in, i.e. the ratio of $D(t)$ to $D(0)$ does not increase overall with increasing t, then $K(t) \to 0$. If $K(t)$ tends to a finite positive value, then in the system dynamical chaos is realized, characterized by the quantity

$$K = \lim_{t \to \infty} K(t) \tag{9.9}$$

which is called the Kolmogorov entropy.

Further information on the physics of chaos and the corresponding mathematical representations are available in the literature [9.26–33].

9.2 Biology and Dynamical Chaos

What has been said above about dynamical chaos has a direct relevance to biology, to biological evolution. This is clearly evidenced by the following positions.

The biosphere, biocenosis, a species, a population, and an organism are all open systems, i.e. systems which exchange matter and energy with the surrounding world. Disequilibrium, openness of a living system is the principal condition of its existence. Equilibrium would have meant death.

Living systems are not only non-equilibrium, they are also quite far from equilibrium. It is exactly for this reason that in the course of evolution and individual development self-organization is realized, and new information is created. In an open system close to equilibrium, these processes would have been impossible. The emergence of new, unpredictable information always occurs from dynamical chaos, and from it alone. New information is contained in each new species and in each new individual that appears as a result of sexual reproduction.

"Evolution is chaos with feedback", Joseph Ford said [9.18]. What is meant here is dynamical chaos. Feedback comprises evolutionary changes in the biosphere which, in turn, exert an influence on the subsequent course of evolution.

One of the basic and indubitable positions of theoretical biology is the magnification principle. Small fluctuations – mutations – become fixed and multiply, and are therefore indefinitely magnified. A sort of strange attractor is realized, which magnifies the initial fluctuation; the "Butterfly Effect" occurs, which magnifies small uncertainties into large-scale changes in weather. However, there is also a difference between evolutionary biology and meteorology, which reduces to the difference between the time scales. In spite of all efforts, it appears to be impossible to forecast the weather for any prolonged period of time. Meteorologists believe that more often than not, tomorrow's weather will not differ from today's weather. What the weather will be like the day after tomorrow is not that clear. The result of evolution (the present-day state of the biosphere) will remain intact for a long period of time, but we cannot say how the biosphere will change in the more remote future. We can only state that if evolution were started anew, it would have led to completely different results. The evolutionary tree with its branches, bifurcations, is on the whole completely unpredictable.

It must be stressed that even the state of the solar system is unpredictable, but for a much longer period of time (billions of years). The solar system emerged from dynamical chaos, its stationary state is fixed for a very large period of time, but small perturbations must lead, sooner or later, to a change of this state. Chaotic motion of the strange attractor type can be maintained in a stationary state for a long time. An example is the red spot of the Jupiter, which, according to present-day data, is turbulence fixed for a prolonged period of time.

Stationarity and stability are different notions. Accordingly, dynamical chaos can also be globally stable. The strange attractor is stable on the global scale in the sense that the trajectory stays inside a given region of phase space. At the same time, the local behaviour of the phase curve is unpredictable.

We have already mentioned Kant's cosmogony as the first fundamental study in history in the area of synergetics or the physics of dissipative systems. Cosmogony, the theory of the origin of the Solar system, was followed by the theory of biological evolution, Darwin's theory, which solves synergetics problems too. The ordered development of the biosphere occurs from chaos, from random variability, from random interactions between organisms and the environment. Chaos gives rise to self-organization and creates biological order.

Speciation and the appearance of higher taxa imply sharp changes in the system, which are similar to phase transitions. This is shown in Figs. 9.12 and 9.13 [9.34]. Figure 9.12 shows the transformation of the original taxon to a new one without branching of the evolutionary tree. Figure 9.13 shows the occurrence of a branching, divergence, which changes the symmetry of the system. As Vorontsov has shown, some progeny of South-American hamsters climbed trees and others preferred the water basin. The symmetry decreased. A transition

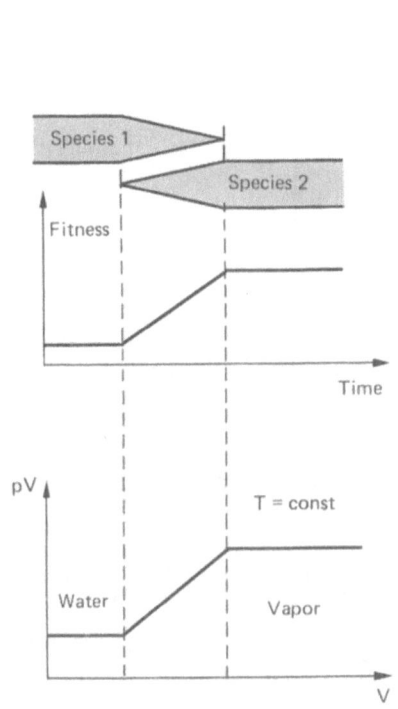

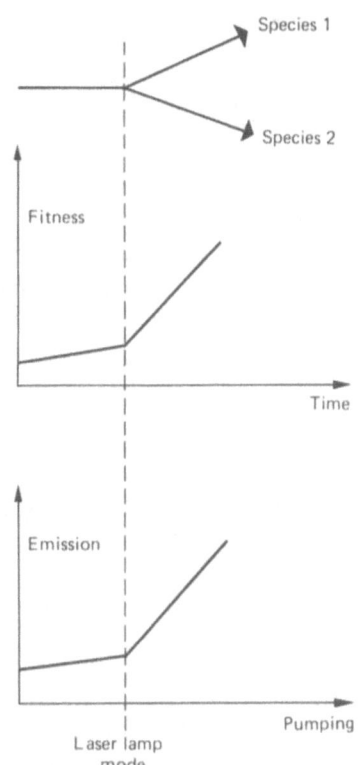

Fig. 9.12. The evolutionary transformation of a species   **Fig. 9.13.** The divergence of a species

without branching is similar to a first-order phase transition; a transition with branching is similar to a second-order phase transition.

It is clear that both types of transition, i.e. bifurcation, are governed by the instability of the intermediate state and by the discrete stability of the resultant states.

Conrad deals with the chaotic mechanisms of biological adaptation in an article called "What Is the Use of Chaos?" [9.35]. Adaptability is the ability of the system to keep functioning under indefinite or unknown environmental conditions. The possible functional role of dynamical chaos is described in Table 9.1.

Let us define more exactly, following Arnold [9.36,37], the notions of bifurcation and catastrophe. As we have seen, bifurcations imply various qualitative changes or metamorphoses of various objects, with a change in the parameters on which they depend. The catastrophes introduced by René Thom are discontinuous changes arising in the form of a sudden response of the system to a smooth change of external conditions. This is a special case of bifurcation. The system emerges in a jump-like manner from the steady-state regime because of the loss of stability and of random perturbations. The system moves on to a

Table 9.1. Possible functional roles of chaos (chaotic mechanisms of adaptability)

1. Search (diversity generation)
   genetic
   behavioural
2. Defence (diversity preservation)
   immunological
   behavioural
   populational
3. Maintenance (disentrainment processes)
   neural and other cellular networks
   age structure of populations
4. Cross-level effects
   interaction between population dynamics and gene structure
5. Dissipation of disturbance (qualitative insensitivity to initial conditions)

new stationary regime of motion. This is a transition from one attractor to another, or crossover from one region to another, in the model of the strange attractor.

It is obvious that what has been said bears a direct relation to speciation and macroevolution.

A real system of any interacting objects in living nature is always a combination of chaos with order. In the case of the Solar system, the order determined by the action of laws of mechanics prevails for a very long period of time, since the dissipation that leads to chaos and an unpredictable behaviour is relatively very small. In biological evolution the situation is incomparably more complicated. Order is specified by heredity and chaos by variability, i.e. in the long run it is inherent in the phosphodiester bond in DNA. The ratio of these two factors determines the directed, irreversible character of evolution, which is far from being reduced to a random shuffling of possible mutations in each generation.

The directionality of evolution stems from constraints on variability that arise from the fact that a certain structure of the preceding organisms, and a certain range of its possible changes, exist (see Chap. 3). The very fact of inheritance dictates a certain order. In this sense evolution resembles functional iteration. Feigenbaum [9.38] writes that since one and the same function is calculated repeatedly, it may be expected that several types of self-consistent behaviour will appear, determined not by a particular form of the function being iterated but by the very fact of its numerous calculations. The driving natural selection is similar in this sense to successive iterations. One might think that peculiarities of this kind underlie convergence phenomena, which are of no adaptive significance. Thus, for example, the leaves of trees are characterized by a limited number of forms, irrespective of their special function. The leaf of the plane tree has the same form as the maple leaf.

Biogeocenosis, as well as the biosphere as a whole, is active kinetic media (AKM) or excitable media (which is the same thing) [9.39,40]. In such media distributed sources of energy or energy-rich substances (for example, ATP in molecular biological systems) exist, which have decreased entropy. In AKM one can, as a rule, pick out elementary volumes containing open point systems which are far from equilibrium. Between such adjacent volumes a linkage, an interaction due to transfer processes is realized. Originally chaotic AKM lose stability at bifurcation points. As a result, structuring occurs in space and time and, in particular, auto-oscillations and autowaves appear. The properties of AKM provide synchronization of these periodic processes [9.39–42]. The appearance of new inhomogeneous states implies a decrease in symmetry: it is an analogue of a nonequilibrium second-order phase transition. The loss of stability is a necessary condition for generation of new information and thus for creation of new forms. The development and complication of the system are possible if the system possesses the required variability. Symmetry-breaking occurs in morphogenesis, in speciation and also in macroevolution, in taxonogenesis.

The model of the interaction between two species – prey and predator – was first constructed by Volterra in his classic work [9.43] (see also [9.41,42]). This is a point, undistributed model of auto-oscillations, which can and must be supplemented by taking account of the spatial migration of species described by the corresponding diffusion terms in kinetic equations. This, of course, complicates the problem.

Directly associated with the problem of speciation is the problem of "gene propagation" put forward by Kolmogorov, Petrovsky, and Piskunov in 1937 ([9.44], see also [9.39]). They considered a distributed genetic system; the spread of the dominant gene in a biallelic system is treated as the movement of the perturbation front in one direction described by the equation

$$\frac{\partial x}{\partial t} = D\frac{\partial^2 x}{\partial r^2} + sx(1 - x)^2 \; , \tag{9.10}$$

where x is the concentration of the dominant gene A; r is the coordinate in space; D is the diffusion coefficient; s is the selection coefficient; the recessive gene a is characterized by fitness 1, the gene A by fitness $1 + s$, with $0 < s \ll 1$. More complex situations are possible, which correspond to the threshold propagation of the wave.

The waves of life mentioned earlier are auto-oscillations of an active kinetic system.

From what has just been said, it follows that there are serious grounds for treating speciation and taxonogenesis as non-equilibrium phase transitions. This analogy is very instructive. Branching bifurcation, which is what is usually meant in a consideration of speciation, is similar to a second-order phase transition, as has been said earlier.

Mutational changes of chromosomes (molecular drive; see Chap. 7) and also point mutations (molecular drift; see Chap. 6) play the role of fluctuations forming in AKM, which has become unstable, embryos of a new phase, i.e. of a diverging species. The corresponding proposition of physical theory is as follows [9.45]: "The transition of the metastable phase to a stable phase occurs by way of the fluctuational appearance of small collections of a new phase, embryos, in a homogeneous medium. As a result of the energetically unfavourable effect of formation of an interface, it proves unstable and disappears again with insufficiently large sizes of the embryo. Only embryos with size a, beginning with a certain size (with a given state of the metastable phase) a_{cr}, are stable; this size will be called critical and embryos of this size will accordingly be referred to as critical."

In speciation and taxonogenesis critical sizes of "embryos" of new species are required; otherwise they will not be able to multiply or be stabilized, thus "hopeful monsters" will die. The critical size of an "embryo" has in this case the meaning of the critical size of a new population N_{cr}, which appears unstable when $N < N_{cr}$. The value of N_{cr} is determined by a number of factors, which of course cannot be reduced to the energetics of the interface. What has been said is valid for first and second-order phase transitions. The critical sizes of a species are necessary for its stable existence; listed in the Red Data Book are animals and plants whose existence is in danger because of small population numbers. Model theoretical calculations pertaining to the role of embryos of new species have been performed by Polezhajev and Sidelnikov [9.46].

The proposition of the similarity of speciation and macroevolution to phase transitions is of thermodynamical nature; the time factor, as we shall see in Sects. 9.3 and 9.4, does not appear in stability problems. Of no less importance for evolution is kinetics, particularly the problem of "embryos", which belongs entirely to this branch of physics.

The existence of relict species, which have changed little or not at all over many millions of years (the crossoptergian *Latimeria*, more than 100 million years; ginkgo trees, about 200 million years) is a phenomenon similar to vitrification. This is kinetic stability, which is directly associated with stabilizing selection. It is legitimate to speak of the virtification of a species.

9.3 Instabilities in Mendelian Systems

Stability problems arise even in elementary problems of population genetics. Let us consider the behaviour of diallelic diploid organisms [9.47]. We have two alleles A and a. The probability of A is equal to p and the probability of a is $q = 1 - p$. Crossing gives rise to homozygotes AA and aa and a heterozygote Aa (for an nth-generation with N births, see Table 9.2 [9.48]). The quantities w_{AA}, w_{Aa} and w_{aa} in Table 9.2 are the fitness coefficients for the corresponding genotypes, which determine their survival. For the $(n + 1)$st generation the total number of A-alleles is equal to

$$2w_{AA}Np_n^2 + 2w_{Aa}Np_nq_n \, , \tag{9.11}$$

since A appears twice in AA and once in Aa. The total number of alleles in the $(n + 1)$st generation is given by

Table 9.2. Zygotes and surviving adult organisms

Genotype	Number of zygotes	Number of surviving adults in $(n + 1)$st generation
AA	Np_n^2	$w_{AA}Np_n^2$
Aa	$2Np_nq_n$	$2w_{Aa}Np_nq_n$
aa	Nq_n^2	$w_{aa}Nq_n^2$

$$2N(w_{AA}p_n^2 + 2w_{Aa}p_nq_n + w_{aa}q_n^2) . \tag{9.12}$$

Hence, the fraction of alleles A in the $(n + 1)$st generation is

$$p_{n+1} = \frac{w_{AA}p_n^2 + w_{Aa}p_nq_n}{w_{AA}p_n^2 + 2w_{Aa}p_nq_n + w_{aa}q_n^2} . \tag{9.13}$$

The change in the probability per generation is given by

$$\Delta p = p_{n+1} - p_n = p_nq_n\frac{(w_{AA} - w_{Aa})p_n + (w_{Aa} - w_{aa})q_n}{w_{AA}p_n^2 + 2w_{Aa}p_nq_n + w_{aa}q_n^2} . \tag{9.14}$$

If $w_{AA} - w_{Aa}$ and $w_{Aa} - w_{aa}$ are small as compared to w_{Aa}, then we can approximately replace the difference equation (9.14) by a differential equation [9.49]:

$$\frac{dp}{dt} = pq\frac{(w_{AA} - w_{Aa})p + (w_{Aa} - w_{aa})q}{w_{AA}p^2 + 2w_{Aa}pq + w_{aa}q^2} , \tag{9.15}$$

where the time t is measured by the number of generations. We seek stationary solutions of Eq. (9.15) corresponding to $\dot{p} = 0$. Two of them are obvious:

$$p_1 = 0 , \quad q_2 = 1 - p_2 = 0 , \quad \text{i.e. } p_2 = 1 . \tag{9.16}$$

In some cases a third root is possible, whose value depends on the quantities w_{AA}, w_{Aa} and w_{aa}. Equation (9.15) may be rewritten in the form

$$\frac{dp}{dt} = p(1 - p)\frac{p(w_{AA} - 2w_{Aa} + w_{aa}) + (w_{Aa} - w_{aa})}{W} , \tag{9.17}$$

where

$$W = w_{AA}p^2 + 2w_{Aa}p(1 - p) + w_{aa}(1 - p)^2 .$$

At $\dot{p} = 0$ the third root is represented by the expression

$$p_3 = \frac{w_{aa} - w_{Aa}}{w_{AA} - 2w_{Aa} + w_{aa}} . \tag{9.18}$$

Nei [9.50] considers concrete cases. They are presented in Table 9.3. The asterisk indicates an identical zeroth solution. The corresponding dependences of the

Table 9.3. Behaviour of a diallelic system

Type	Fitness coefficients			Stationary state		
	w_{AA}	w_{Aa}	w_{aa}	p_1	p_2	p_3
1. Semidominant	1	$1 - s/2$	$1 - s$	0	0	$\underline{1}$
2. Dominant	1	$1 - s$	1	0	$\underline{1}$	$\underline{1}$
3. Recessive	1	$1 - s$	$1 - s$	0	$\underline{1}$	0*
4. Overdominant, largely of the heterozygote	$1 - s_1$	1	$1 - s_2$	0	1	$\underline{s_2/(s_1 + s_2)}$
5. Overdominant, largely of the homozygote	1	$1 - s$	1	$\underline{0}$	$\underline{1}$	1/2

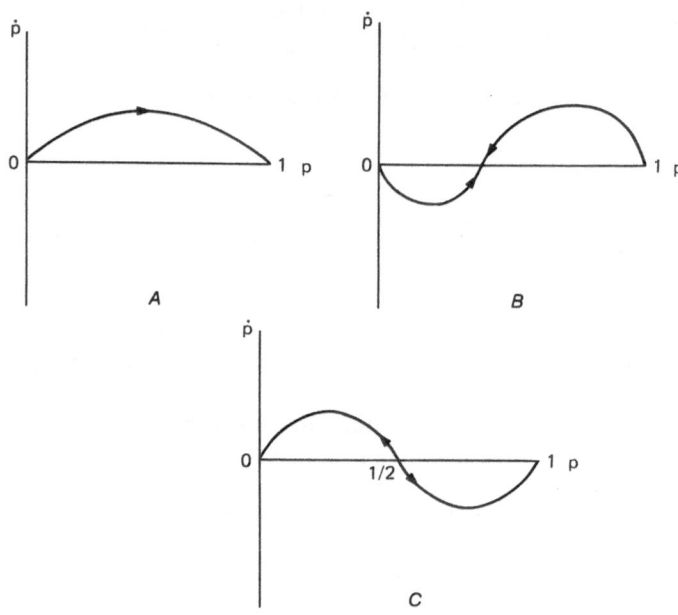

Fig. 9.14. The behaviour of the diallele system (see Table 9.3). (*A*) Types 1, 2, 3. (*B*) Type 4. (*C*) Type 5

derivatives $\dot{p}$ on p are given in Fig. 9.14. In Table 9.3 the stable stationary states are underlined. Of greatest interest is the last case, which describes the separation of "two phases" with an unstable intermediate state, $p_3 = 1/2$. The equation $\dot{p} = 0$ assumes in this case the following form:

$$\frac{dp}{dt} = p(1 - p)\frac{2p - 1}{1 - 2sp(1 - p)} = 0 \ . \tag{9.19}$$

This case is not encountered often, however; usually heterozygotes have advantages over homozygotes.

In Eq. (9.17), however, no account is taken of mutations that transform allele A to allele a and vice versa with the respective frequencies μ and v. A more general equation, which is considered by Belintsev and Volkenstein [9.51,52] has the form

$$\frac{dp}{dt} = p(1-p)\frac{p(w_{AA} - 2w_{Aa} + w_{aa}) + (w_{Aa} - w_{aa})}{W} - \mu p + v(1-p) . \quad (9.20)$$

The form of the equation does not depend on whether selection precedes mutations or vice versa. We rewrite Eq. (9.20) in the form

$$\frac{dp}{dt} = [W(p)]^{-1}(\alpha_0 - \alpha_1 p - \alpha_2 p^2 - \alpha_3 p^3) . \quad (9.21)$$

Here

$$\alpha_0 = vw_{Aa} , \quad \alpha_1 = \mu w_{aa} - \beta , \quad \alpha_2 = \beta\frac{3\gamma - 1}{\gamma}$$

$$\alpha_3 = \beta\frac{1 - 2\gamma}{\gamma} , \quad \beta = w_{Aa} - w_{aa} .$$

The quantity γ expresses the degree of dominance of allele A in the heterozygote Aa:

$$0 < \gamma = \frac{w_{Aa} - w_{aa}}{w_{AA} - w_{aa}} < 1 .$$

The above expressions for the coefficients α hold if

$$\beta \ll w_{AA}, w_{Aa}, w_{aa} , \quad \mu, v \ll 1 , \quad v \ll \mu .$$

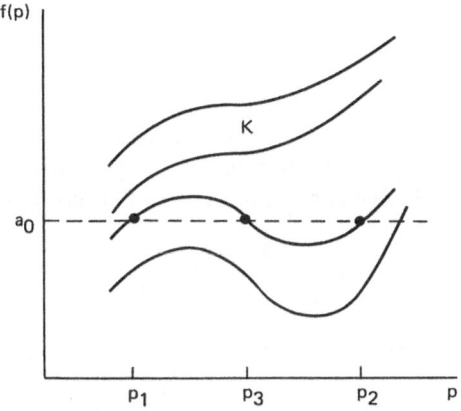

Fig. 9.15. Stationary states of p. See Eq. (9.22)

For the stationary state $\dot{p} = 0$ there is a corresponding cubic equation

$$\alpha_0 = f(p) = \alpha_1 p + \alpha_2 p^2 + \alpha_3 p^3 \qquad (9.22)$$

which has three roots. In the cases considered above, the presence of the terms μp and $v(1 - p)$ in (9.20) changes almost nothing since $\mu, v \ll 1$. If $\mu, v \ll w_{Aa} - w_{aa}$, $w_{AA} - w_{Aa}$, addition of mutational terms does not give anything new either. But the differences in fitness may be very small. In the general case, however, interesting situations arise. A change in conditions of existence of a population corresponds to a change in the parameters β, μ, v, which determine, according to (9.21), the pace of evolution and, according to (9.22), its final outcome. Figure 9.15 presents a diagram of the stationary states p (9.22) at different values of β in the cases of $\gamma < 1/3$ ($\alpha_2 < 0$, $\alpha_3 > 0$) and $\beta < \mu w_{AA}$ ($\alpha_1 > 0$). The cubic equation (9.22) has three material positive roots, which coincide at the critical point K, in which

$$p_K = \frac{3\gamma - 1}{3(2\gamma - 1)} \; ,$$

$$\left(\frac{v}{\mu}\right)_K = \frac{(1 - 3\gamma)^3}{9(1 - 2\gamma)(1 - 3\gamma + 3\gamma^2)} \qquad (9.23)$$

$$\left(\frac{\mu}{\beta}\right)_K = \frac{1}{w_{AA}} \frac{(1 - 3\gamma + 3\gamma^2)}{3\gamma(1 - 2\gamma)} \; .$$

It is easy to see that no critical point is possible for either of the cases indicated in Table 9.3; in particular, the condition $0 < \gamma < 1$ is not fulfilled. However, in the case

$$w_{AA} = 1 \; , \quad w_{Aa} = 1 - s_1 \; , \quad w_{aa} = 1 - s_2$$

the critical point is possible if

$$\frac{s_2}{s_1} > \frac{3}{2} \; .$$

The curves in Fig. 9.15 are similar to the van der Waals curves for a real gas. If we compare α_0, α_1, p with pressure P, temperature RT and concentration $1/V$, respectively, Eq. (9.22) will correspond to an equation of state in virial form:

$$P = \frac{RT}{V} - \frac{C_2}{V^2} + \frac{C_3}{V^3} \; . \qquad (9.24)$$

For $\beta > \beta_k$ we obtain three steady-state values p_1, p_2, p_3, with p_3 being unstable and p_1, p_2 stable. Like a real gas, a population may be in two locally stable stationary states. The transition from one state to another is analogous to an equilibrium first-order phase transition. This is a situation similar to the one considered by Schlögl ([9.53], see also [9.42]) for a chemical system with an autocatalytic reaction. A phase transition is caused by sufficiently large fluctuations of the population size and composition.

Let us now assume that the population is continuously spread within its habitat. The population dynamics is described by an equation, which includes the migration of individuals:

$$\frac{\partial p}{\partial t} = \varphi(p) + D\Delta p \qquad (9.25)$$

where the diffusion coefficient D is expressed in terms of the mean distance r to which the individuals migrate during their lifetime:

$$D = \frac{r^2}{2} \qquad (9.26)$$

where Δ is the Laplace operator and $\varphi(p)$ is the right-hand side of Eq. (9.25). Diffusion does not upset the stability of spatially homogeneous stationary states (the solution for an undistributed, point system). However, here the appearance of a stable, spatially inhomogenous distribution $p(r)$ is possible. Two spatially separated phases can arise, each of which corresponds to a stable state. According to (9.25), the condition for their stable coexistence is found by minimizing the functional

$$\psi = \int dr\, \mathscr{L} \qquad (9.27)$$

where we have introduced a Lagrangian, which is equal to

$$\mathscr{L} = \frac{D}{2}(\nabla p)^2 - \int \varphi(p)\, dp \ . \qquad (9.28)$$

Minimization of (9.27) with the total volume of the system being constant, when the volume of the transition layer is much less than the volume of each of the phases, gives

$$\varphi(p_1) = \varphi(p_2) = 0 \ , \quad \int_{p_1}^{p_2} \varphi(p)\, dp = 0 \ . \qquad (9.29)$$

The first equality in (9.29) is the condition of stationarity of the solutions p_1 and p_2, and the second is analogous to the Maxwell condition for the van der Waals gas.

The spatial separation of a population is the first stage on the allopatric path of speciation. Further existence of the allopatric form leads, once again, to the accumulation in it of new characters and to biological isolation. The geographical separation of the population is similar to spatial separation in the case of an equilibrium first-order phase transition.

The structure of the population, which evolves in accordance with Eq.

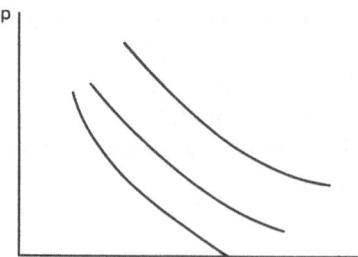

Fig. 9.16. A second-order transition in evolution

(9.20), may also undergo transitions similar to second-order phase transitions, if with certain bifurcation values of the parameters the stability of the stationary states against small fluctuations is upset. Such a transition occurs when a_1 turns to zero, i.e. at $\beta = \mu w_{AA}$. At $v = 0$ ($\alpha_0 = 0$) at the point $\beta = \mu w_{AA}$ the stationary dependence $\rho(\alpha_1)$ shows a kink characteristic of a second-order phase transition (Fig. 9.16). The quantities p, α_0 and $\alpha_1 = \mu w_{AA} - \beta$ appear to be analogues of the magnetization, the magnetic field strength and the deviation from the Curie point $(T - T_C)/T_C$ for a ferromagnetic-paramagnetic transition. It can be shown, accordingly, that near the transition point, the relaxation time for small fluctuations in the stable stationary state and the sensitivity of the system to a change in the parameters sharply increase. In a distributed system with additional introduction of random external influences, upon approach to the transition point the exchange fluctuations and the correlation radius of fluctuations increase. The transition to a new organization through the point of stability loss occurs in a fluctuational manner; ever increasing volumes are encompassed by correlated fluctuations, which result from small external influences. In such situations the population assumes lability, which may lead to rapid progressive evolution, say to the emergence of a new adaptation zone (aromorphosis according to Severtsov). Such "uncovering of evolutionary reserves" occurs with a low intensity of selective elimination ($\beta = \mu w_{AA}$). During evolution the stages of continuous development under a steady-state regime alternate with transient stages, which are similar to phase transitions.

Thus, in the simplest biallele system, which obeys Mendel's laws, there are possible discrete stable states, between which transitions similar to phase transitions take place. The statics of these states and the dynamics of the transitions are determined by the mutation frequencies μ, v and by relative survival w_i, i.e. variability and natural selection.

Berg in his book *Nomogenesis*, which has been mentioned earlier, was the first to formulate a very important position: "The birth and death of individuals, species, and ideas are a catastrophic process. The appearance of all these categories is preceded by a long hidden period of development, which occurs according to certain laws, and then a leap occurs – saltus, which manifests itself in their emergence, propagation on the Earth's surface and conquering a place 'under the sun'. The process of transition of a gas to a liquid is a jumpwise process" [9.54].

Berg maintained that speciation is similar to a phase transition. Can we say that what has been said above proves this proposition? Of course, the emergence of a new species cannot be the result of the mutation of a single allele gene. This is a process which involves a multitude of genes. The mutation of one gene can, however, play a role in speciation if this gene is regulatory and responsible for the behaviour of an assembly of other genes. Such a situation is realized in homoetic mutations.

Two questions arise in this connection. First, in what measure are these conceptions applicable to a multi-allele system? Second, to what extent are Mendel's laws and population genetics valid? The equations of population genetics have already been used by us for restructuring a genome in speciation?

Calculations for a multi-allele system encounter a number of difficulties and have not yet been carried out. One might think, however, that the nonlinearities, equations of high power, are characteristic of such a system too. If a phase transition occurs in a biallele system, then it is the more probable that it will occur in a system of many genes.

It is necessary to emphasize that the model treatment made above is purely thermodynamic; we are speaking here of stationary states and of their stability, but not of the rate of transition from one state to another, not of the kinetics; speciation is probably similar to a phase transition, but it can occur either slowly or rapidly, depending on internal and external conditions. The gradualism-punctualism controversy is thus removed – the rate of speciation may be high or low; this process may proceed in bursts or gradually. However, the concept of punctualism is more profound and instructive.

As regards the role of Mendel's inheritance in speciation and macroevolution, it can be revealed only when the molecular-genetic foundations of these processes are established. This was discussed in Chap. 7; the phenomena of molecular drive, which have presumably played a determining role in speciation and macroevolution, are of the non-Mendelian nature.

9.4 Speciation and Bifurcations

Let us consider the ordinary patterns of species divergence: gradual (Fig. 9.17a) and punctual (Fig. 9.18a), and let us try to treat these patterns physicomathematically [9.55]. Do these figures play only a purely illustrative role or can they express real regularities?

It is obvious that the branching point in both cases implies a bifurcation point. Gradual divergence may be described by a series of non-linear equations. One of the simplest equations contains a cubic term:

$$\frac{dx}{dt} = x^3 - \lambda x \ . \tag{9.30}$$

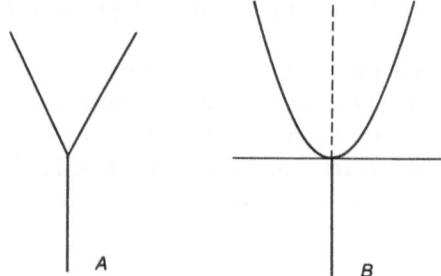

Fig. 9.17. Schematic representation of the gradual divergence of species (A) and its mathematical description (B)

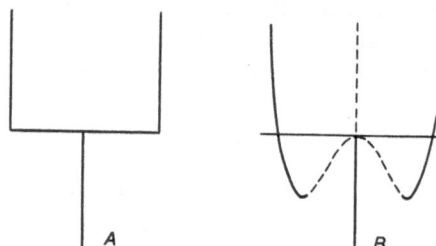

Fig. 9.18. Schematic representation of the punctuated divergence of species (A) and its mathematical description (B)

Here x is a certain morphological quantitative character and λ is a parameter which changes during evolution. The average value of x, corresponding to the ancestral species, is assumed to be $x = 0$; positive and negative values of x are deviations from $x = 0$. The stationary solutions of (9.30) with

$$\dot{x}_{st} = x_{st}^3 - \lambda x_{st} = 0 \tag{9.31}$$

correspond to the following values [9.21]:

$$x_{st} = 0 \; , \quad x_{st} = \pm \lambda^{1/2} \; .$$

The first solution is stable when $\lambda \leqslant 0$, the second and third solutions are stable when $\lambda > 0$. In Figs. 9.17b and 9.18b the stable solutions are indicated by solid lines and the unstable solutions by broken lines.

The parabolic divergence sketched in Fig. 9.17b is not obligatory, of course. But the simple solution of Eq. (9.30) conveys the main features of the gradual divergence of species.

Punctuated speciation (Fig. 9.18a) may be represented by the more complex fifth-order equation [9.55]

$$\frac{dx}{dt} = x[\lambda - x^2(x^2 - a^2)] \, , \tag{9.32}$$

where a is an additional parameter. The stationary solutions corresponding to $x = 0$ are as follows:

$$x_{st} = 0 \, , \quad x_{st} = \pm \left[\frac{a^2}{2} \pm \left(\frac{a^4}{4} + \lambda \right)^{1/2} \right]^{1/2} \, . \tag{9.33}$$

The solutions are shown in Fig. 9.18b. The situation is similar to the one described by the van der Waals equation for a real gas. When the solution that is stable with $\lambda < 0$ attains the value $\lambda = 0$, the system chooses a stable solution corresponding either to $x_{st} = -a$ or $x_{st} = a$. Subsequent changes imply an increase in x along the fourth-order parabola, which is steeper than in the case of gradual speciation shown in Fig. 9.17b.

We see that the simplified schemes in Figs. 9.17a and 9.18a may have a real physico-mathematical meaning.

Of course, these positions cannot be regarded as a true physical theory of speciation. They only describe the general meaning of gradual and punctual patterns. Construction of the theory requires an independent derivation of non-linear equations (9.30) and (9.32) (or other equations giving similar results), on the basis of the molecular conceptions of speciation.

These conceptions are contained in the molecular drive concept (Chap. 7). But a theory that encounters formidable difficulties has not yet been worked out.

However, it has become possible to describe mathematically the possibilities of gradual and punctual speciation as a result of bifurcations in non-linear, non-equilibrium open systems. The other arguments given in Chap. 2 provide evidence for punctualism. Of course, the phenomenological approach described above is purely thermodynamical, and we have no answer to the questions referring to the rate of evolution; the phase transition in speciation may be either rapid or slow.

In the only figure in *The Origin of Species* (given on page 115), a multitude of branches emerge from the branching points, but all except two are rapidly terminated. Biologists usually portray binary divergence, but they state that this is done only for the sake of simplicity; a single species can be a source of more than two new species.

As has already been shown, phylogenetic development is described in terms of the theory of dynamical systems, in terms of synergetics. The evolution of taxa may be represented by a gradual change of the stationary states of a dynamical system. At certain stages of evolution the system's parameters assume critical values at which the stationary states lose stability and experience bifurcations. Are they binary or multiple [9.56]?

Figure 9.19 shows all bifurcations of stationary states. Just as in Figs. 9.17b and 9.18b, the broken lines represent unstable states. There may exist other stable states in the supercritical region (Fig. 9.19b and c). However, these states

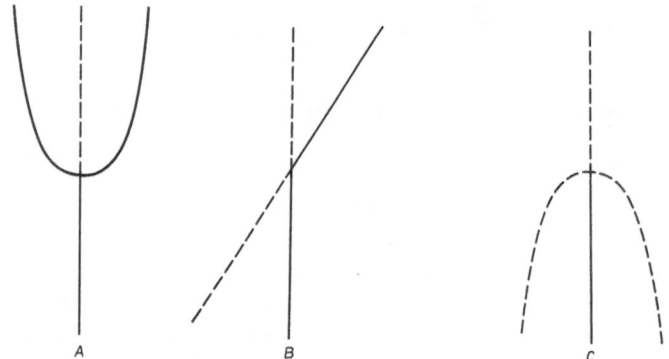

Fig. 9.19. Bifurcations of stationary states

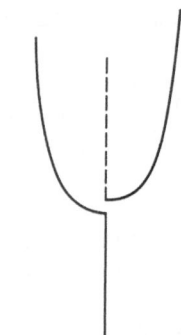

Fig. 9.20. Schematic representation of the binary divergence after symmetry breaking

are remote and we shall not consider them here. We shall confine ourselves to
the simple case of binary gradual divergence described by Eq. (9.30). For a
system with two parameters

$$\dot{x} = x(\lambda - x^2) + \varepsilon \; . \tag{9.34}$$

This type of bifurcation is a so-called assembly [9.57]. A slight shift in the
second parameter ε transforms the bifurcation diagram to the one shown in Fig.
9.20. A continuous transition is possible only to one of the two new stable states:
to the right or the left one, depending on the sign of ε. This means that a state
may be chosen by the system not because of the suitable initial conditions, but
due to a perturbation of the dynamical system.

Multiple branching is possible only for a degenerate bifurcation. Figure
9.21 shows a bifurcation in which four new stable states ($x_{st} = \pm \lambda^{1/2}/a, \pm \lambda^{1/2}/c$)
and two unstable states ($x_{st} = \pm \lambda^{1/2}/b$) arise simultaneously at the critical point
of the original branch. This diagram corresponds to a triple assembly:

$$\dot{x} = x(\lambda - ax^2)(\lambda - bx^2)(\lambda - cx^2) + \varepsilon \; . \tag{9.35}$$

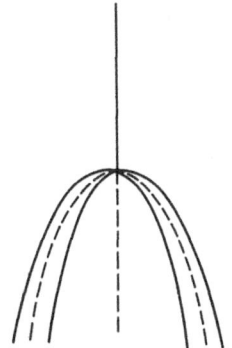

Fig. 9.21. Triple assembly

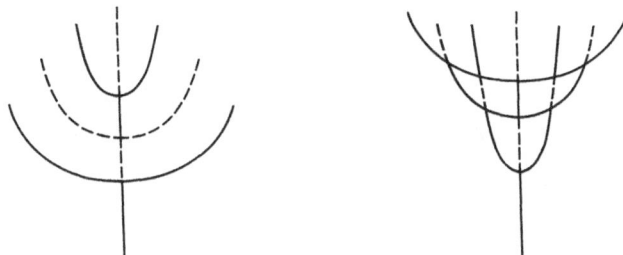

Fig. 9.22. Examples of triple assembly

Degeneracy is always created by certain special conditions that increase symmetry. Small perturbations of the dynamics can break symmetry and lift degeneracy. For example, the elementary systems (9.35) are readily split by small shifts:

$$\dot{x} = x(\lambda - \alpha - ax^2)(\lambda - \beta - bx^2)(\lambda - \gamma - cx^2) \ . \qquad (9.36)$$

After such splitting of the degenerate bifurcation the continuous transition, which maintains stability, leads only to the two remotest branches, i.e. to two maximally differing states. Some examples of the corresponding diagrams are presented in Fig. 9.22. An additional perturbation by the additive parameter ε retains the continuous path only to one of the two stable states.

In these dynamical models account was taken of only one degree of freedom. In a more general case, the interaction of various degrees of freedom can produce secondary bifurcations of the stable branches. This means that the splitting of the degenerate system transforms multiple branching to successive binary divergences. The simplest model of such a situation is described by the following equations:

$$\dot{x} = x(\lambda - x^2 - \alpha y^2)$$
$$\dot{y} = y(\lambda - \delta - y^2 - \alpha x^2) \qquad (9.37)$$

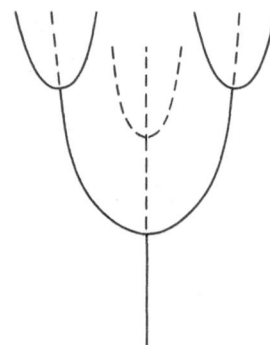

Fig. 9.23. Two-dimensional splitting into successive binary bifurcations

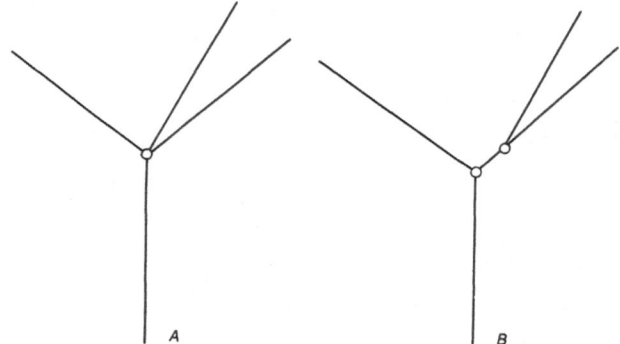

Fig. 9.24. Degenerate (*A*) and real (*B*) splitting of the evolutionary tree

where δ is the splitting; $\alpha < 1$. The corresponding diagram is shown in Fig. 9.23. Similar but more complicated examples of secondary bifurcations in the theory of dissapative systems have been explored by Belintsev et al. [9.58]. These model calculations allow one to presume that only binary bifurcations are actually realized in evolutionary trees. If a multiple branching of the type shown in Fig. 9.24a is sketched, this means that the true diagram must contain two closely spaced points of binary divergence (Fig. 9.24b).

Based on the conceptions of speciation and macroevolution as non-equilibrium phase transitions, we come to the conclusions that the probability of multiple branchings of the evolutionary tree is extremely low. Multiple branchings are considered to be highly improbable in the concept of punctualism.

9.5 Motion in the Fitness Landscape

In all the preceding treatment the rate of evolution has not been discussed. The qualitative considerations regarding the constraints on natural selection described in Chap. 3 provide serious grounds for regarding evolution as directed and, hence, accelerated. This is what gives, in general, an answer to the question put earlier: How could the biosphere have been produced for 3.5×10^9 years of its existence?

Note that the evolution was accelerated after the transition from pro-karyotes to eukaryotes, from unicellular to multicellular organisms with their sexual multiplication. It is this transition that determined the directedness of the subsequent evolution on the basis of fixation of a multitude of characters of the already formed multi-cellular organisms.

The necessity of the construction of the corresponding physico-mathematical models is obvious.

Let us first consider the model proposed by Wright [9.59,60]. There is a so-called fitness landscape in which a population is distributed. This is a multi-dimensional space, on the co-ordinates of which quantitative characters are laid off and one axis represents fitness. The landscape is a rugged terrain on which valleys alternate with peaks corresponding to fitness maxima. These peaks represent ecological niches.

We have already encountered such a fitness landscape in dealing with molecular, pre-biotic evolution in the Eigen theory (Chap. 8). However, the landscape figuring in this theory is different from a combination of valleys and peaks; it is a mountainous territory, the movement in which occurs along mountain ridges without loss of height. The relationship between the Eigen molecular theory and the theory of evolution of organisms has been considered above.

In the simplest case, which allows one to arrive at qualitative conclusions, the model developed by Wright may be limited to a single quantitative character z, which arises as a result of the joint action of a number of genes. Then the dependence of mean fitness W on the average value of $\bar{z}$ will be represented by a curve with several maxima. Let us assume that there are only two such peaks, i.e. ecological niches (Fig. 9.25).

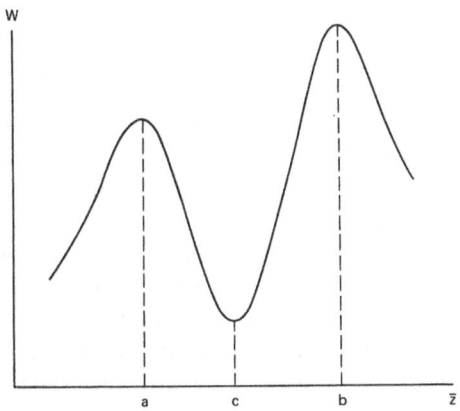

Fig. 9.25. Simple evolutionary landscape

Lande has studied the behaviour of a population in such a landscape in his work called "Expected time of the random genetic drift of a population between stable phenotypic states" [9.61]. It is immediately clear that as a result of a permanently existing random genetic drift in such a model, the population will exist for a long period of time in its original niche, while drifting near to the corresponding peak. The task is to explore the dynamics of a possible movement to another niche, to another peak. Such a transition, which represents evolutionary development, may be depicted as the descent from a peak into a valley and a subsequent ascent to a new peak. If it is much higher than the original one, the situation becomes stabilized. The fitness space under consideration is different from a mountainous country that appears in the Eigen theory.

In Lande's work, natural selection and random drift are modelled with the aid of the Kolmogorov diffusion equation. Calculation shows that with appreciable selection the population spends a long period of time drifting near to the original peak. For the mean phenotype, the expected time of the shift between the two peaks increases in an approximately exponential manner with the effective population size N_e. Compared to this time, the expected time of existence of intermediate forms upon transition from one peak to another proves to be very short. It may be lower than the level of resolution in the palaeontological record. This short time also depends on the population size, not exponentially but logarithmically.

Lande notes the agreement between these theoretical results and the concept of punctualism. A prolonged time of residence at the initial (and final) peak corresponds to stasis and a short time of migration through a valley corresponds to speciation.

Developing the same ideas, Newman, Cohen and Kipnis carried out further calculations [9.62]. The transitions between the peaks appear to be rapid and to proceed in one direction even with small random variations, which are first directed opposite to selection.

The simplest model of evolution of an average genetic trait $\bar{z}(t)$ is described by the equation

$$d\bar{z}(t) = \frac{dW(\bar{z}(t))}{dt} dt + \alpha \, dG(t) \ . \tag{9.38}$$

The function $W(\bar{z}(t))$ represents a one-dimensional phenotypic landscape. If $W > 0$ at a given $\bar{z}(t)$, selection pushes $\bar{z}$ to large values. The quantity α provides a measure of the drift, random variations; $G(t)$ is a random process of the type of ordinary Brownian motion.

Let us assume that there are two peaks – a small peak a and large peak b. As $\bar{z}$ increases W increases up to $\bar{z} = a$, descends into a valley at $\bar{z} = c$, ascends to $\bar{z} = b$, and then falls off (Fig. 9.25). We assume the time to be discrete; in palaeontology $\bar{z}(t)$ is observed over a finite number of eras n, $0 < t_1 < t_2 \cdots < t_n$. There is a natural time scale τ_α associated with the transition from a to b. The scale increases sharply with decreasing α. Let us assume that for small α, the intervals of observation t_i are proportional to τ_α.

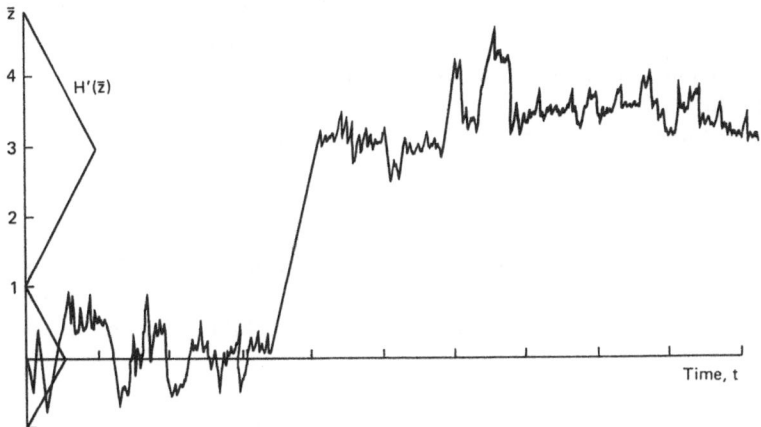

Fig. 9.26. Results of calculations according to Eq. (9.39). See text for explanation [9.62]

With $\alpha \to 0$ the observed values will be fixed at a during a certain period of time, after which they will instantaneously jump over to b and will stay there for an indefinite period of time. For small α the transitions $a \to c$ are hardly probable and occur as a result of random variations. Conversely, the movement $c \to b$ is created by selection. We write

$$\bar{z}(t + 1) - \bar{z}(t) = H(\bar{z}(t)) + sg_t \, , \tag{9.39}$$

where g_t specifies the sign of the random variable; with equal probability $g_t = \pm 1$. Figure 9.26 shows the results of calculations at $H = 0.01$ or -0.01, $s = 0.07$, $a = 0$, $b = 3$, $c = 1$ (in dimensionless units [9.62]). These results are in full agreement with what was said above. We see that state a is unstable. It may be asserted that introduction of a rare mutant that has advantages in a population in which it was previously absent can cause a sharp transition from the unstable state to a more stable state, $a \to b$, practically by-passing c. It should be emphasized that this means just an increase of small fluctuations up to a macroscopic level.

Newman and his co-workers [9.62] believe that they have confirmed the concept of punctualism. Gradual evolution is possible only in a changing landscape with the fitness peaks shifting gradually in a single direction. Punctuation arises upon formation of a critical embryo due to a random drift, i.e. a geographical region in which the transitions $a \to b$ to a higher peak is realized.

The work of Newman and his co-workers [9.62] has been recognized as complete and final evidence in favour of punctualism [9.63]. We cannot agree with this, however.

The fitness landscape proposed by Wright and the works of Lande [9.61] and Newman and his co-workers [9.62] are really consistent with the concept of punctualism. This is not surprising since punctualism was originally implied in these works; the discrete fitness peaks themselves are punctual. The model is quite reasonable, but it has a major shortcoming. It is presumed that ecological niches exist independently of an evolving population, which finds its way into the niche (ascends to a peak), just as a billiard ball falls into the pocket. The concept of punctualism, however, is based on the limited adaptiveness of the emerging species. Moreover, the populations themselves are important participants in ecosystems which they form.

The conceptions discussed above are in agreement with the fundamental idea of dynamic chaos. However, this idea is not developed here. The very appearance of peaks, ecological niches, is a consequence of the non-linearities of the corresponding relations, a consequence of instabilities and bifurcations.

9.6 Directionality of Evolution and Synergetics

In Chap. 3 we considered the biological factors that determine the directedness of evolution. They reduce to constraints on natural selection, which are governed by the established structure of evolving organisms and by limited opportunities of its change in both phylogeny and ontogeny. From rather general considerations it follows that the constraints that create the directedness of evolution must accelerate it strongly.

In Chap. 8, devoted to the Eigen theory, we considered the directedness of evolution of macromolecules. Directedness in this case is also determined by constraints produced by the succession of mutating molecular structures. As we have seen, the emergence of new species (quasispecies) of macromolecules has the character of non-equilibrium phase transitions which are similar to second-order phase transitions. Such behaviour is realizable only in dissipative or synergetic systems, i.e. in open far-from-equilibrium systems.

The directedness of the evolution of organisms may be represented by a mathematical model directly associated with the Lande model described in the preceding section. The Lande model proceeds from the presumption of the normal distribution of quantitative characters with a fixed dispersion matrix. Thus, in the Lande model evolution is experienced only by average phenotypes governed by gradients of average fitness as functions of average values of quantitative phenotypic characters. In fact, mutations and natural selection must presumably change not only average values, but also the distribution function. Therefore, there are good reasons for regarding as more adequate those models of evolution which are based on the Fisher-Eigen equation for the distribution function $p(z, t)$ in the space of genotypic characters. The corresponding analysis has been carried out by Livshits and Volkenstein [9.64].

The Fisher-Eigen equation for the problem under consideration (see [9.65–67]) has the form

$$\frac{\partial p}{\partial t} = (W(z) - \overline{W})p + \frac{\partial}{\partial z} D \frac{\partial p}{\partial z} \ . \tag{9.40}$$

The first term on the right-hand side describes natural selection; $W(z)$ is the selective value of characters z or the fitness of phenotype z;

$$\overline{W} = \int W(z)p\,dz \tag{9.41}$$

is the average fitness of the population. The quantity $\overline{W}$ determines the rate of growth of the entire population. The last term on the right-hand side describes the random mutational change of phenotypic characters as an effective diffusion in the phenotypic landscape. It is thus presumed that this random process is continuous [9.65], i.e. that among mutations random transitions with a maximally small change of characters predominate.

The mathematical model of evolution based on the Fisher-Eigen equation has been studied by Ebeling and his co-workers [9.66,67]. They derived a general expression for the solution of this non-linear equation in terms of the eigenfunctions and eigenvalues of the associated Schrödinger equation, and also, by analogy with the corresponding quantum-mechanical problem, studied the specific features of evolutionary dynamics in the random fitness landscape. We will be interested in the evolution of the distribution function $p(z,t)$ concentrated near the maximum of the fitness function $W(z)$.

Suppose that at the moment t under consideration the distribution is normal, Gaussian:

$$p(z,t) = \frac{1}{(2\pi v(t))^{1/2}} \exp\left[-\frac{(z-\bar{z}(t))^2}{2v(t)}\right]. \tag{9.42}$$

For

$$W(z) = W_m - \frac{\varepsilon}{2}z^2$$

(which may be regarded as a general case near the fitness maximum) the Gaussian distribution (9.42) is the exact solution at all t, in which case the average value $\bar{z}$ and the dispersion v obey the equations

$$\frac{d\bar{z}(t)}{dt} = -\varepsilon v\bar{z} \tag{9.43}$$

$$\frac{dv(t)}{dt} = 2D - \varepsilon v \;. \tag{9.44}$$

The average phenotype drifts to the fitness maximum $\bar{z} \to 0$ and the dispersion to a stable equilibrium value $v_{st} = (2D/\varepsilon)^{1/2}$.

The one-dimensional model (9.40) gives a general notion of the character of evolutionary changes of the average phenotype and dispersions, but it does not describe phenomena that are determined by the interaction of characters. Let us now consider the n-dimensional Fisher-Eigen equation (9.40) with a quadratic fitness function

$$W(z) = W_m - \frac{1}{2} \sum_{i,j=1}^{n} \varepsilon_{ij} z_i z_j \qquad (9.45)$$

and with the matrix D_{ij} which is non-diagonal in the general case [9.64].

From the viewpoint of the model under discussion, evolution in the multi-dimensional phenotypic landscape does not reduce to evolution of phenotypic characters taken separately. The evolutionary interaction of different characters z_i, z_j is determined by their relationship under natural selection and in a mutational process due to the corresponding off-diagonal elements of the matrices ε_{ij} and D_{ij}. This relationship may be due to the combined nature of the quantitative traits involved here, and to differences in the intensity of selection and mutations of individual elementary characters.

The reduction of (9.40) and (9.45) for the Gaussian distributions

$$p = (2\pi)^{-n/2} \beta^{1/2} \left[-\frac{1}{2} \sum_{i,j} \beta_{ij}(z_i - \bar{z}_i)(z_j - \bar{z}_j) \right] , \qquad (9.46)$$

where β is the determinant of the matrix D_{ij}, leads to the equations

$$\frac{d\bar{z}_i}{dt} = -\sum_{j,k} \beta_{ij}^{-1} \varepsilon_{jk} \bar{z}_k \qquad (9.47)$$

$$\frac{d\beta_{ij}}{dt} = \varepsilon_{ij} - 2 \sum_{k,l} \beta_{ik} D_{kl} \beta_{lj} . \qquad (9.48)$$

If ε_{ij} is a positively determined matrix (i.e. if the fitness $W(z)$ is at a maximum at $z = 0$), then the average phenotype evolves to the optimal $\bar{z} \to 0$. In fact, since the matrix β_{ij} is also positively determined, then the inverse matrix β_{ij}^{-1} and the matrix product

$$\sum_{k} \beta_{ik}^{-1} \varepsilon_{kj}$$

are also positively determined. Thus, all the eigennumbers of the problem (9.47), which is linear with respect to $\bar{z}$, are negative, which indicates that $\bar{z}$ tends monotonically towards zero. This inference coincides with the predictions of the one-dimensional model. At the same time, Eqs. (9.47) and (9.48) demonstrate the specific features of evolution in the space of many characters. The direction of evolution at each particular moment of time is determined not only by the fitness gradient but also by he correlations between the characters. According to the distribution (9.45), the correlation between the ith and jth characters is given by

$$C_{ij} = \overline{(z_i - \bar{z}_i)(z_j - \bar{z}_j)} = \beta_{ij}^{-1} . \qquad (9.49)$$

At $i = j$ this expression gives the dispersion

$$v_i = \beta_{ii}^{-1} .$$

The correlations of phenotypic characters, which are not supported by the corresponding degree of correlation of the mutational process or selection, may disappear with time. Let us consider, for example, a two-dimensional case (z_1, z_2) and let us assume that neither fitness nor mutations "mix" these characters, i.e. $\varepsilon_{12} = 0$, $D_{12} = 0$. Then, from (9.49) it follows that

$$\frac{dC_{12}}{dt} = -C_{12}(\varepsilon_{11}v_1 + \varepsilon_{22}v_2) \; . \tag{9.50}$$

The correlation decreases since the dispersions v_1, v_2 and the coefficients ε_{11}, ε_{22} are positive.

Does the disappearance of the correlations mean that the evolution of the corresponding characters is totally independent? Does the limitation of the range of variation of individual characters influence the rate of evolution? This is the key question. Qualitatively it seems to be fairly evident that the constraints on natural selection imposed by the structure and dynamics of organisms must accelerate evolution. The mathematical modelling of this situation has been described by Livshits and Volkenstein [9.64].

Such a constraint can be modelled in Eq. (9.40) by the conditions $\varepsilon_{11} \gg \varepsilon_{22} > 0$; $\varepsilon_{12} = 0$. In the absence of the correlation of characters 1 and 2, the distribution function is thought to be the product

$$p(z_1, z_2, t) = p_1(z, t)p_2(z_2, t) \; , \tag{9.51}$$

in which case p_1 and p_2 evolve separately in accordance with Eqs. (9.47) and (9.48) with the corresponding coefficients. The first character attains its optimal value during a time of the order of $(v_1\varepsilon_{11})^{-1}$. The range of variation of the first character is limited by a small value of the maximum dispersion:

$$v_1^{st} = (2D/\varepsilon_{11})^{1/2} \; . \tag{9.52}$$

Irrespective of this, the second character tends to its optimum at a rate of the order of $v_2\varepsilon_{22}$, and its dispersion tends towards

$$v_2^{st} = (2D/\varepsilon_{22})^{1/2} \; . \tag{9.53}$$

At first glance, the independence indicated contradicts the expected behaviour: the limitation of the possibilities, among which we have to look for the optimum, must speed up the search. As a matter of fact, there is no contradication here. The limitation of the range of variation of individual characters has an appreciable effect on the probability of finding the system in a given volume (dz_1, dz_2) near the optimum $(0, 0)$:

$$p(0, 0, t)\, dz_1\, dz_2 = \frac{dz_1}{(2\pi v_1^{st})^{1/2}} \frac{dz_2}{(2\pi v_1^{st})^{1/2}} \exp\left[-\frac{\bar{z}_2^2(t)}{2v_2} \right] \sim (\varepsilon_{11})^{1/4} \; . \tag{9.54}$$

The narrower the "corridor" of variation of the character $\Delta z_1 \sim (\varepsilon_{11})^{-1/4}$, the higher the probability amplitude. This effect persists in the limiting case of random walk $W(z) = 0$, $\overline{W}(0) = 0$. The "corridor" that restricts diffusion spreading in other directions does not permit the probability amplitude to drop to zero. In the multidimensional landscape this effect may be enormous. Indeed, upon diffusion in the n-dimensional space without restrictions, the amplitude decreases as

$$p_{(n)} \sim (Dt)^{-n/2} \tag{9.55}$$

and upon one-dimensional diffusion inside the "corridor" the tubes of the cross-section are given by

$$p_{(1)} \sim S_{n-1}^{-1} (Dt)^{-1/2} \ . \tag{9.56}$$

An increase in the probability is measured by the ratio

$$\frac{p_{(1)}}{p_{(n)}} \sim \frac{(Dt)^{(n-1)/2}}{S_{n-1}} \to \infty \ . \tag{9.57}$$

It has been shown that the presence of constraints on selection sharply accelerates evolution. As we know, there was enough time and, as we see, there had to be enough time. The interaction of characters ignored in this calculation may either increase or decrease the ratio (9.57), but it cannot alter the general inference.

The problem of the effect of the constraints and of the relationship between the characters has also been explored by Burger [9.68] on the basis of the Lande mathematical model. The basic conclusions of this work, i.e. the existence of a magnitude of dispersion optimal for evolutionary dynamics, and the limitation imposed on the dispersion by the possibility of evolution along the "corridor," follow from the specific features of the model.

The situation in the case of organisms is different from the one dealt with in the Eigen theory. It is not easy here to introduce the sequence space that would correlate in some manner with the fitness landscape. Since the genome is very large and is composed, to a considerable extent, of non-coding DNA, and because the population is small, the mutants located in the sequence space are only rarely found to be near the peaks of the landscape. The landscape itself is discrete and contains certain islands of high fitness, which are only rarely connected with one another by "mountainous ridges". We are dealing with diversified geography in the systems being compared. As a matter of fact, biological evolution is governed predominantly by molecular drive rather than by molecular drift. In other words, in contrast to systems that appear in the Eigen theory, changes in the genomes of organisms which determine speciation and macroevolution consist of significant changes in chromosomes.

In the theory of organismic evolution we have to consider the sequence space in the broad sense of the word, with account taken of both the chromosomal changes and the fitness space, which characterize phenotypes. Conversely, the Eigen theory does not consider phenotypes at all, and selection acts directly on genotypes, i.e. on the primary structures of macromolecules which take part in hypercycles. No degeneracy is present here, except for the closeness of the properties of macromolecules that form a single quasispecies.

Naturally, the tendency arises to apply the ideas of the Eigen theory (the quasispecies, the hypercycle, the sequence space) to the evolution of organisms as well. But this is a matter for future work. Nonetheless there is every ground for believing that Eigen's ideas will have a beneficial influence on further development of investigations in this most important area of theoretical biology.

Let us return to the fitness landscape. We have considered a model in which the distribution and dispersion of characters evolves. The limitations and correlations of characters may have a decisive effect on the direction and rate of evolutionary changes. In principle, theory must take into account the dependence of the fitness landscape itself on evolving populations. The processes of occupation of fitness peaks are co-operative in the sense that the curvature and height of ascents depend on the number of "mountain climbers" that have reached the peak.

As has been pointed out earlier, we can invert the landscape by replacing the peaks with energy wells. In this case, evolution may be likened to currents of water flowing downhill. Sooner or later the water finds its way to the deepest wells. What has been said above is tantamount to taking account of the fact that the running-down of the water itself produces new wells and depressions.

Of course, changes in environmental conditions, both biotic and abiotic, have a direct effect on the fitness landscape.

Another approach to the problem of the directedness and, hence, the rate of evolution, which has not yet been realized, is possible on the basis of the study of Markovian processes with stochastic matrices, whose many off-diagonal elements are equal to zero and which undergo change in the successive evolutionary stages.

In concluding this chapter, let us consider the general problems of the theory of dissipative structures or synergetics. The most important sections of this new and very topical area of physics are tied up with cosmology and meteorology, with non-linear optics and with all the branches of biology. Two paths have led to the development of synergetics. Prigogine began constructing the thermodynamics of irreversible processes with a consideration of open non-equilibrium systems that are near equilibrium. These systems are described by linear equations and neither periodic processes nor phase transitions occur in them. Further, Prigogine turned his attention to open systems that are far from equilibrium, which he termed dissipative. The works of Prigogine and his collaborators were fundamental, giving rise to a new branch of science (see [9.21,69–74]). Haken was led to this area, which he called synergetics, by non-linear optics. The transition from incoherent to coherent laser radiation is a non-equilibrium phase transition, the emergence of order from chaos [9.75,76]. Synergetics is just the science of chaos and self-organization due to instabilities, of non-equilibrium phase transitions. Haken defined synergetics as the theory of the emergence of new qualities at the macroscopic level. Such emergence implies a non-equilibrium phase transition realizable due to the collective, co-operative interactions of the system's elements. These processes are possible only in dissipative systems. Thus, the notions of synergetics and dissipative systems are not identical, but they characterize identical systems and their identical behaviour. The advent of this area of physics implies the switchover to new theoretical thinking, which has led, for the first time, to the true union of biology and physics. Let us recall once again that the first synergetics study was *The Origin of Species* by Darwin, who disclosed the mechanism of the transition from chaotic variability to the ordered development of the biosphere.

Synergetics is directly associated with cybernetics and with information theory. This is to be discussed in the next chapter.

10. Informational Aspects of Evolution

10.1 Order, Information and Entropy

We have already dealt with the emergence of order from chaos; the latter implies the complete absence of the former. In the natural sciences we need a quantitative measure of order.

As is well known, the quantitative measure of *disorder* is entropy. The Boltzmann formula for entropy is

$$S = k \ln P \qquad (10.1)$$

where Boltzmann's constant $k = 1.38 \times 10^{-16}$ erg·deg^{-1}; P is the statistical weight of the state under consideration, i.e. the number of ways it can be realized. At the limit, in the case of complete order, the state of the system is unambiguous, being realized in only a single way, i.e. $P = 1$ and, hence, $S = 0$. Such is the equilibrium state of any pure substance at the absolute zero of temperature (0 K) (the Nernst theorem). On the other hand, a given system is characterized by maximum disorder exactly in a choatic state realized in a maximum number of ways: $P = P_{max}$, $S = S_{max}$. For an isolated system, at a temperature different from the absolute zero of temperature, this is just the equilibrium state under given external conditions.

We may introduce, as a measure of order, the decrease of entropy as compared to its value in the most disordered, chaotic state. This is entropy with the minus sign, negentropy. It should be stressed that the term "negentropy" has a conditional meaning. Schrödinger wrote that an organism "feeds on negentropy" [10.1]. This is a figurative expression; negentropy means the efflux of entropy from an open system into the surroundings.

We give here an elementary example of the decrease of entropy. When a mixture of two gases is separated into two different gases, the entropy decreases by an amount

$$\Delta S = k N_A \left(n_1 \ln \frac{n_1 + n_2}{n_1} + n_2 \ln \frac{n_1 + n_2}{n_2} \right) \qquad (10.2)$$

where n_1 and n_2 are the numbers of moles of the first and second gases; N_A is the Avogadro number. Of course, such a separation is possible only if appropriate work is done; otherwise, the second law of thermodynamics would be violated. Upon separation a certain order has arisen. Upon mixing, the order disappears and the entropy increases by the same amount ΔS.

Here we are speaking of molecular chaos in the ordinary sense of the word. This is not deterministic, dynamic chaos, but rather equilibrium chaos. It conveys no information, in contrast to deterministic chaos (see Sect. 10.7).

Conversely, any ordered state implies the presence of certain information (the first gas is in one half of the vessel and the second gas in the other half; these two halves are separated by a partition).

Information, i.e. a set of data on ordering, may serve as its measure. But we can speak of a measure only if it can be fixed. Data are information if they can be transmitted and memorized.

The word "memorizing" has a broad meaning here. This is the fixation of existing order for a more or less prolonged period of time, fixation that can be realized in various ways. In biology we deal with memorizing at the genetic and, hence, the phenotypic level, and also with memorizing determined by the functioning of the nervous system, beginning with conditional reflexes and ending with the higher nervous activity.

When man dies, only information about him is left, the memory of him in the mind of those who knew him, photographs, portraits, letters. He leaves the results of his work, be it machines made by him or books he has written. He also leaves genes in the genomes of his children, grandchildren and remote progeny. Finally, if the deceased is not burnt in the crematorium, his skeleton is left, the examination of which can provide some information. But the latter provides the least information, though science is capable of restoring his face from his skull.

Let us recall the basic propositions of canonical information theory, which is basically a theory of communications (cf. [10.2,3]). The amount of information obtained as a result of the choice of one of P equally probable possibilities is expressed by the formula

$$I = \log_2 P \ . \tag{10.3}$$

The unit of information, the bit, is obtained at $P = 2$, say in coin tossing. If the possibilities are not equally probable, then the Shannon formula is valid. This is

$$I = -\sum_{i=1}^{M} p_i \log_2 p_i \tag{10.4}$$

where p_i is the probability of realization of the ith event out of M possible ones. The probabilities p_i are normalized:

$$\sum_{i=1}^{M} p_i = 1 \ . \tag{10.5}$$

For example, the appearance of the letter "O" in a Russian text is equal to 0.090 and the probability of occurrence of the letter "Φ" is 0.002 [10.2].

The transformation of formula (10.4) to formula (10.3) is simple. If all p_i are identical, then they are equal to $1/M = 1/P$. We obtain (10.3).

As has been said, the increase of order (implying the increase of information) is expressed by a decrease in entropy. Let us consider the relationship between S and I.

Suppose that we have a system consisting of N molecules, which may be in M different states. If in the system N_1 molecules are in state 1, N_2 molecules in state 2, etc., N_M molecules in state M, the statistical weight P is equal to

$$P = \frac{N!}{N_1! N_2! \dots N_M!} \tag{10.6}$$

in which case

$$\sum_{i=1}^{M} N_i = N \ . \tag{10.7}$$

According to (10.1), the entropy of such a state of the system as a whole is given by

$$S = k \ln P \cong k \left(N \ln N - \sum_{i=1}^{M} N_i \ln N_i \right) \tag{10.8}$$

or, denoting the fraction of molecules in state i,

$$p_i = N_i/N \tag{10.9}$$

$$S = -k \sum_{i=1}^{M} p_i \ln p_i = -k \ln 2 \sum_{i=1}^{M} p_i \log_2 p_i \ . \tag{10.10}$$

Comparing (10.10) with (10.4), we obtain

$$S = k \ln 2I \ . \tag{10.11}$$

As a matter of fact, the situations considered for molecules and letters are quite similar.

The value of $k \ln 2 = 10^{-16}$ erg$\cdot$deg^{-1} is very small. It is clear that the receipt of any information is associated with a decrease in entropy, which is paid for by an increase in entropy in another part of the system [10.4]. Otherwise, the second law of thermodynamics would be violated. But does formula (10.11) imply the quantitative equivalency of the real memorized, i.e. macroscopic information and entropy?

This question has been analyzed in detail by Romanovsky et al. [10.5]. The equivalency (10.11) exists only for micro-information, say for information about the instantaneous values of the co-ordinates and momenta of gas molecules present in a certain vessel. Such information cannot be memorized without special devices performing a considerable amount of work (say devices for instantaneous photographing of molecules). Here the microscopic information is transformed to macroscopic information. The entropy "cost" of macro-information is many orders of magnitude higher than the value determined by formula (10.11).

This can be illustrated by a simple example. One bit of macro-information is obtained in coin tossing. The entropy changes when the energy is consumed on tossing, and due to the release of heat when the coin hits the floor. This entropy is many orders of magnitude greater than $k \ln 2$. The coin is macroscopic and may be infinitely large.

Information is always macroscopic in the long run. Microinformation is entropy. True information is created on the invariable condition that entropy is released, the amount of effluent entropy being many times greater than the information received.

We see that the relationship between entropy and information is not simple at all. Certain details which seem to be essential are contained in a work by Ebeling and Volkenstein [10.6].

From a physical point of view, information is a quantity which is transferred from the source to the receiver and which reduces the uncertainty of the system's state. Information is a binary relationship between two systems, i.e. the source and the receiver, and is always associated with the energy and entropy fluxes between these systems.

The transfer of information may be treated as a special form of entropy transfer. There are other forms of entropy transfer, such as heat transfer, which have nothing to do with information transfer. Inside a certain body, the entropy has no form; it assumes a form only when transferred from one body to another.

Informational processes require non-equilibrium conditions which are realized as a result of the efflux of entropy into the surroundings. Such processes are possible only in systems supplied with valuable energy, such as electric energy for computers, food for animals, and sunlight for plants. The second law of thermodynamics pertains to the transfer of overall entropy, a small part of which is informational entropy according to Shannon (formula (10.4)). Only this part can be memorized.

Entropy is a measure of the value of energy transferred between bodies in the form of heat or work. For a system with fixed energy E, and fixed values of other extensive parameters X, the entropy is at a maximum in the equilibrium state. A decrease in entropy determines the value of energy, which may be expressed, for example, by a dimensionless quantity:

$$W(E, X, t) = 1 - \exp\{[S(E, X, t) - S_{eq}(E, X)]/k\} \ . \tag{10.12}$$

In the language of statistical mechanics this means that the value of entropy is determined by the part of phase space occupied by the system. In an isolated system the value of energy satisfies the conditions

$$W(E, X, t) \geqslant 0 \ , \quad \frac{\partial}{\partial t} W(E, X, t) \leqslant 0 \ . \tag{10.13}$$

Let us recall some basic propositions of canonical information theory (CIT). This theory is concerned only with the amount of information in accordance with formulas (10.3) and (10.4), ignoring the sense, content or value of information. Accordingly, it does not consider the reception of information. The receptor considered in canonical information theory is only capable of distinguishing one letter from another, it does not understand the text of a message.

The CIT introduces the notion of redundant information. Redundant information is repetitive information. It is used in technology to suppress the constantly existing noise, which hinders the exact transmission of a message. For example, a sinking ship repeatedly sends a distress signal for help, SOS, in order to increase the probability of its reception.

Any letter text contains redundant information; if the language is known, a number of letters may be omitted in the text without loss of its meaning (cf. [10.3]). The measure of redundancy is

$$R = 1 - I/I_0 \, , \tag{10.14}$$

where I is the amount of information per letter of the text, the language structure of which is known in advance; I_0 corresponds to the assumption of the equal probability of appearance of all letters. It is obvious that $I < I_0$; knowledge of the language implies the presence of prior information. For example, the calculation of I for the Russian language according to the Shannon formula (11.4) gives $I = 4.35$ bits, and according to formula (10.2) we have $I_0 = \log_2 32 = 5$ bits (the Russian alphabet contains 32 letters) [10.2,3]. As has repeatedly been said, new information is created as a result of "the accidental choice remembered", which was first pointed out by Quastler [10.7]. All the three words in this definition are essential. Memorizing implies the fixation of the results of a certain process, in the course of which a non-deterministic, random choice is made. Such is, for example, the gain or loss in a roulette game. A system that generates new information must be multi-stationary. Unless the roulette wheel is stopped there is no information, there is only entropy. When the ball stops in one of the many possible positions, information arises (see also [10.3,5]).

Memorizing, i.e. reception, requires a special treatment, which is ignored by canonical information theory. In biology we deal with two kinds of memory: genetic memory, and memory in the direct sense of the word, about which the Russian poet N.S. Gumilev wrote:

> Memory, thou with the giant's hand
> Direct life as the horse by the bridle.

The appearance of a new organism as a result of the union of the male and female sex cells is the creation of new information. This process leads to non-deterministic results; it is not known in advance that progeny will be produced by a given pair and one cannot predict what recombination of genomes is realized in the zygote. New information is also created in speciation and macroevolution. The creative activity of man in any area implies the generation of new information (cf. [10.3]). Recall the words of Blok given earlier.

In connection with this, let us dwell on the old problem of the creative role of natural selection. Does selection create something fundamentally new or is it only a sieve which does not pass non-adaptive traits? From what has already been said, it follows that selection plays a creative role, producing new information. For this to be done chaos is needed, which is what occurs during evolution. However, the emergence of dynamic order from chaos does not always imply the generation of new information. For example, the Bénard structure in a nonequilibrium viscous liquid (see [10.3], page 125) is unambiguous; there is no random choice here. But each new species implies new information.

Painters sometimes treat creative work as the elimination of redundancy. On one of the best pages of "Anna Karenina" Tolstoy wrote about the painter Mikhailov:

"But, while introducing corrections, he did not change the figure but only rejected what concealed it. He removed, as it were, the coats because of which it was not seen."

This description is metaphoric. A poet, a painter, a sculptor, and natural selection are all concerned with the generation of new information rather than with the removal of coatings or with sifting.

Certain constraints are always imposed on a work of art. This refers literally to the canvas on which a picture has been painted. A sonet always limits the possibilities of a poet. It imposes restrictions but leaves enough room for imagination. The same character in the informational aspect is exhibited by constraints on natural selection (see Chap. 3).

Information is only something which produces information (Weizsäcker, see [10.8]). Information is a measure of structure; without structure there is no information, without information there is no structure. A structure is the information memorized. Accordingly, aesthetics cannot exist without information. At the same time, as was said above, information is a real physical notion. Polanyi maintained that "all objects which convey information are irreducible to the laws of physics and chemistry" [10.9]. In essence, this statement implies the non-physical nature of any historical branch of science, be it cosmology or biology. As a matter of fact, history consists of the transmission and memorization of information. Polanyi was mistaken, because he could not know the physics of dissipative systems or synergetics (see Sect. 10.7). It may be asserted that information theory is a legitimate part of physics.

10.2 Information and Evolution

The ideas of information theory are of universal importance since everything different from homogeneous grey chaos contains information. The diverse phenomena in nature and society are associated with the creation, transmission, coding and reception of information. Accordingly, informational approaches allow us to tackle from a different angle a number of problems concerning the natural sciences and the humanities.

It should be stressed once again that the study of informational aspects of phenomena can by no means be regarded as fundamentally different from other physical approaches, say from estimates of the corresponding balance of energy and free energy. An informational message is material, being associated with the change of both energy and entropy. As we have seen, the minimal entropy cost of one bit of information is $k \ln 2$ and the minimal energy cost is $kT \ln 2$ [10.3]. The actual changes in entropy and energy are many orders of magnitude greater.

Information theory is a part of cybernetics, a science concerned with the study of "control and communication in an animal and a machine" (Wiener [10.10]). A living organism is a controllable and controlling machine; control is accomplished by feedback at the genetic and metabolic levels. As has been repeatedly said before, in contrast to machines built by man, in which the control and linkage have a mechanical, electrical and/or magnetic character, an organism is a chemical machine. The signals carrying information are molecules and ions; the sources, transducers and receptors of signals are also built up of molecules. The functioning of a living system reduces to molecular interactions.

We can estimate the amount of information in an organism and in its constituent parts. According to the estimate made by Blumenfeld [10.11], the biggest contribution to the information of the human organism is made by proteins; this contribution is about 10^{26} bits. Other contributions are smaller by many orders of magnitude. This estimate does not make a distinction between basic and redundant information. Schrödinger called an organism an aperiodic crystal [10.1]. What is meant here is the condensed, solid state of the organism and the absence of crystalline periodicity in its structure. But at the same time this implies a large amount of non-redundant information. The synergetic, non-equilibrium ordering of the organism and cell is different from the equilibrium ordering of a crystal and this distinction is first of all expressed in a much greater content of non-redundant, i.e. valuable information (see below).

Cybernetics and information theory were first applied to the problems of biological evolution in the works of Schmalhausen, the founder of the theory of stabilizing selection [10.12].

Schmalhausen treats evolution as a process controlled by the interaction of an elementary evolving system (a population) with the biogeocenosis. The biogeocenosis, which also incorporates the population considered, is a regulatory mechanism. The flow of information is directed from the biogeocenosis to the population. An organism develops from a zygote which has received inherited genetic information from the parents. This information is then transformed in the processes of individual development. As a result, a phenotype arises, which expresses the properties of the genotype at the individual level. The direct flow of information about the events in the biogeocenosis is transmitted through the multiplication of approbated phenotypes.

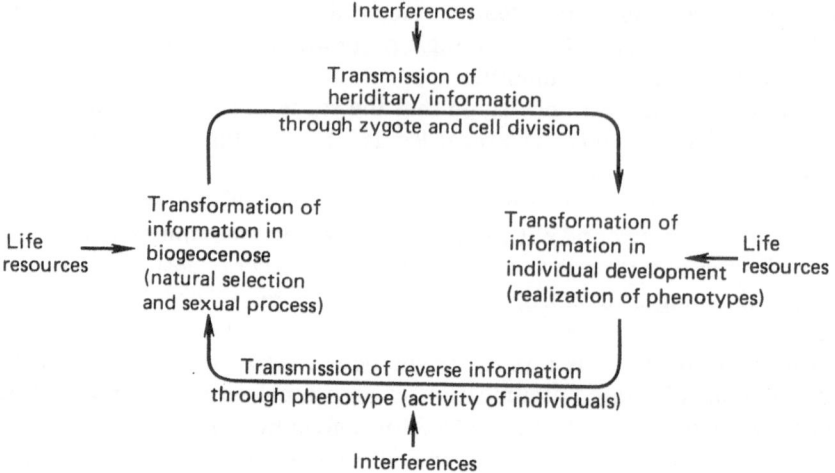

Fig. 10.1. Scheme of the regulatory mechanism of evolution according to Schmalhausen

The reverse flow of information is directed from the population to the bio-geocenosis through phenotypes. "Natural selection is a legitimate expression of the results of a complex control system in the biogeocenosis. Selection is based on a comparative evaluation of phenotypes inside the population ... and serves as a principal mechanism of transformation of backward information in the biogeocenosis".

The general scheme of the regulating mechanism of evolution according to Schmalhausen is shown in Fig. 10.1. The transfer of information by means of certain feedbacks is accompanied by noise. The transformation of information in an individual development occurs by means of living systems. The transformation of information in the biogeocenosis is associated with external factors. The mechanism of heriditary transfer is incorporated into individuals, whereas changes in the biogeocenosis are related to the population as a whole. The population changes and evolves, and the information about these changes is transmitted through the multiplication of individuals.

The analysis carried out by Schmalhausen is qualitative. This is a kind of translation from the ordinary biological language of evolutionary theory into the language of cybernetics. Such a translation does not directly provide new information but contributes to a better understanding of the regulatory processes in evolution.

One of the first attempts to apply information theory to evolution was undertaken by Gatlin [10.13]. She has made estimates of the informational content of DNA. DNA is built up of four types of nucleotides: A, T, G, and C. If their appearance is a given chain unit were equally probable, then according to (11.3) the amount of information per nucleotide would be as follows:

$$I_1 = \log_2 4 = 2 \text{ bits} .$$

With account taken of unequal probability, the Shannon formula should be used. For the bacteria *Micrococcus lisodeiticus*, $p(C) = p(G) = 0.355$, $p(A) = p(T) = 0.145$. We have

$$I_2 = - \sum_{i=1}^{4} p_i \log_2 p_i = 1.87 \text{ bits} .$$

·Like a linguistic text, the DNA text is a Markov chain; there are stochastic matrices p_{ij}, which express the probability of appearance of a nucleotide j in the chain unit that follows the nucleotide i. The probabilities of doublets ij for different i and j differ. Accordingly, we have

$$I_3 = - \sum_{i,j=1}^{16} p_{ij} \log_2 p_{ij} \tag{10.15}$$

with

$$\sum_{i,j=1}^{16} p_{ij} = 1 . \tag{10.16}$$

The differences between p_{ij} and $p_i p_j$, however, are not great. Nonetheless, Gatlin, while studying the values of $I_1, I_2, I_3 \ldots$ for DNA tried to establish the evolutionary significance of the relationships between these quantities. This attempt boils down to the determination of the redundancy of information (cf. [10.3]). However, Gatlin's work had no consequences for science for several reasons. Firstly, the amount of information is a quantity which in itself provides little progress in the solution of biological problems. What use can be made of the fact that $I_1 = 2$ bits for DNA? Secondly, the determination of the composition, i.e. values of p_i, for the total DNA of an organism carries little information either, since in eukaryotes only a small fraction of DNA participates in the biosynthesis of protein, i.e. in the formation of the structure of the organism. This is also valid for p_{ij}, p_{ijk}, etc. Thirdly, the data on the nucleotide sequences in DNA required for finding Markov chains were quite insufficient when Gatlin published her work in 1972; the data available today are also insufficient for that purpose. Gatlin should, however, receive credit for stating the problem of quantitative informational approaches to evolution, of attracting attention to this area.

Fifteen years later, Wicken published his book under the remarkable title *Evolution, Thermodynamics and Information. Extending the Darwin Program* [10.14]. Unfortunately, the author has insufficient knowledge of the three areas of science presented in the title. The book mostly contains methodological statements, which are, as a rule, erroneous. The author states that he is "an ontological anti-reductionist and a methodological reductionist". The meaning of these words is obscure. An example is the following statement: "The origin has at its disposal more than pure physics and chemistry. It also has natural selection ... Organization provides boundary conditions for the operation of physico-chemical processes that cannot be deduced from the principles that control these processes." These statements are quite unfounded and, as shown in Chap. 1, they are simply erroneous. The author maintains the "a message or a structure cannot have entropy." In the chapter devoted to biological organization it is said that "organization includes a function while the physical sciences do not deal with functions." These statements have nothing to do with science.

As useless as the book cited above is the book *Evolution as Entropy* by Brooks and Wiley [10.15]. The basic proposition in this work is that speciation is controlled by the stochastic premises of the second law of thermodynamics. One may only regret that in 43 years since the publication of Schrödinger's work, a book has appeared whose authors do not understand the role of the second law of thermodynamics in living nature.

In a collection of works which was edited by Weber, Depew and Smith [10.16] there are no such embarrassing errors, but neither are there constructive ideas that could make it possible to go a step forward in the understanding of the thermodynamic and informational aspects of evolution. In this respect the monograph written by Ebeling and Feistel [10.17] is incomparably more informative.

In all the books mentioned above [10.13–17] the authors are concerned only with the amount of information and, hence, with entropy. But, by confining oneself to these concepts alone, one can hardly say anything about evolution.

Of prime importance to informational approaches in biology are the words written by Schmalhausen in the draft of his unfinished book "*Cybernetics as the Science of the Self-development of Living Things*" ([10.12], pages 194, 218, 219): "... in the world of living things the quality or value of information is often of decisive importance. In all cases when information is compared and selected, this is done on the basis of its quality evaluation. Along the lines of feedback a comparison is always made of the actual result of a certain action with that coded in the program. This invariably implies, first of all, an evaluation according to the quality of information.... Thus, it is the qualitative estimation of information that becomes most essential in biology ... No appropriate methods have yet been worked out for estimation of the quality of information. In special cases, perhaps the quality of information may also be expressed quantitatively".

10.3 The Value of Information

The amount or quantity of information is a certain measure of order, which can be received and memorized. The treatment of reception in canonical information theory is, however, limited to this proposition. On the other hand, the semantics, content, meaning or value of information (in what follows use will be made of the last term) cannot be defined at all without considering reception. No value of information exists, irrespective of its reception (cf. [10.3,18]). First of all it is necessary to find out what the reception of information means.

The reception of information is a non-equilibrium, irreversible process of transition of the receptor system from a relatively unstable state to a more stable state. The process is irreversible since the information received is not returned. The process is non-equilibrium since upon reception the flow of information is not balanced by the reverse flow from the receptor to the source. Finally, the receptor system must be in an unstable state, otherwise it will not perceive information, and under the influence of the information the system must pass to a new, more stable state. In other words, a "goal" is needed. The presence of a "purpose" implies nothing other than instability; the achievement of the goal is the transition to a more stable state.

In order to receive information, the system must possess a certain level of reception, i.e. the receiving act cannot be realized if the system is not properly prepared, that is, if it has no reserve of preliminary information – a glossary. It is impossible to read a book written in a language unknown to the reader.

Thus, the value of information may be associated with the achievement of a "purpose" in the sense indicated. The value of information is determined by the consequences of its reception.

The problem of the origin of valuable information is very important to biology. It can be expressed by the formula

$$V = \log_2(P/P_0) \tag{10.17}$$

where P_0 and P are respectively the probabilities of achieving a purpose before and after the information is received. This expression is similar to that proposed in the monographs of Bongard [10.19] and Kharkevitch [10.20]. It is convenient for cases where the probability of achieving a purpose is low. Stratanovitch associates the value of information with a decrease in the "expenses" of achieving the goal [10.21]. The definition proposed by Stratanovitch is convenient to use if the probability of achieving the purpose is not low, and there are many ways of accomplishing it.

The probability of random generation or receipt of valuable information in biology is low. Therefore, use may be made of the formula given above. If after the information is received the purpose is achieved with a probability P close to unity, $V = -\log_2 P_0$, i.e. it coincides with the maximum amount of information:

$$V = -\log_2 p_0 = \log_2 P_0 = I \ .$$

According to (10.17), the value of information may be positive, negative (in the case of disinformation) or zero (if the receipt of information does not affect the achievement of the purpose).

The problem of information value as applied to biology and, in particular, to evolutionary theory has been discussed in a number of works [10.22,23] (see also [10.3,5,18]).

The amount of valuable information depends on the accuracy of specification of the purpose: it is smaller, the more accurate is this specification. In biology, the purpose is often formulated only approximately, say "the purpose is to live". If we replace the word "to live" by "to survive", the purpose will be clarified to a certain extent but it still remains uncertain.

The problem of the purpose in biology is quite clear if we are speaking of the higher nervous activity. In an artificial system, such as a slot machine, the purpose is given from outside; the machine itself is incapable of pursuing its objectives. Could a system exist that would be capable of striving towards the realization of a goal it has formulated itself? This is the fundamental question for the problem of the origin of life. We may say that the objective of an elementary evolving system (a population) is to achieve a state favouring life and reproduction.

Since the purpose implies instability, it may be said that a cloud of cosmic dust has as its purpose the creation of a star.

A population attains a greater stability by way of adaptation to an appropriate ecological niche. The choice of the niche is specified unambiguously; here we deal with the "accidental choice remembered," i.e. new information is generated.

The purpose of DNA consists, roughly speaking, of the synthesis of proteins which provide the existence of cells and organisms; a biosynthetic system attains a more stable state, while producing proteins. However, in this case there is no choice, and the synthesis of a particular protein in cells of an individual of a particular species cannot be regarded as the creation of new information. The synthesis is determined by the original structure of the genome. New information is produced as a result of mutations and recombinations of genomes.

Thus, while considering the value of biological, in particular genetic, information, we can use the notion of the purpose.

The value of information received is especially high for trigger systems, which are characterized by high instability, i.e. which sharply change their state under very slight influences. For example, the receipt of one bit of information such as the change of traffic lights from red to green causes the change of direction for the entire flow of transport. One bit on tossing a coin may mean very much if we cast lots. In such situations the receptor system is unstable because it possesses a large reserve of specific information. In biology we have to deal with trigger systems at all levels of organization.

Let us consider the properties of a dynamical system with its inner purposes, i.e. with instabilities. Suppose the system is described by the equation

$$\frac{dx_i}{dt} = F_i(x_1, x_2, \ldots, x_m) \tag{10.18}$$

where x is a vector in m-dimensional phase space ($i = 1, 1, \ldots, m$). The F_i functions satisfy three conditions: (1) there is multistationarity, i.e. a certain number of stationary states, among which a choice can be made; (2) there are regions of instability in phase space, which is necessary for generation of new information; (3) the system is open and far from equilibrium, i.e. dissipative, which is required for memorizing the information. Up to a certain moment t_c the system is unstable; at moment t_c it is found to be in a region of attraction of one of N attractors – stationary states. In this way a choice is made.

Let us divide the system into v subsystems and let us consider the state of each of them for $t < t_c$, the subsystems being regarded as isolated.

It is possible that at a moment preceding t_c, each subsystem is already in the region of attraction of the jth stationary state ($j = 1, 2, \ldots, N$), with j being different for different subsystems. Each subsystem already has information but the system (10.18) has no information. Under these conditions the purpose of each subsystem may be formulated as the necessity of retaining the information generated. The purpose of the subsystem that has chosen the jth state, is to bring the entire system into the same state. The purposes of different subsystems may be antagonistic since the choice of a stationary state by the entire system rules out a different choice.

In biology such a purpose is equivalent to the survival of a population. If the purpose is formulated, we can specify the value of information created in each of the subsystems. The information value depends on the state of the entire system at a given moment; it varies with time. It is thus important whether the possibility exists of dividing the system into subsystems which possess information at the moment when the system as a whole has no information at all.

These concepts can be illustrated by a model of code choice proposed by Chernavskaya and Chernavsky (see [10.5]). The model is expressed by the formula

$$\frac{dx_i}{dt} = ax_i - b \sum_{j \neq i}^{N} x_i x_j \; ; \quad i, j = 1, 2, \ldots, N \; , \tag{10.19}$$

where N is a large number of various possible codes; x_i is the number of individuals with an ith code. The linear term represents self-reproduction and the quadratic term represents an antagonistic interaction. The model is similar to the Eigen-Schuster hypercycle (see Sect. 8.2), but it allows for the choice of one code out of a series of equivalent variants, i.e. in a case when the code chosen has no *a priori* advantages.

The system (10.19) has the following properties:

1. There are N stationary states corresponding to the survival of pure populations ($x_i \to \infty$, $x_{j \neq i} \to 0$, $i, j = 1, 2, \ldots, N$).
2. The system is dissipative and the stationary states are stable.
3. There are two stable points: $x_i = 0$ and $x_1 = x_2 = \cdots = x_N = \bar{x} = a/Nb$.

At large values of N, a/Nb is close to zero.

There is no instability region here. However, if we take into account the external noise, then the region adjoining $x_i = \bar{x}$ corresponds to instability. The elements of the system (10.19) are individuals, whose goal is to survive and preserve the code they had.

Suppose that $x_1 = x_2 = \cdots = x^0$ at the initial moment $t = 0$. The *a priori* probability of survival is given by

$$P_{in}^0 = \sum_{i=1}^N p_i^0 w_i^0 = \sum_{i=1}^N \frac{1}{N} \frac{x^0}{X^0} = \frac{x^0}{X^0} = \frac{1}{N} \ . \tag{10.20}$$

Here $w_i^0 = N^{-1}$ is the probability of choice of the ith code,

$$p_i = x^0/X^0 \equiv x^0/\sum x_i^0 = x^0/Nx^0 \equiv N^{-1} \ .$$

The *a posteriori* probability of survival after receiving the instruction to choose the kth variant of the code can be found by putting $w_i = \delta_{ik}$ in (10.20),

$$P_f^0 = p_k^0 = \frac{x_k^0}{X^0} = \frac{1}{N}$$

and the value of information is equal to

$$V = \log_2 P_f^0/P_{in}^0 = 0 \ . \tag{10.21}$$

In this case the receipt of information does not affect the achievement of the goal, which is natural under symmetric conditions.

The original symmetric state is unstable; with the passing of time it is destroyed and the system tends towards one of the stable states in which only one variant of code predominates (say, the kth variant). At the moment t (such that at $\gg 1$) the number of individuals of type k is given by

$$x_k(t) \cong x_0 e^{at} \ . \tag{10.22}$$

The number of individuals of type i is

$$x_i(t) = x_0 e^{-at} = x_0(t) \ll x_k(t) \ ; \quad i \neq k \ . \tag{10.23}$$

The *a priori* probability of survival is here given by

$$P_{in}(t) = \sum_{i \neq k}^N P_i(t)w_i + P_k(t)w_k = N^{-1} \frac{x_k(t)}{X(t)} + \frac{\bar{x}}{X(t)} \ . \tag{10.24}$$

Here

$$X(t) = \sum_{i=1}^N x_i(t) = x_k(t) + (N-1)\bar{x}(t) \ .$$

The *a posteriori* probability after the kth code is chosen is

$$P_f(t) = P_k(t) = x_k(t)/X(t) = [1 + N\varepsilon(t)]^{-1} \ , \tag{10.25}$$

where $\varepsilon = \bar{x}(t)/X(t) \ll 1$. The value of this information is

$$V_k(t) = \log_2 \frac{N}{1 + N\varepsilon(t)} = \log_2 N - \log_2[1 + N\varepsilon(t)] \ , \tag{10.26}$$

where $\varepsilon(t)$ tends to zero with time and the quantity $V_k(t)$ increases to the maximum value of information about the code:

$$V_k(t) \to \log_2 N - N\varepsilon(t) = I_{\max} - N\varepsilon(t) \ . \qquad (10.27)$$

This example shows that the value of information is important in evolutionary processes. This refers to all sciences dealing with developing systems, to linguistics, economics, etc. The random generation of valuable information has a low probability, but it occurs during development. The problem of low probability drops out if the information is not valuable at the moment of generation, but its value progressively increases as a result of the interactions taking place within the system.

10.4 Information Value, Indispensability and Complexity

The fraction of information with a zero value can be established even with an obscure specification of the purpose. An example of such information is redundant information, which does not change the probability of achieving the purpose. The value of redundant information is thus zero. In a linguistic text 50% of letters are redundant; having omitted these letters, we still can achieve the purpose, i.e. we can understand the text correctly. Redundant information may, however, assume value if it is efficiently used to eliminate the "noise".

On the other hand, non-redundant information is valuable. Thus, the measure of information value may be the degree of non-redundancy, indispensability of a message or of its element.

Neglecting the different probabilities of occurrence of this or that latter in a text written in the Russian language, we obtain, as has been said above, the amount of information per letter:

$$I_0 = \log_2 32 = 5 \text{ bits} \ .$$

With all these probabilities being taken into account we obtain in accordance with the Shannon formula:

$$I_1 = 4.35 \text{ bits} \ .$$

With account taken of the probabilities of pair combinations,

$$I_2 = 3.52 \text{ bits} \ .$$

With account taken of the probabilities of triple combinations,

$$I_3 = 3.01 \text{ bits}$$

and so on (see [10.2,3]). The redundancy increases in the following sequence:

$$R_n = 1 - I_n/I_0 \ . \tag{10.28}$$

We obtain

$$R_0 = 0 \ , \quad R_1 = 0.13 \ , \quad R_2 = 0.30 \ , \quad R_3 = 0.40\ldots \ .$$

The value of the non-redundant fraction of the elements of a message also increases. We replace the decrease of the amount of information per letter by a proportional decrease of the number of letters N in the message:

$$N_n/N = I_n/I_0 = 1 - R_n \ . \tag{10.29}$$

The relative value of information is expressed by the formula

$$V_n = I_0/I_n = (1 - R_n)^{-1} \ . \tag{10.30}$$

For the Russian language we have

$$V_0 = 1 \ , \quad V_1 = 1.15 \ , \quad V_2 = 1.42 \ , \quad V_3 = 1.66\ldots \ .$$

We will identify the value of information with the indispensability in application to biology. How does the indispensability, non-redundancy change during biological development and evolution?

In ontogeny there is no choice; it is predetermined (if we distract from somatic mutations that arise after the zygote is formed). Thus, no new information is generated in ontogeny but the information programmed into the genome is gradually revealed, and recoded. The essential feature of ontogeny is that the recoding of information, while not changing the quality of information, implies at the same time an increase in indispensability at the levels of cells and tissues, which become more and more specialized. This is evidenced, in particular, by experiments carried out by Spiegelmann.

During ontogeny the totipotency of cells is replaced by unipotency, the presumptive embryo is converted to a determined one. The irreplaceability, i.e. the value of information, is increased.

The entire path of evolution from prokaryotes to eukaryotes, from unicellular organisms to multicellular ones and from reptiles to birds to mammals, etc., implies an increase in the irreplaceability of the organismal systems at the levels of cells, tissues and organs. In this sense, evolution is progressive. Primitive organisms, such as the sponge, while being dispersed into separate cells, recover upon fusion of these cells. More complex organisms cannot do this.

Saunders and Ho [10.24] suggest the use of the increase of complexity as a characteristic of progressive evolution. The fact that a mammal is more complex than blue-green algae seems to be evident. It is necessary, however, to introduce a more rigorous definition of the complexity concept if we wish to carry out at least a semiquantitative investigation.

The definition of complexity that may be used with respect to biology is given in the works of Kolmogorov [10.25] and Chaitin [10.26]. Earlier John von Neumann defined the complexity of a system by using the number of its constituent parts [10.27]. He pointed out that with a high degree of complexity of an object its description may be more complex than the object itself.

According to Kolmogorov, the complexity of a certain object may be treated as a message determined by a minimal, most economical program coding this message, which is expressed in bits. The complexity thus defined is equivalent to randomness. A random sequence of zeros and units has a degree of complexity coinciding with the length of this sequence expressed in bits. For example, the message

$$0100010110010111$$

has no abbreviated program, its length being 16 bits. On the other hand, the sequence 0101010101010101 has an abbreviated program $(01)^8$.

It should be stressed that the complexity defined in this way is relative; it depends on the level of reception. For the physiologist the amount of information contained in the bovine brain is hundreds and thousands of bits, and for the butcher it does not exceed 5 bits since the brain is one of the approximately 30 parts of the bovine body consumed for food, and $\log_2 30 = 4.91$ bits.

Evidently, the complexity increases in the hierarchy of taxa in the sequence from the kingdom to an individual. Each successive taxon is characterized by an ever increasing specialization and, accordingly, by an ever increasing length of the program describing the system. At the same time, the irreplaceability increases in the sequence from the kingdom to an individual.

At the kingdom level any animals are interchangeable, e.g. cockroach and lion and man; at the phylum level, a man and cobra; at the class level, man and marsupial; at the order level, man and baboon; at the superfamily level, man and chimpanzee; at the genus level, modern man *Homo sapiens* and his ancestor *Homo habilis*. At the species level all human beings are equivalent, be they Charles Darwin or Lysenko. Finally, at the individual level, all human beings are different from one another; they are not interchangeable.

Does the complexity always increase during evolution? Numerous counter-examples may be cited. When an organism evolves from free to parasitic existence, the complexity increases. All helminths are devoid of organs of vision. The crustacean *Sacculina carcini*, a parasite that lives in crabs, has a system of branching appendages which cover all the internal organs of the host. The appendages enter a sac which contains the sexual organs. Other organs – the intestine, organs of sense and excretory organs – are missing. These organs are present in the larvae of *Sacculina carcini*. The female of the echiurid worm *Bonellia viridis* has macroscopic sizes; it is a complex multicellular organism, which has diverse functions. On the other hand, the male is microscopic, lives in the sexual ducts of the female and is capable only of impregnation. Evolutionary simplification is also encountered in vertebrates. In four families of deep-sea fish (the anglerfish *Caulophrinidae*, *Ceratidea*, *Neoceratidae*, *Linophrinidae*) special relationships between the sexes are also encountered. For example, in *Ceratis holboelli* the female is about 1 m in length and the male 15 mm. The male penetrates into the female's skin, after which its jaws, eyes and intestine are reduced so that eventually it is transformed into an appendage which produces semen.

Evidently we should speak not of the increase of complexity but of the evolutionary increase of irreplaceability, the value of information. Complexity and irreplaceability are not identical notions. Complexity refers to structure and irreplaceability implies also function. Complexity characterizes a system as a whole, irreplaceability only a part of it. Thus, the notion of irreplaceability proves more elaborate.

Since evolution occurs by way of divergence, specialization and adaptation to new ecological niches, the irreplaceability must really increase. This is what implies the growth of perfection of a biological system (perfection in the sense of optimal adaptation to particular ecological niches). *Sacculina carcini* or the *Bonellia* male are no less perfect in this respect than freely existing organisms.

An increase of irreplaceability during evolution has also been established at the molecular level. Associated with this are a number of complex problems of molecular evolution discussed in Chaps. 6, 7 and 8.

10.5 The Value of Amino-Acid Residues

We have defined the value of information as the degree of irreplaceability and, hence, as the degree of non-redundancy. This definition is reasonable, though tentative. Let us consider protein "texts" from this viewpoint. Comparing the primary structures of various homologous proteins, we can establish the interchangeability of amino-acid residues in proteins, i.e. the functional similarity of amino-acid residues (FSA). Based on the data reported in Dayhoff's Atlas [10.28], Bachinsky and Ratner [10.29,30] determined the values of N_{ij}, which are the sums of the positions of amino-acid residues in proteins; one protein contains a residue i and the other, which is homologous to the former, contains a residue j. The quantities N_{ij} characterize the replaceability of residue i by residue j. It is obvious that N_{ij} depend on the frequencies of occurrence of these residues N_i and N_j. Ruling out this dependence, we can determine FSA as follows:

$$\text{FSA} = \frac{2N_{ij}}{N_i + N_j} \,. \tag{10.31}$$

Table 10.1 presents the values of N_i, N_{ij} and FSA obtained by Bachinsky and Ratner [10.29,30]. The values of N_{ij} are given in the left lower part and those of FSA in the right upper part of the table.

Let us determine the tentative values of codons expressed in terms of the FSA of the corresponding amino-acid residues [10.31]. Let us consider, as an example, the codon AAA corresponding to Lys. The mutations and the corre-

Table 10.1. Value of N_i, the matrix N_{ij} and FSA

	N_i	1	2	3	4	5	6	7	8	9	10	11	12	13	14	15	16	17	18	19	20	21
1. Ala	455		29	27	17	9	13	16	18	17	10	29	12	6	12	8	4	7	3	6	4	1
2. Ser	430	129		31	10	7	16	25	13	18	7	25	13	12	12	4	7	5	4	5	4	3
3. Thr	345	109	122		16	9	19	17	17	14	12	15	15	13	9	4	6	6	4	7	6	0
4. Val	338	69	40	55		23	10	7	9	5	35	13	12	7	9	8	10	5	6	11	2	0
5. Leu	332	37	25	29	78		6	6	6	2	24	5	10	5	4	16	7	5	1	15	4	1
6. Lys	308	50	58	62	32	19		17	17	14	5	11	18	25	9	1	5	7	0	5	0	4
7. Asn	287	60	32	54	21	20	50		15	27	3	20	15	11	6	4	7	14	1	3	2	1
8. Glu	284	66	48	52	27	20	51	41		31	5	14	25	8	7	1	5	4	2	3	0	1
9. Asp	262	61	63	44	14	7	39	75	86		3	18	12	6	8	0	5	6	1	2	0	0
10. Ile	236	35	24	36	99	67	13	8	14	7		5	8	7	4	10	11	4	1	10	4	1
11. Gly	232	99	84	43	37	15	31	52	35	44	11		12	8	11	6	5	4	2	2	5	2
12. Gln	230	41	43	44	35	27	49	39	66	29	19	27		15	10	2	4	12	2	4	3	0
13. Arg	227	21	40	36	19	14	66	27	20	15	16	18	35		5	5	8	7	1	6	3	2
14. Pro	188	39	38	25	23	11	23	15	17	17	8	23	21	10		5	3	3	1	1	1	3
15. Phe	159	26	13	10	20	40	3	10	3	1	19	12	3	9	5		28	9	1	5	13	2
16. Tyr	141	12	20	15	23	17	12	14	10	10	21	10	7	14	5	42		9	1	5	6	0
17. His	139	20	15	15	11	11	16	29	8	13	7	8	23	13	5	43	13		2	2	6	0
18. Cys	104	9	12	9	14	3	1	2	4	1	2	6	3	5	1	1	1	2		2	0	0
19. Met	99	17	13	16	25	32	11	5	6	4	16	3	7	9	1	7	6	2	2		4	0
20. Trp	48	9	9	4	3	8	0	3	0	0	5	3	4	4	1	13	6	6	0	3		0
21. Term	33	2	7	3	0	1	6	1	3	2	1	2	0	3	3	2	0	0	0	0	0	

sponding values of FSA are as follows:

CAA	Lys-Gln	18	ACA	Lys-Thr	19	AAC	Lys-Asn	17
GAA	Lys-Glu	17	AGA	Lys-Arg	25	AAG	Lys-Lys	100
UAA	Lys-Term	4	AUA	Lys-Ile	5	AAU	Lys-Asn	17

The replacement by an identical residue is characterized by an arbitrarily large value of FSA (100). We sum up these 9 numbers and divide the sum by 300. We obtain a tentative value which is the larger, the greater is the replaceability of the residue. In the case of AAA, we see that $q = 0.74$. The value of information, i.e. the degree of irreplaceability of the residue, is the greater, the lower is that value of q. Let us define the value of the codon as

$$v = 1/(q + 0.50) \qquad (10.32)$$

so that the values of v are of the order of unity. For the codon AAA we have $v = 0.81$. Table 10.2 gives the values v of all codons xyz.

The differences in values of v for jointly degenerate codons are determined by the differences in the results of their mutations.

Averaging v over all jointly degenerate codons, we find the tentative values of amino-acid residues in proteins which are characterized by their degrees of irreplaceability. Roughly speaking, the value of a residue is the higher, the more seldom it occurs in proteins; the most valuable residues are Trp and Met, and the least valuable are Val and Ala. In turn, the occurrence of residues N_i is in

Table 10.2. Values of codons in relative units

x		A		G		C		U		z
	y									
A		Lys	0.81	Arg	0.70	Thr	0.55	Ile	0.64	A
		Lys	0.81	Arg	0.70	Thr	0.57	Met	1.25	G
		Asn	0.79	Ser	0.80	Thr	0.55	Ile	0.65	C
		Asn	0.79	Ser	0.80	Thr	0.55	Ile	0.65	U
G		Glu	0.86	Gly	0.57	Ala	0.52	Val	0.53	A
		Glu	0.86	Gly	0.57	Ala	0.52	Val	0.55	G
		Asp	0.94	Gly	0.55	Ala	0.52	Val	0.54	C
		Asp	0.94	Gly	0.55	Ala	0.52	Val	0.54	U
C		Gln	0.76	Arg	0.51	Pro	0.60	Leu	0.49	A
		Gln	0.76	Arg	0.51	Pro	0.60	Leu	0.50	G
		His	0.77	Arg	0.61	Pro	0.61	Leu	0.57	C
		His	0.77	Arg	0.61	Pro	0.61	Leu	0.57	U
U		Term	—	Term	—	Ser	0.56	Leu	0.68	A
		Term	—	Trp	1.82	Ser	0.54	Leu	0.69	G
		Tyr	0.98	Cys	1.12	Ser	0.56	Phe	0.86	C
		Tyr	0.98	Cys	1.12	Ser	0.56	Phe	0.86	U

Table 10.3. Values of amino-acid residues,
their frequencies and degeneracy of codons

	v	N_i (%)	n
1. Trp	1.83	0.99	1
2. Met	1.25	2.04	1
3. Cys	1.12	2.14	2
4. Tyr	0.98	2.91	2
5. His	0.94	2.83	2
6. Gln	0.86	3.28	2
7. Phe	0.86	4.75	2
8. Lys	0.81	6.35	2
9. Asn	0.79	5.93	2
10. Asp	0.77	5.40	2
11. Glu	0.76	5.65	2
12. Ile	0.65	4.87	3
13. Ser	0.64	8.90	6
14. Arg	0.61	3.89	6
15. Pro	0.60	4.69	4
16. Leu	0.58	6.87	6
17. Gly	0.56	7.12	4
18. Thr	0.55	4.78	4
19. Val	0.54	6.97	4
20. Ala	0.52	9.40	4

crude agreement with the number of codons n, this being an indirect confirmation of the neutral theory of evolution. These data are shown in Table 10.3.

As has been noted earlier, biological development, both ontogeny and phylogeny, implies an increase of the complexity and specificity of the system. An increase in the amount of information during development is quite evident, since the number of various types of proteins, cells, tissues and organs increases.

It may be presumed that the value of information defined as the degree of irreplaceability of an element of a message also increases during evolution and individual development. The divergence of species, implying a decrease in symmetry and an increase in specificity, increases the degree of irreplaceability. The same situation is realized in embryogenesis, which is illustrated by the replacement of the presumptive parts of the embryo with determined ones.

The irreplaceability of the elements of the system implies in itself the reduced fitness of the system to environmental conditions; the loss of a single irreplaceable element can lead to the death of the whole organism since this element cannot be replaced by another one. Such a situation does not arise if the elements are interchangeable, like, for example, the cells of sponge. However, the increase of the complexity of the system due to an increase in the degree of irreplaceability of its elements creates higher autonomy of the organism, i.e. its lesser dependence on the environment. Presumably, nature finds a certain optimum.

Woese [10.32] treats the specific features of a developing biological system

Table 10.4. Differences in total values of amino-acid residues in
cytochrome c

Mammals	Difference*	Birds and turtle	Difference**
Man	0.00	Penguin	0.00
Rhesus	−0.10	Hen	−0.05
Donkey	−0.34	Emu	−0.30
Horse	−0.43	Duck	−0.30
Pig	−0.58	Pigeon	−0.58
Rabbit	−0.66	Turtle	−0.80
Whale	−0.88		
Kangaroo	−0.88		
Dog	−1.06		
Elephant	−1.22		
Bat	−1.25		

* Compared to humans
** Compared to penguins

as the properties of a machine with a high degree of complexity. A machine is
built so that it can perform functions that are not specified *a priori*, at least in
details. An example is the functioning of the immune system.

One might think that the value of information, i.e. the irreplaceability of
the elements of a message, must increase during evolution at any level, beginning with the molecular level. Reichert and co-workers [10.33] have shown that
the primary structure of cytochrome c is "refined" in phylogeny. The regularity
of the system is enhanced, this regularity being disclosed by the Fourier analysis
technique, the "noise" being reduced. The value of the message is increased.

Let us employ the values of v for amino-acid residues given in Table 10.3
to calculate the overall values of homologous proteins. These values are to
a certain extent consistent with the course of evolution for cytochrome c (Table
10.4).

Among mammals the highest overall value is exhibited by the human cytochrome c; among birds, the penguin cytochrome c, this being consistent with the
The Island of Penguins by Anatole France. This is in no way associated with the
higher development of the human brain in comparison with the brain of the bat.
This means that *Homo sapiens* has travelled a longer path of evolution. Naturally, over many millions of years the irreplaceable residues have accumulated
in an ever increasing quantity.

There is nothing like this in the case of haemoglobin: there is a sufficiently
random scatter of the overall values of v. This may be accounted for by the
relative youth of haemoglobin as compared to cytochrome c. Therefore, the
fraction of neutral mutations in haemoglobin is greater than in cytochrome c;
haemoglobin has been "edited" to a lesser extent. "Editing" implies the progress
along the path from neutralism to selectionism, figuratively "from Kimura to
Darwin". We may say this since neutral theory does not contradict Darwinism.

Yano and Hasegawa [10.34] have shown that the Shannon informational entropy of the protein sequence increases during evolution; the content of rarely occurring residues increases and the content of those occurring more frequently falls off. The rarely occurring residues possess higher values and those which occur more frequently have lower values (Table 10.3). Both an increase in the content of rarely encountered residues and a decrease in the content of frequently encountered residues imply an increase in the value of the protein chain treated as a message. At the same time, an increase in informational entropy implies a decrease in the amount of information. We can once again see that the amount of information and its value are quite different notions.

The work of Conrad and Volkenstein [10.35], which is devoted to the irreplaceability of amino-acid residues and self-facilitation of evolution, is a further development of Volkenstein's work [10.31].

Earlier Conrad [10.36] had formulated the principle of self-facilitation of evolution. According to this principle, the amenability of genes to evolution is variable and subject to natural selection. This is the ability to adjust to evolution. Two propositions are essential here. Firstly, different versions of protein (isoenzymes) may be identical, being at the same time subject in different degrees to slight changes in fitness as a result of mutations. Secondly, if an increase in fitness calls for the simultaneous appearance of two or more genetic events, the rate of evolution is low. However, if a new state can be achieved as a result of a series of intermediate states, evolution is accelerated. Both these propositions mean that evolution may progress by way of self-acceleration and self-facilitation [10.37]. As a matter of fact, the same reasonings determine the directionality of evolution.

It is evident that there must be a relationship between the evolutionary amenability of protein and the replaceability of its amino-acid residues. Out of two isofunctional proteins the one which possesses a higher degree of average replaceability of residues must be more readily subject to change in step-by-step evolution. There is perhaps a certain relationship between irreplaceability and adaptation. A protein with a higher degree of irreplaceability of residues is not necessarily better adapted. But the evolution of specialized functions is accompanied by an increase in both fitness and irreplaceability. It may be presumed that the ability to incorporate valuable residues is a prerequisite for the evolution of a functional protein.

The self-facilitation principle may be formulated in terms of the interdependence of irreplaceability and replaceability. By increasing the number of less valuable amino-acid residues in protein, we can increase the rate with which it can gain, lose or replace more valuable residues. "Cheap" residues are especially important during the periods of adaptive radiation.

Volkenstein [10.31] has shown that the probability of transitions (the replacement of a purine with the other purine and of a pyrimidine with the other pyrimidine, $A \rightleftharpoons G$ and $C \rightleftharpoons U$ in RNA) is higher than that of transversions (purine $\rightleftharpoons$ pyrimidine, A, $G \rightleftharpoons C$, U). If we estimate these probabilities on the basis of the protein structure, the probability of $A \rightleftharpoons G$ will be appreciably higher than that of $C \rightleftharpoons U$; in haemoglobin $p_{AG} = 0.378$ and $p_{CU} = 0.133$. However, if these probabilities are evaluated in a straightforward way, from the substitutions observed in RNA, the situation will change: $p_{AG} = 0.182$ and $p_{CU} = 0.210$. The higher value of p_{AG} in proteins is associated with the fact that the average values of FSA are much greater for $A \rightleftharpoons G$ than for $C \rightleftharpoons U$ and for transversions. The difference in the probabilities as determined for protein and RNA evidently means that the transitions $C \rightleftharpoons U$ and transversions increase the probability of replacements leading to structurally and functionally related residues.

As has already been pointed out, two kinds of selection exist: directive selection and stabilizing selection. In the former case, silent mutations, which do not change the residue, travel in the direction of improved adaptation. In the latter case, these mutations increase the mutational stability of protein. In both cases the selection of silent mutations, which increase the replaceability of codons, undergoes self-acceleration.

Neutral theory (see Chap. 6) deals with just stabilizing selection, which is predominantly realized at the molecular level.

Cytochrome c and haemoglobin contain a considerable reserve of "cheap" information, which makes possible a gradual selection of more valuable information. An increase in information value in the active sites of protein is accompanied by a decrease in the remaining parts. This process has gone much further in cytochrome c than in haemoglobin; about one-third of the residues in cytochrome c are hardly replaceable.

Three relevant propositions may be formulated in terms of information theory [10.35]. Firstly, the number of mutations and selection acts required to increase the total value of information in protein is reduced if the value of information per residue decreases. The total value of information here is the sum of the irreplaceabilities of the residues.

Secondly, the number of mutations and selection acts required to increase the diversity of residues in protein falls off with increasing diversity per amino-acid residue. The diversity may be represented by the Shannon entropy:

$$D(G_i) = \sum_{i=1}^{N} f(a_i) \log f(a_i) , \qquad (10.33)$$

where G_i is the allele i and $f(a_i)$ is the frequency of the residue a_i.

Thirdly, the number of mutations and selection acts required to increase the diversity and the total value of information in protein increases with increasing uncertainty of the primary structure. The uncertainty is expressed by the formula

$$H(G_i) = \sum_{i=1}^{N} p(G_i) \log p(G_i) , \qquad (10.34)$$

where $p(G_i)$ is the *a posteriori* probability of the allele G_i in a large population.

In order to maintain the rate of evolution, highly replaceable residues, which lower the value of information per residue, must be incorporated to compensate for the indispensable units. The incorporation of such highly replaceable residues means that a larger number of variants of protein will have similar fitness. Therefore, an increase in the uncertainty of the primary structure is consistent with neutral theory.

The general conclusion is that addition of elements with low information value may reduce the cost of acquisition of new evolutionary information.

It is quite natural to wonder about amount of information in protein and the fraction of valuable information. This question is not simple since the amount of information is far from being reduced to the choice of a particular sequence. The primary structure of a protein built up of 150 amino-acid residues contains information equal to

$$I_0 = \log_2 20^{150} = 645 \text{ bits} .$$

However, the spatial structure of protein contains a much larger amount of information determined by all kinds of interactions between non-neighbouring residues.

We have adopted a tentative definition of information value in terms of the degree of its irreplaceability, non-redundancy. Therefore, the measure of information value in protein may be the degree of neutrality of corresponding point substitutions. Let us go back to Table 6.1, which gives the rates of evolutionary substitutions for a number of proteins. It was pointed out that the rate of substitutions, of which the lowest is for histone H4 and the highest for fibrino-peptides, expresses the fraction of the "passive", i.e. replaceable, part of protein. The histone molecule H4 is entirely "active". This is just the information value of protein, of which the largest is for histone H4 and the smallest for fibrino-peptides. If the fraction of valuable information is taken to be proportional to the rate of substitutions, then this fraction will be 100% for histone, 0.11% for fibrinopeptides, and 4.5% for cytochrome c. Of course, these figures are tentative.

Which machine is more complex and contains a larger amount of valuable information – one built by man (say, a car) or a protein? The number of parts of which a car is made is much larger than the number of amino-acid residues in protein. The clearances and tolerances in a good car are reduced to a minimum; the parts are replaceable only by identical ones. On the other hand, most of the residues in the protein molecule are replaceable. Accordingly, the car is more complex than myoglobin and carries a larger amount of valuable information.

10.6 Game Models of Evolution

The concepts of information and goal are the key concepts in the theory of games. Any game of chance consists of the generation of information as a result of the fixation of a random choice, and of the use of that information in order to achieve a goal, i.e. in order to win. A random choice is made by casting a die or by extracting a card from the pack. In strategic games such as chess, which are of a creative nature, each move creates or brings out information. Here the situation is more complicated than in games of chance since, in principle (and only in principle!) a chess game is quite deterministic; a higher mind could have found the perfect move in any position. However, the number of potential moves in the game in progress is so large that neither man nor computer can detect this determination in a finite period of time. This makes possible a free (i.e. random) choice, and new information is generated. This calls for intuition rather than logic.

The game consists of the evaluation of information and, accordingly, of finding optimal moves leading to a win. Information value is defined here as an increase in the probability of achieving this goal, in accordance with formula (10.17).

In conventional card games (bridge, preference, vint) the randomness of a choice is created by reshuffling the pack, but the information generated is used creatively, strategically, on the basis of an evaluation much similar to that made in a chess game.

Of course, game models do not reproduce evolution since there are no players, but they enable one to understand better certain aspects of evolutionary development. A comprehensive analysis of a number of game models has been made by M. Eigen and R. Winkler [10.37]. Let us describe some of these models.

1. *The "Random-Walk" Game.* There is a square board divided into $n \times n$ squares. The game is played by two persons. One of the players has n^2 white draughtsmen, the second has n^2 black draughtsmen. Each player places $n^2/2$ of his draughtsmen on his half of the board, leaving $n^2/2$ draughtsmen in reserve. The board is thus initially occupied by $n^2/2$ black and $n^2/2$ white draughtsmen. A coin is cast. If heads turn up, the "white" player replaces a black draughtsman in an arbitrary position by a white one; if tails turn up, the replacement is reversed. The state of the system fluctuates randomly; any occupation of the board has an equal probability – $0:n^2$ and $n^2/2:n^2/2$ and $n^2:0$. We may introduce, however, an additional rule, postulating the interaction between the draughtsmen: any of one player's pieces surrounded by his opponent's draughtsmen are replaced by pieces of his opponent's colour. The game is terminated if one of the species of draughtsmen "dies out". This is a model of phase transition as a result of co-operative interactions. If no additional rule is introduced, no new information is generated; if the interaction is involved in the game, the generation of new information becomes possible.

2. *The "Equilibrium" Game* (this model was proposed by Paul and Tatiana Ehrenfest, 1907). Suppose we have a chessboard on which black and white draughtsmen are placed arbitrarily, with all the squares being occupied. We have two octahedral dice, the faces of one carrying the numbers of horizontal

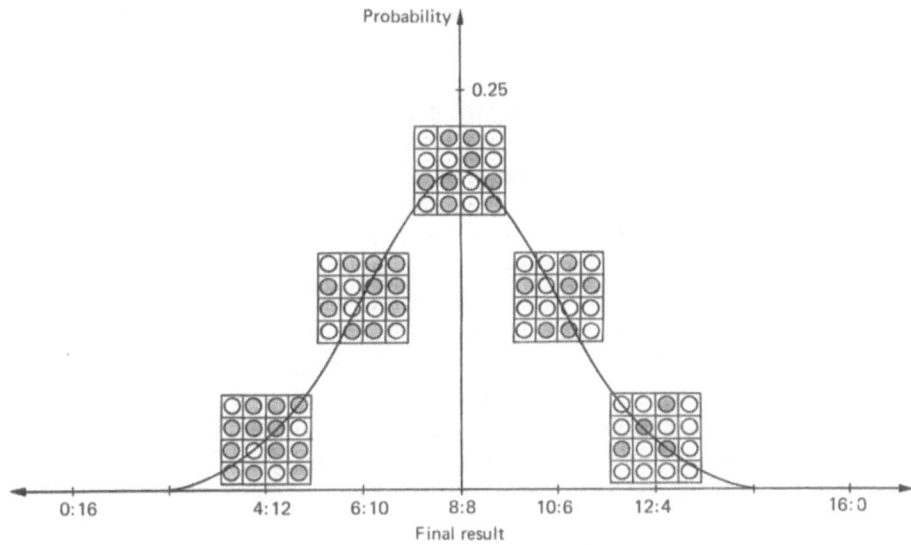

Fig. 10.2. Distribution of probabilities in the "Equilibrium" game [10.37]

rows from 1 to 8, and the faces of the other the letters from a to h indicating the vertical rows. The rules of the game are as follows: having thrown both dice, we replace a draughtsman, whose coordinates have turned up, by one of the other color. Nearly equal numbers of black and white pieces will appear on the board after a sufficient number of throws. The probability distribution is Gaussian with a maximum at $N/2 = 32$ (Fig. 10.2). No information is produced in the system. As soon as a deviation from equilibrium appears, the probability of its reduction rises in direct proportion.

Now we change the rules of the game: a draughtsman whose co-ordinates are given by a throw of the dice is not replaced by a piece of the other colour; we take another draughtsman of the same colour. In this case, the even distribution is unstable; if the initial state was represented by 32 white and 32 black pieces, then after nearly 64 throws draughtsmen of the same colour will be left on the board. Information is generated due to a chance excess of the population. The game simulates the "survival of the survivor" but not Darwinian evolution.

3. *The "Beads" Game* (Eigen). Inspired by the ideas developed by Hermann Hesse in his famous novel *The Glass Bead Game (Das Glasperlenspiel)*, Eigen proposed a game of his own to illustrate the theory he wanted to advance. We have four types of glass beads, say blue, yellow, red and green, placed on the squares of a chessboard. We throw two octahedral dice. With odd throws the "death" of a glass bead occurs, and it is removed from the board; with even throws a bead is "born," and a similar bead is placed on the square that was freed on the previous throw. The system fluctuates randomly and no information is created. We now introduce selective advantages into the game. After an

even throw of two octahedral dice we throw an ordinary cubic die and double the blue bead if any number of points is obtained on this die; we double the red if 1, 2, 3, 4 or 5 (but not 6) points are obtained; we double the yellow if 1, 2, 3 or 4 points are obtained; and we double the green if we obtain 1, 2 or 3 points. This game is a selection; in the long run the board will be entirely occupied by blue beads. We have obtained a peculiar model of Darwinian evolution, which has no mutations, however. For mutations to be taken into account, we introduce a modified variant of the game. We have only two types of beads: blue and yellow. The blue beads are doubled 4 times faster than the yellow ones. We ascribe to blue beads a certain probability of erroneous reproduction, namely the replacement of a blue bead with a yellow one. In contrast, the yellow beads multiply without errors. Calculations for the error probabilities of 0, 25, 50 and 75 percent show that in the last two cases the blue beads die out; with an error of 25 percent their number exceeds, on an average, the number of yellow beads, but the yellow beads do not die out; with precise reproduction the blue beads replace the yellow ones.

4. *The "Life" Game* (Conway). A large board is divided into squares. Each square may be either empty or occupied by a chip. The state depends on the occupation of the neighbouring squares. Each square has 8 neighbouring squares. The initial layout of the chips is specified by the player. The sequence of moves simulates the succession of generations. There are three possibilities: (a) Survival. A chip survives in the next generation if two or three neighboring squares are occupied. (b) Death. A chip is removed from the board if more than three or less than two neighboring squares are occupied; the first case corresponds to overpopulation, and the second to evolution. (c) Birth. An empty square is occupied by a chip if three and only three neighboring squares are occupied.

As a result, death, an unlimited growth or stationary behaviour with stable or oscillating configurations become possible. Some examples are shown in Fig. 10.3.

5. *Chess.* This game is an analogue of evolution in certain respects.

In the initial position the chessplayer can make a choice out of 20 possible moves. In fact, any person who can play chess does not turn over in his mind all these moves; he limits himself to five or six possibilities. As the fame progresses the number of possible moves increases, but the choice becomes narrower and narrower. After each move there is generated new information, a new "ecological situation". The opponent's move plays the role of a mutation.

The similarity between a chess game and evolution is determined by two circumstances. In a chess game the rules of movement are specified and cannot be changed during the game – this is not the croquet from "Alice's Adventures in Wonderland"! In evolution the main types of biologically functional molecules, primarily proteins and nucleic acids, are specified. Evolution is irreversible and so is a chess game – the moves cannot be reversed. Evolution is directional, canalized; once it takes a certain path, the further events are specified.

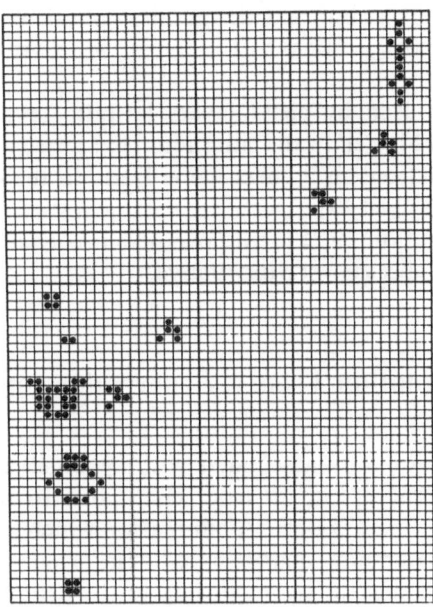

Fig. 10.3. The game of "Life". The Gun configuration of the oscillator is the complex on the left-hand side. After 30 generations it reacquires its original form. In the meantime it emits, on average, four Gliders, stable configurations which propagate across the net. When a Glider encounters an Eater (*top right*) it is absorbed. The Eater is a 15th generation oscillator which is unaltered in the process of consuming a Glider

The same is valid in chess; the beginning chosen specifies the general direction of the further progress of the game.

Chess serves as a source of another interesting analogy. According to the theory developed by W. Steinitz, the game should be played positionally, with accumulation of small advantages. When they are sufficient, the chessplayer must look for a combinational, decisive way leading to a win. The non-triviality of this theory, which has been analyzed in detail by E. Lasker is as follows: if the positional advantages are not used at the proper moment, they disappear.

Evidently, positional play may be likened to macroevolution, which is not accompanied by a radical change in the ecological situation (speciation). A combination may be compared to a punctuated change, with rapid creation of new information. In chess, as well as in evolution, either a combination of gradualism and punctualism or (according to Severtsov) a combination of idio-adaptations and aromorphoses is realized.

The great chess thinker Emmanuel Lasker, who held the title of world chess champion for 27 years, treated a chess game as a model of the "life struggle", "struggle for existence". In his time the synthetic theory of evolution was only in its infancy, and information theory did not exist at all. I think the analogy discussed would have been of interest to Lasker.

Let us mention in passing that the greatest physicist of the 20th century, Einstein, devoted a special article to Lasker, in which he compared him with Spinoza [10.39].

The terminology used by chessplayers is quite revealing: M. Botvinnik writes, "Eiwe adapted himself to the situations that arose in a game ... Analysis of adaptational evaluations in a chess game leads to the creation of a perfect chess-playing machine" [10.40].

Of course, in evolution there are no partners involved in a game. Nonetheless, the analogy with chess is very instructive.

10.7 Information Theory, Synergetics, and Biology

Information theory is directly associated with synergetics. The generation of information and of structure arises from chaos due to the instability of the preceding state. Thus is history born. A bifurcation which generates information implies a decrease in symmetry; chaos devoid of information is more symmetrical than self-organized new states. According to the definition given by Haken [10.41], synergetics is the theory of appearance of new qualities at a macroscopic level. Such appearance occurs as a result of self-organization due to co-operative interactions of the system's elements. But this is what implies the creation of information.

Biology is concerned with the study of historical processes, i.e. the birth and death of new information. Information is the determining notion for the very existence of life. Without the origin, transmission and reception of information, one cannot speak of life.

However, as we have seen, the cannonical information theory (i.e. the theory of communications), which does not take account of semantics, provides almost nothing of value to biology. Therefore, the reception and remembrance (in particular, genetic) of information are of decisive importance to life. It is exactly these problems that need to be solved in the further development of information theory and, hence, of synergetics. Memorizing is a temporal non-equilibrium process characterized by a large relaxation time. In this sense, the vitrification of a liquid may serve as the memorizing model and, hence, as the reception model, with the reservation that it is microscopic information, i.e., entropy, that is fixed in glass. Macroscopic information at another level, of course, is contained in any article made of glass. Art creates information, science extracts it from the surrounding world.

Self-organization in living nature may be treated as the attainment of certain objectives through a message, i.e. attractors. As has been already pointed out, the drive towards a goal and its achievement implies the transition from an unstable to a stable state. Processes of this kind are schematically shown in Fig. 10.4 [10.41]. Various situations are possible here: the attainment of a single attractor through a single message; the attainment of a single attractor through two messages; the attainment of two attractors through a single message. In the last case divergence occurs, a branching bifurcation. All these processes imply the self-organization of the system and are made possible due to fluctuations inside the system, i.e. due to changes in the initial conditions. In such models semantics is created, information value, as the "relative importance" of individual attractors [10.41].

The problem of information value, which is inextricably tied up with reception, is far from being solved. As we have seen, interesting results can be obtained with the aid of the tentative definition of information value as the indispensability, non-redundancy of information. However, the transition from static information theory, in which time does not figure, to dynamic information theory, which includes reception and memorizing and, hence, time and semantics, has not yet been realized in physics.

Biological information begins with instabilities upon creation of a collective state. It becomes meaningful only with respect to the surroundings, since life is the property of a dissipative system which interacts with the environment. In biological systems high levels of co-ordination between their individual parts are realized. In this sense, the behaviour of a biological system is coherent. This implies compression, the compaction of the amount (but not the value!) of information, which is similar to compression that occurs upon transition from incoherent to coherent laser radiation (Haken [10.41,42]). An integrated description, which is possible for a highly co-ordinated system, requires a much smaller number of bits than the description of the system's elements. It is clear that the multicellular slime mold *Dictyostelium discoideum*, which emerged as a result of the self-organization of a population of amoebae, can be described by means of a much smaller amount of information than is required for the population. In the long run, the answer to the question of whether an organism is alive or dead contains only one bit of information. Haken has constructed a mathematical description of a self-organizing system in terms of information theory [10.42].

It should be stressed that, while speaking of compression, we are dealing with two aspects. The first aspect is the objective characterization of a self-organized system; instead of a very large amount of information about all individual subsystems, we are dealing with information about integrity, about order parameters, which are of macroscopic nature. The second aspect is associated with reception of available information, i.e. with semantics. We have already given the example of the bovine brain, which contains thousands of bits of information for the biologist and only five for the butcher. We are dealing here with the different levels of reception and, hence, with different values.

One bit obtained in the solution of the life-death dilemma expresses static information. The situation is quite unambiguous. One cannot be slightly dead, just as it is impossible to be slightly pregnant. But while considering the dynamics of life and death, we encounter a much more complicated picture. Suppose we are speaking of a human being dying of a serious disease. The information about the number of days left for him to live consists of many bits.

In concluding this chapter and the book, let us formulate once again the basic positions that underlie the physical approaches to biological evolution.

Living organisms at all levels, beginning with a living cell and ending with the biosphere as a whole, are dissipative systems, i.e. open far-from-equilibrium systems. Such systems may be in the state of dynamic deterministic chaos, capable of bifurcations as a result of instabilities. Such bifurcations are speciation, macroevolution and the emergence of various ontogenetic stages. These bifurcations have the character of non-equilibrium phase transitions.

The theory of dissipative systems, synergetics, is a part of physics concerned with historical processes. Like cosmology, biology is a historical science. Synergetics as a part of physics closely tied up with information theory was developed in the second half of our century. The lack of knowledge and misunderstanding of synergetics is responsible for contrasting physics with biology, for the so-called anti-reductionism (Mayr and Polanyi). It is precisely in synergetics that physics is united with biology and it is this union of the two sciences that makes possible the unravelling of the deep-seated meaning of evolutionary processes. Synergetics involves the bringing together of the natural sciences and literature and art, the branches of creative activities producing new information, which is what occurs in biology.

The other physical approaches are based on molecular biology and molecular biophysics. This is a different level of investigation in evolutionary biology, which is practically (and only practically!) complementary to the level of synergetics. In this area the most important achievements are the neutral theory of molecular evolution developed by Kimura (Chap. 6) and the theory of evolution of macromolecules worked out by Eigen (Chap. 8). The relationship between these theories is considered in Chap. 9. In the Eigen theory, the above-indicated practical complementarity is overcome – the theory is directly associated with synergetics and information theory; speciation in the world of macromolecules has the character of a second-order phase transition.

The molecular theory of organismic evolution has not yet been united with synergetic approaches and its development is beset with formidable difficulties. Of great importance here is the concept of molecular drive (Chap. 7), but no physico-mathematical description of evolutionary processes (bifurcational speciation and macroevolution) has yet been derived from it.

The key problem of evolutionary theory is the relationship between genotype and phenotype studied at different levels, beginning with the structure of proteins and nucleic acids and ending with palaeontology. As we have seen, this problem is missing in systems that are studied in the Eigen theory. The solution of this problem requires considerable effort and a large amount of time. The physical approaches realized in the study of ontogeny and phylogeny are of prime importance here.

The theory of evolution is itself a living system, which is vigorously developing today in conjunction with molecular biology, synergetics, and information theory, i.e. with physics. And the theory of evolution developed by Darwin has not only retained its validity, it has also acquired a new, deeper meaning.

References

Chapter 1

1.1   Mayr, E. (1968): *Animal Species and Evolution* (Harward University Press, New York)
1.2   Poljansky, G. (1982): *Proc. of VI Internat. Congress on Protozoology. Special Congress Volume of Acta Protozoologica.* Part I, pp. 23–30
1.3   Kretschmer, E. (1977): *Körper-bau und Charakter* 26 Aufl (Springer, Heidelberg)
1.4   Pavlov, I. (1973): *Twenty Years Experience of Objective Studies of the Highest Nervous Activity (Behavior) of Animals* (Nauka, Moscow) (in Russian)
1.5   Prigogine, I. (1980): *From Being to Becoming* (W. Freeman and Co., San Francisco)
1.6   Shklovsky, I. (1976): *Universe, Life, Intellect* (Nauka, Moscow) (in Russian)
1.7   Ebeling, W., Feistel R. (1982): *Physik der Selbstorganisation und Evolution* (Akademie-Verlag, Berlin)
1.8   Schmalhausen, I. (1969): *Problems of Darwinism* (Nauka, Moscow) (in Russian)
1.9   Lamarck, J. (1809): *Philosophie Zoologique* (Paris) (English translation: Hugh Elliot, *The Zoological Philosophy*, 1914, Macmillan, London)
1.10 Karyakin, Yu. (1989): Dostoevsky and the Eve of the 21st Century (Sovetsky Pisatel', Moscow) (in Russian)
1.11 Lamarck, J.-B. (1820): *Systéme analytique des connaissances positives de l'homme* (Paris)
1.12 Bljakher, L. (1971): *The Problem of Inheritance of Acquired Characters* (Nauka, Moscow) (in Russian)
1.13 Grant, V. (1977): *Organismic Evolution* (W. Freeman, San Francisco)
1.14 Popper, K. (1959): *The Logic of Scientific Discovery* (London)
1.15 Popper, K. (1980): *New Scientist*, 611
1.16 Monod, J. (1975): In *Problems of Scientific Revolution. Progress and Obstacles to Progress in Science*, ed. R. Harré (Oxford)
1.17 Berg, L. (1969): *Nomogenesis or Evolution Determined by Law* (MIT Press, Cambridge, Mass.)
1.18 Lyubishchev, A. (1982): *Problems of Form, Systematics and Evolution of Organisms* (Nauka, Moscow) (in Russian)
1.19 Jacob, F., Wollman, E. (1961): *Sexuality and Genetics of Bacteria* (Academic Press, New York)
1.20 Schmalhausen, I. (1949): *Factors of Evolution. The Theory of Stabilizing Selection* (Blakiston, Philadelphia)
1.21 Schmalhausen, I. (1983): *Ways and Laws of the Evolutionary Process* (Nauka, Moscow) (in Russian)
1.22 Darwin, C. (1872): *The Origin of Species.* 6th edn. London: John Murray (Reprinted 1958, The New American Library, New York) Chap. 14
1.23 Cavalli-Sforza, L., Piazza, A., Menozzi, P., Mountain, J. (1988): *Proc. Nat. Acad. Sci. USA* **85**, 6002–6
1.24 Lewin, R. (1988): *Science* **242**, 514
1.25 Illitch-Switych, V. (1971): *Comparison of Nostratic Languages* (Moscow) (in Russian)
1.26 Harding, R., Sokal, R. (1988): *Proc. Nat. Acad. Sci. USA* **85**, 9370–72
1.26a Barton, N., Jones, J. (1990): *Nature* **346**, 415–6
1.27 Volkenstein, M. (1983): *General Biophysics* (Academic Press, New York)

1.28 Volkenstein, M. (1994): *Biophysics* (Mir Publishers and The American Institute of Physics, Moscow and New York; translated by Artavaz Beknazarov), 2nd ed.
1.29 Schrödinger, E. (1945): *What Is Life?* (Cambridge University Press, New York and London)
1.30 Quastler, H. (1964): *The Emergence of Biological Organization* (Yale University Press, New Haven, Conn.)
1.31 Volkenstein, M. (1989): *Voprosy Filosofii* **8**, 20–33.
1.32 Wigner, E. (1987): *Symmetries and Reflexions* (Indiana University Press, Bloomington, In.)
1.33 Elsasser, W. (1958): *The Physical Foundation of Biology* (Pergamon Press, Oxford)
1.34 Polanyi, M. (1968): *Science* **160**, 1308–13
1.35 Küppers, B.-O. ed. (1990): *Leben = Physik + Chemie?* (Piper, München and Zürich)
1.36 Mayr, E. (1982): *The Growth of Biological Thought. Diversity, Evolution and Inheritance* (Harvard University Press, Cambridge, Mass.)
1.37 Simpson, G. (1964): *This View of Life* (Harcourt, Brace and World, New York)
1.38 Skvortsov, A. (1988): *Priroda* **1**, 17–9
1.39 Bohr, N. (1955): *The Unity of Knowledge* (New York, pp. 17–62)
1.40 Volkenstein, M. (1988): *Uspekhi Fizicheskikh Nauk* **154**, 279–97
1.41 Bohr, N. (1933): *Nature* **131**, 421–3, 457–9
1.42 Bohr, N. (1938): *Congressi di fisica, radiologia e biologia sperimentale* (Bologna, pp. 6–15)
1.43 Volkenstein, M. (1987): *Molek. Biol.* **21**, 630–41
1.44 Bohr, N. (1937): *Atomphysik og Menneskelig Erkendelse* (København)
1.45 Heitler, W. (1951): In *Albert Einstein: Philosopher-Scientist*, ed. Schlipp-Tudor (New York, pp. 196, 197)
1.46 Goethe, W., *Faust*, part I, lines 1936–41
1.47 Bohr, N. (1960): In *Symposia of the Society for Experimental Biology* 14 (Cambridge, pp. 1–5)
1.48 Bohr, N. (1960): *The Connexion Between the Sciences. Address at the International Congress of Pharmaceutical Sciences* (Copenhagen)
1.49 Volkenstein, M. (1972): *Crossroads of Science* (Nauka, Moscow) (in Russian)
1.50 Bohr, N. (1963): *Naturwissenschaften* **50**, 21–5
1.51 Ginsburg, V. (1967): In *Niels Bohr: Life and Works*, ed. B. Kuznetsov (Nauka, Moscow) (in Russian)
1.52 Volkenstein, M. (1990): *Entropie und Information* (Harri Deutsch. Thun., Frankfurt/Main)
1.53 Bauer, E. (1935): *Theoretical Biology* (Leningrad) (in Russian)
1.54 Küppers, B.-O. (1990): *Information and the Origin of Life* (The MIT Press, Cambridge, Mass.)
1.55 Monod, J. (1970): *Le Hasard et la Necessité* (Seuil, Paris)
1.56 Weil, H. (1954): In *Columbia University in the City of New York. Bicentennial Celebration* (New York)
1.57 Futuyma, D.J. (1988): *Evolution* **42**, 217–26
1.58 *Philosophical Encyclopedic Dictionary* (1989) (Moscow) (in Russian)

Chapter 2

2.1 Darwin, C. (1933): *Diary of the Voyage of H.M.S. "Beagle"* (Cambridge University Press, Cambridge)
2.2 *Biological Encyclopedic Dictionary* (1986) (Moscow) (in Russian)
2.3 Mayr, E. (1978): In *Evolution. A Scientific American Book* (W. Freeman, San Francisco)
2.4 Simpson, G. (1980): *Splendid Isolation* (Yale University Press, New Haven)
2.5 Schmalhausen, I. (1969): *The Problems of Darwinism* (Nauka, Leningrad) (in Russian)
2.6 Mayr, E. (1982): *The Growth of Biological Thought* (Harvard University Press, Cambridge, Mass.)
2.7 Mayr, E. (1970): *Populations, Species and Evolution* (Harvard University Press, Cambridge, Mass.)

2.8   Simpson, G. (1961): *Principles of Animal Taxonomy* (Columbia University Press, New York)
2.9   Dobzhansky, Th. (1937): *Genetics and the Origin of Species* (Columbia University Press, New York)
2.10   Génermont, J. (1979): *Les Méchanismes de l'Evolution* (Dunod Université, Paris)
2.11   Poljansky, G. (1982): *Acta Protozoologica. Special Volume, Proceedings of VI International Congress of Protozoology*, Part I, pp. 23–30
2.12   Takhtajan, A. (1984): Preface to the Russian edition of the book by V. Grant, *Organismic Evolution* (Mir Publishers, Moscow) (in Russian)
2.13   Vrba, E. (1980): *South African J. Sci.* **76**, 61–84
2.14   Lewin, R. (1980): *Science* **210**, 883–7
2.15   Grant, V. (1981): *Plant Speciation* (Columbia University Press, New York)
2.16   Darwin, C. (1958): *The Origin of Species* (The New American Library, New York)
2.17   Volkenstein, M. (1994): *Biophysics* (Mir Publishers and The American Institute of Physics, Moscow and New York; translated by Artavaz Beknazarov), 2nd ed.
2.18   Solbrig, O., Solbrig, T. (1979): *Introduction to Population Biology and Evolution* (Addison-Wesley Publishing Co., Reading, Mass.)
2.19   Darwin, C. (1985): *Variations of Animals and Plants under Domestication* (John Murray, London)
2.20   Belyaev, D. (1989): In *Vavilov's Legacy in Contemporary* Biology (Nauka, Moscow) (in Russian)
2.21   Darwin, Ch. (1958): *The Origin of Species*. Chapter 6 (The New American Library, New York)
2.22   Gould, S. (1982): *The Panda's Thumb*. Chapter 4 (W. W. Norton, New York)
2.23   Gould, S. (1980): *Paleobiology* **6**(1), 119–30
2.24   Gould, S., Eldredge, N. (1977): *Paleobiology* **3**, 115–51
2.25   Goldschmidt, R.B. (1940): *The Material Basis of Evolution* (Yale University Press, New Haven, Conn.)
2.26   Eldredge, N., Gould, S.J. (1972): "Punctuated equilibria: an alternative to phyletic gradualism." In *Models in Paleobiology*, ed. T.J.M. Schopf, pp. 82–115 (W. Freeman, Cooper and Co., San Francisco)
2.27   Darwin, Ch. (1958): *The Origin of Species*. Chapter 10 (The New American Library, New York)
2.28   Stanley, S.M. (1979): *Macroevolution: Pattern and Process* (W. Freeman and Co., San Francisco)
2.29   Williamson, P.G. (1981): *Nature* **293**, 437–43; **294**, 214–5
2.30   Rhodes, F. (1983): *Nature* **305**, 269–72
2.31   Cronin, J.E., Boaz, N.T., Stringer, C.B., Rak, Y. (1981): "Tempo and mode in hominid evolution." Nature **292**, 113–22
2.32   Avise, J. (1977): *Proc. Nat. Acad. Sci. USA* **74**, 5083–7
2.33   Sheldon, P. (1987): *Nature* **330**, 561–3
2.34   Maynard Smith, J. (1987): *Nature* **330**, 516–9
2.35   Eldredge, N., Gould, S.J. (1988): *Nature* **332**, 211–2
2.36   Dawkins, R. (1988): *The Blind Watchmaker* (Penguin Books, London)
2.37   Severtsov, A. (1967): *The Main Directions of the Evolutionary Process* (University Press, Moscow) (in Russian)
2.38   Simpson, G.G. (1944): *Tempo and Mode in Evolution* (Columbia University Press, New York)
2.39   Mayr, E. (1954): "Change of genetic environment and evolution." In *Evolution as a Process*, eds. J. Huxley, A. Hardy, and E. Ford (Allen and Unwin, London, pp. 157–80)
2.40   Schindewolf, O. (1936): *Paläontologie, Entwicklungslehre, und Genetik* (Borntraeger, Berlin)
2.41   Darwin, C. (1958): *The Origin of Species*. Chapter XIII
2.42   Wolken, J. (1975): *Photoprocesses, Photoreceptors and Evolution* (Academic Press, New York)
2.43   Haldane, J. (1966): *The Cause of Evolution* (Cornell University Press, Ithaca, N.Y.)
2.44   Valentine, J. (1981): In *Evolution* (Mir Publishers, Moscow) (in Russian)
2.45   Bolshakov, V. et al (1979): *Doklady Akad. Nauk SSSR* **249**, 1462–6

2.46 Timofeeff-Ressovsky, N., Vorontsov, N., Jablokov, A. (1975): *Kurzer Grundriss der Evolutionstheorie* (G. Fischer Verlag, Jeny)
2.47 Darwin, C. (1958): *The Origin of Species* (The New American Library, New York, p. 170)

Chapter 3

3.1 Calow, P. (1984): *Evolutionary Principles* (Blackie, Glasgow)
3.2 Simpson, G. (1951): *Horses: the Story of the Horse Family in the Modern World and through Sixty Million Years of Evolution* (Oxford University Press, Oxford)
3.3 Grant, V. (1977): *Organismic Evolution* (W. Freeman and Co., San Francisco)
3.4 Darwin, Ch. (1958): *The Origin of Species*. Chapter 1 (The New American Library, New York)
3.5 Timofeeff-Ressovsky, N., Vorontsov, N., Jablokov, A. (1975): *Kurzer Grundriss der Evolutionstheorie* (G. Fischer Verlag, Jena)
3.6 Maynard Smith, J., Burian, P., Kauffman, S., Alberch, P., Campbell, J., Goodwin, B., Lande, R., Raup, D., Wolpert, L. (1985): *Quart. Rev. Biol.* **60**, 265–87
3.7 Spurway, H. (1949): *La Ricerca Scientifica, Suppl.*, 3–9
3.8 Vavilov, N. (1967): *The Law of Homologeous Series in Hereditary Variability* (Nauka, Leningrad) (in Russian)
3.9 Waddington, C.H. (1957): *The Strategy of the Genes* (Allen and Unwin, London; Macmillan Co., New York)
3.10 Raff, R., Kaufman, T. (1983): *Embryos, Genes and Evolution* (Macmillan Publishing Co. and Collier Macmillan Publishers, New York and London)
3.11 Dawid, I. et al. (1982): "Genomic Change and Morphogenetic Evolution." In *Evolution and Development*, ed. J. Bonner, pp. 19–39 (Springer-Verlag, Berlin, Heidelberg and New York).
3.12 Gerhart, J. et al. (1982). "The cellular basis of morphogenetic change." In *Evolution and Development*, ed. J.T. Bonner, pp. 86–114 (Springer-Verlag, Berlin, Heidelberg and New York)
3.13 Bonner, J.T., ed. (1982): *Evolution and Development* (Springer-Verlag, Berlin, Heidelberg and New York)
3.14 Wolpert, L. (1969): *J. Theor. Biol.* **25**, 1–47
3.15 Wolpert, L. (1978): *Scientific American* **239**, No. 4, October, 124–37
3.16 Gurvich, A. (1977): *Selected Works* (Meditsina, Moscow) (in Russian)
3.17 Kaufman, T., Wakimoto, B. (1982): "Genes that control high level development switches." In *Evolution and Development*, ed. J.T. Bonner, pp. 189–205 (Springer-Verlag, Berlin, Heidelberg and New York)
3.18 Katz, M. (1982): "Ontogenetic mechanisms: the middle ground of evolution." In *Evolution and Development*, ed. J.T. Bonner, pp. 207–12 (Springer-Verlag, Berlin, Heidelberg and New York)
3.19 Horn, H. et al. (1982): "Adaptive aspects of development." In *Evolution and Development*, ed. J.T. Bonner, pp. 215–35 (Springer-Verlag, Berlin, Heidelberg and New York)
3.20 Volkenstein, M. (1983): *General Biophysics* (Academic Press, New York)
3.21 Volkenstein, M. (1994): *Biophysics*, 2nd edn. (Mir Publishers and The American Institute of Physics, Moscow and New York; translated by Artavaz Beknazarov)
3.22 Barth, L. (1953): *Embryology* (The Dryden Press, New York)
3.23 Wagner, R., Mitchell, H. (1955): *Genetics and Metabolism* (John Willey and Sons, Inc., and Chapman and Hall, Ltd., New York and London)
3.24 Khesin, R. et al. (1963): *Biokhimiya* **27**, 761, 1092, 1962; **28**, 1070
3.25 Jacob, F. (1982). *The Possible and the Actual* (Pantheon, New York)
3.26 Garcia-Bellido, A., P. Lawrence, and G. Morata (1979): *Scientific American* **241**, 1, 203–20
3.27 Struhl, G. (1984): *Nature* **310**, 10–11
3.28 Gould, S. (1980): *Paleobiology* **6**, 119–30
3.29 Gould, S. (1982): *Science* **216**, 380–7

3.30   Von Baer, K.E. In *The Origin of Species*, Chapter 14
3.31   Haeckel, E. (1866): *Generelle Morphologie der Organismen* (Reimer, Berlin)
3.32   Vorontsov, N. (1984): *Priroda* **8**, 75–87
3.33   Gould, S. (1977): *Ontogeny and Phylogeny* (The Belknap Press of Harvard University Press, Cambridge, Mass., London)
3.34   Garstang, W. (1951): *Larval Forms with Other Zoological Verses* (Basil Blackwell, Oxford)
3.35   Berg, L. (1969): *Nomogenesis or Evolution Determined by Law* (MIT Press, Cambridge, Mass.)
3.36   Turing, A. (1952): *Phil. Trans. Roy. Soc. London* **237B**, 32
3.37   Crick, F. (1970): *Nature* **225**, 420–4
3.38   Meinhardt, H. (1982): *Models of Biological Pattern Formation* (Academic Press, New York)
3.39   Gierer, A., Meinhardt, H. (1974): *J. Cell Sci.* **15**, 321
3.40   Belintsev, B. (1983): *Uspekhi Fiz. Nauk* **141**, 55–101
3.41   Belintsev, B. (1991): *Physical Foundations of Biological Morphogenesis* (Nauka, Moscow) (in Russian)
3.42   Webster, G., Wolpert, L. (1966): *J. Embryol. Exp. Morphol.* **16**, 91
3.43   Gierer, A., Meinhardt, H. (1972): *Kybernetik* **12**, 30
3.44   Schaller, H., Bodenmüller, H. (1981): *Naturwissenschaften* **68**, 252–6
3.45   Fulton, A. (1985): *The Cytoskeleton. Cellular Architecture and Choreography* (Chapman and Hill, London and New York)
3.46   Sussman, M. (1973): *Developmental Biology* (Prentice-Hall, Inc., Englewood Cliffs, New Jersey)
3.47   Zavalishina, L., Belousov, L., Ostroumova, T. (1980): *Ontogenesis* **11**, 492–9
3.48   Ashall, F., Puck, T. (1984): *Proc. Nat. Acad. Sci. USA* **81**, 5145–9
3.49   Volkenstein, M. (1962): *Doklady Akad. Nauk SSSR* **146**, 1426–9
3.50   Vorobjev, V., Kukhareva, L. (1965): *Doklady Akad. Nauk SSSR* **165**, 435–9
3.51   Lyubishchev, A. (1982): *Problems of Form, Systematics and Evolution of Organisms* (Nauka, Moscow) (in Russian)
3.52   Darwin, C. (1875): *The Variation of Animals and Plants under Domestication* (John Murray, London)
3.53   Gould, S. (1982): "Change in developmental timing as a mechanism of macroevolution." In *Evolution and Development*, ed. J. Bonner (Springer-Verlag, Berlin, Heidelberg and New York)
3.54   Mayr, E. (1985): *The Growth of Biological Thought* (The Belknap Press of Harvard University Press, Cambridge, Mass., and London)

Chapter 4

4.1   Timofeeff-Ressovsky, N., Vorontsov, N., Yablokov, A. (1975): *Kurzer Grundriss der Evolutionstheorie* (G. Fischer Verlag, Jena)
4.2   Schmalhausen, I. (1983): *Selected Works. Ways and Laws of the Evolutionary Process* (Nauka, Moscow) (in Russian)
4.3   Popper, K. (1959): *The Logic of Scientific Discovery* (Routledge, London)
4.4   Lewontin, R.C. (1979): "Adaptation." In *A Scientific American* Book "Evolution" (W.H. Freeman, San Francisco)
4.5   Grant, V. (1977): *Organismic Evolution* (W.H. Freeman, San Francisco)
4.6   Lack, D. (1947): *Darwin's Finches* (Cambridge University Press, Cambridge)
4.7   Schmalhausen, I. (1949): *Factors of Evolution. The Theory of Stabilizing Selection* (Blakiston, Philadelphia)
4.8   Severtsov, A. (1967): *The Main Directions of the Evolutionary Process* (Moscow University Press, Moscow) (in Russian)
4.9   Luria, S., Delbrück, M. (1954): *Genetics* **39**, 169
4.10   Cairns, J., Overbaugh, J., Miller, S. (1988): *Nature* **335**, 142–5
4.11   Ryan, F. (1952): *Nature* **169**, 882–3
4.12   Shapiro, J. (1984): *Molec. Genet.* **194**, 79–90

4.13   Hall, B. (1988): *Genetics* **120**, 887–97
4.14   *Nature* **337**, 119–20 (1988)
4.15   Symonds, N. (1989): *Nature* **337**, 119–20
4.16   Lenski, R. et al. (1989): Nature **337**, 123–4
4.17   Volkenstein, V. (1931): *An Essay on Contemporary Aesthetics* (Akademia, Moscow) (in Russian)
4.18   Alexandrov, V. (1977): *Cells, Macromolecules, and Temperature* (Springer-Verlag, Berlin)
4.19   Alexandrov, V. (1985): *Reactivity of Cells and Proteins* (Nauka, Leningrad) (in Russian)
4.20   Volkenstein, M. (1994): *Biophysics* (Mir Publishers and the American Institute of Physics, Moscow and New York; translated by Artavaz Beknazarov), 2nd ed.
4.21   Hochachka, P., Somero, G. (1973): *Strategies of Biochemical Adaptation* (W. Saunders and Co., Philadelphia, London and Toronto)
4.22   Hochachka, P., Somero, G. (1984): *Biochemical Adaptation* (Princeton University Press, Princeton, New Jersey)
4.23   Conrad, M. (1983): *Adaptability. The Significance of Variability from Molecules to Ecosystems* (Plenum Press, New York)
4.24   Gould, S., Lewontin, R. (1979): *Proc. Roy. Soc. London* **B205**, 581–98
4.25   Chetverikov, S. (1965): *Bulletin of the Moscow Society of Naturalists. Ser. Biol.* **70**(4), 33–74
4.26   Gould, S. (1980): *Paleobiology* **6**(1), 119–30
4.27   Gould, S., Vrba, E. (1982): *Paleobiology* **8**, 4–15
4.28   Vrba, E. (1983): *Science* **222**, 387–9

Chapter 5

5.1   Volkenstein, M. (1994): *Biophysics*, 2nd edn. (Mir Publishers and The American Institute of Physics, translated by Artavaz Beknazarov, Moscow and New York)
5.2   Volkenstein, M. (1963): *Configurational Statistics of Polymeric Chains* (Interscience Publishers, New York)
5.3   Birstein, T., Ptitsyn, O. (1966): *Conformations of Macromolecules* (Interscience Publishers, New York)
5.4   Flory, P. (1969): *Statistical Mechanics of Chain Molecules* (Interscience Publishers, New York)
5.5   Lifshitz, I., Grosberg, A., Khokhlov, A. (1978): *Rev. Mod. Phys.* **50**, 683–713
5.6   Grosberg, A., Khokhlov, A. (1989): *Statistical Physics of Macromolecules* (Nauka, Moscow) (in Russian)
5.7   Golovanov, I., Sobolev, V., Volkenstein, M. (1991): *Biopolymers* (Nauka, Moscow) (in Russian)
5.8   Levitt, M., Chothia, C. (1976): *Nature* **261**, 552–8
5.9   Chothia, C. (1984): *Ann. Rev. Biochem.* **53**, 537–72
5.10   Ptitsyn, O. (1985): *J. Biosciences* (India) **8**, 1–13
5.11   Chou, K., Maggiore, G., Némethy, G., Scheraga, H. (1988): *Proc. Nat. Acad. Sci. USA* **85**, 4295–9
5.12   Golovanov, I., Tsygankova, I., Nautshitel, V., Volkenstein, M. (1989): *Doklady Akad. Nauk SSSR* **309**, 633–7; *Molek. Biol.* **25**, 133–43 (1991)
5.13   Schrödinger, E. (1945): *What is Life?* (Cambridge University Press, Cambridge)
5.14   Frauenfelder, H., Parak, F., Young, R. (1988): *Ann. Rev. Biophys., Biophys. Chem.* **17**, 451
5.15   Iben, E. et al. (1989): *Phys. Rev. Letters* **62**, 1916–9
5.16   Frauenfelder, H., Steinbach, P., Young, R. (1989): *Chemica Scripta* **29A**
5.17   Frauenfelder, H. (1989): *Nature* **338**, 623–4
5.18   Frauenfelder, H. (1989): *Internat. J. Quantum Chem.* **35**, 711–5
5.19   Goldansky, V., Krupiansky, Y., Flerov, V. (1983): *Doklady Akad. Nauk SSSR* **272**, 978

5.20 Volkenstein, M., Ptitsyn, O. (1955): *Doklady Akad. Nauk SSSR* **103**, 795; *J. Technical Phys.* **26**, 2204 (1956)
5.21 Volkenstein, M., Sharonov, Y. (1961): *Vysokomolek. Soedin.* **3**, 1740–5; (1962) **4**, 917–21
5.22 Polyakov, K. et al. (1987): *Kristallografia* **32**, 918–26
5.23 Volkenstein, M. (1971): *Izv. Akad. Nauk SSSR. Ser. Biol.* 805–14
5.24 Volkenstein, M. (1982): *Vestnik Akad. Nauk SSSR* **10**, 56–63
5.25 Volkenstein, M., Golovanov, I., Sobolev, V. (1982): *Molecular Orbitals in Enzymology* (Nauka, Moscow) (in Russian)
5.26 Volkenstein, M. (1982): *Physics and Biology* (Academic Press, New York)
5.27 Volkenstein, M. (1981): *J. Theor. Biol.* **89**, 45–51
5.28 Volkenstein, M. (1972): *J. Theor. Biol.* **34**, 193–5
5.29 Shaitan, K., Rubin, A. (1985): *Biofizika* **30**, 517–26
5.30 Onuchi, J., Wolynes, P. (1988): *J. Phys. Chem.* **92**, 6495–503
5.31 Krakoviak, M. et al. (1975): *Doklady Akad. Nauk SSSR* **224**, 873–6
5.32 Krakoviak, M. et al. (1975): *Izv. Akad. Nauk SSSR. Ser. Fiz.* **39**, 2354–8
5.33 Krakoviak, M. et al. (1978): *Vysokomolek. Soedin.* **1320**, 129–32
5.34 Volkenstein, M., Sharonov, Y. (1977): *Priroda* **5**, 30–41
5.35 Sharonov, Y., Pismensky, V., Yarmola, E. (1988): *Molek. Biol.* **22**, 1272–83; 1491–506
5.36 Sharonov, Y., Pismensky, V., Yarmola, E. (1988): *FEBS Letters* **235**, 63–6
5.37 Dreinsicke, D., Schulz, G. (1988): *J. Mol. Biol.* **203**, 1021–8
5.38 Sinev, M., Razgulyaev, O., Timchenko, A., Ptitsyn, O. (1989): *Eur. J. Biochem.* **180**, 61–6
5.39 Williams, J., Clegg, H., Hutch, J. (1961): *J. Mol. Biol.* **3**, 532–40
5.40 Finkelstein, A., Ptitsyn, O. (1971): *J. Mol Biol.* **62**, 613–24
5.41 Poroikov, V., Esipova, N., Tumanian, V. (1976): *Biofizika* **21**, 397–400
5.42 Ptitsyn, O. (1984): *Molek. Biol.* **18**, 574–90; (1985) *J. Mol. Structures* **123**, 45–65
5.43 Ptitsyn, O., Finkelstein, A. (1980): *Quart. Rev. Biophys.* **13**, 339–86
5.44 Chan, H.S., Dill, K. (1989): *J. Chem. Phys.* **90**, 492–509
5.45 Chan, H.S., Dill, K. (1990): *Proc. Nat. Acad. Sci. USA* **87**, 6388–92
5.46 Chan, H.S., Dill, K. (1989): *Macromolecules* **22**, 4559–73
5.47 Lau, K.F., Dill, K. (1989): *Macromolecules* **22**, 3986–97; (1990): *Proc. Nat. Acad. Sci. USA* **87**, 638–42
5.48 Stitgter, D., Alonso, D., Dill, K. (1991): *Proc. Nat. Acad. Sci. USA* (in press)
5.49 Shakhnovich, E., Gutin, A. (1989): *Formation of a Unique Structure in Polypeptide Chains. A Theoretical Investigation with the Aid of the Replica Approach.* Preprint (Pushchino) (in Russian).
5.50 Shakhnovich, E., Gutin, A. (1989): *J. Phys. A: Math. Gen.* **22**, 1647–59
5.51 Shakhnovich, E., Gutin, A. (1989): *J. Phys. France* **50**, 1843–50
5.52 Shakhnovich, E., Gutin, A. (1991): *J. Theor. Biol.* (in press)
5.53 Shakhnovich, E., Gutin, A. (1990): *Nature* **346**, 773
5.54 Lesk, A., Chothia, C. (1980): *J. Mol. Biol.* **136**, 225–70
5.55 Bashford, D., Chothia, C., Lesk, A. (1987): *J. Mol. Biol.* **196**, 199–216
5.56 Spivak, V. (1986): *Molek. Biol.* **20**, 789–97
5.57 Imai, K. et al. (1989): *Protein Sequences and Data Analysis* **2**, 81–6
5.58 Dickerson, R. (1980): *Scientific American* **242**, 3, 98–111
5.59 Chipens, G., Polevaja, L., Veretennikova, N., Krikis, A. (1980): *The Structure and Functions of Low-Molecular* Peptides (Zinatne, Riga) (in Russian)
5.60 Crick, F., Barnett, L., Brenner, S., Watts-Tobin, R. (1961): *Nature* **192**, 1227–31
5.61 Volkenstein, M. (1981): *Zhur. Obshch. Biol.* **42**, 680–6
5.62 Anfinsen, C.B. (1959): *The Molecular Basis of Evolution* (J. Wiley and Sons, New York; Chapman and Hall, London)
5.63 Kimura, M. (1983): *The Neutral Theory of Molecular Evolution* (Cambridge University Press, Cambridge)
5.64 Altukhov, Y. (1990): *Genetic Processes in Populations* (Nauka, Moscow) (in Russian)
5.65 Ptitsyn, O. (1981): *FEBS Letters* **131**, 197–203

5.66 Montelione, G., Wutrich, K., Nice, E., Burgess, A., and Scheraga H. (1987): *Proc. Nat. Acad. Sci. USA* **84**, 5226–30

5.67 Gibson, K., Scheraga, H., (1988): In *Structure and Expression, Vol. 1: From Proteins to Ribosomes*, ed. M. Sarma and R. Sarma (Adenine Press, Gulderband, New York)

5.68 Gibson, K., Chin, S., Pincus, M., Clement, E., Scheraga, H. (1986): In *Lecture Notes in Chemistry, Vol. 44: Supercomputer Simulations in Chemistry*, ed. M. Dupuis (Springer-Verlag, Berlin)

5.69 Vasquez, M., Scheraga, H. (1988): *J. Biomol. Structure and Dynamics* **5**, 705, 757

5.70 Montelione, G., Scheraga, H. (1989): *Accounts of Chemical Research* **22**, 70–6

5.71 Ripoll, D., Scheraga, H. (1988): *Biopolymers* **27**, 1283–90

5.72 Scheraga, H. (1989): *Chemica Scripta* **20**, 20–5

5.73 Dickerson, R., Geis, I. (1969): *The Structure and Action of Proteins* (Harper and Row Publishers, New York, Evanston, London)

5.74 Miller, S., Janin, J., Lesk, A., Chothia, C. (1987): *J. Molec. Biol.* **196**, 641–50

5.75 Chothia, C., Lesk, A. (1986): *EMBO Journal* **5**, 823–9

5.76 Chothia, C., Lesk, A. (1987): *Cold Spring Harbor Symposia. Quart. Biol.* **52**, 399–405

5.77 Perutz, M. (1983): *Molec. Biol. Evol.* **1**, 1–28

5.78 Volkenstein, M. (1966): *Biochim. Biophys. Acta* **119**, 421–6

5.79 Volkenstein, M. (1977): *Molecular Biophysics* (Academic Press, New York, San Francisco and London).

5.80 Creighton, T., Darby, N. (1989): *Trends in Biochem. Sciences* **14**, 319–24

5.81 Bode, W., Greyling, H., Huber, R., Otlewski, J., Wilusz, T. (1989): *FEBS Letters* **242**, 285–91

5.82 Creighton, T., Charles, I. (1987): *Cold Spring Harbor Symposia. Quart. Biol.* **52**, 511–9

5.83 Ptitsyn, O., Volkenstein, M. (1986): *J. Biomol. Structure and Dynamics* **4**, 137–56

5.84 Doolittle, R. (1979): In *The Proteins*, Vol. 4 (Academic Press, New York) pp. 2–118

5.85 Jacob, F. (1977): *Science* **196**, 1161–6

5.86 Jacob, F. (1985): In *Evolution from Molecules to Men*, ed. D.S. Bendall (Cambridge University Press, Cambridge)

5.87 Eisenberg, D. et al. (1986): *Proteins* **1**, 16–22

5.88 Ho, S., De Grado, W. (1987): *J. Amer. Chem. Soc.* **109**, 6751–8

5.89 Lear, J., Wasserman, Z., De Grado, W. (1988): *Science* **240**, 1173–81

5.90 Regan, L., De Grado, W. (1988): *Science* **241**, 976–8

5.91 De Grado, W., Wasserman, Z., Lear, J. (1988): *Science* **243**, 622–8

5.92 Matsumara, M., Matthews, B. (1988): *Science* **243**, 792–4

5.93 Warshel, A., Sussman, F., Hwang, J.-K. (1988): *J. Molec. Biol.* **201**, 139–59

5.94 Novotny, J., Rashin, A., Bruccoleri, R. (1988): *Proteins* **4**, 19–30

5.95 Russel, A., Fersht, A. (1987): *Nature* **328**, 496–500

5.96 Egami, F. (1975): *J. Biochem.* **77**, 1165–75

5.97 Williams, R. (1985): *Eur. J. Biochem.* **150**, 231–48

5.98 Vallee, B., Williams, R. (1968): *Proc. Nat. Acad. Sci. USA* **59**, 498–505

5.99 Vallee, B., Williams, R. (1968): *Chemistry in Britain* **4**, 398–505

5.100 Berg, J. (1985): *Cold Spring Harbor Symposia. Quant. Biol.* **52**, 579–85

5.101 Coon, M., White, R. (1980): In *Metal Ion Activation of Dioxygen*, ed. T. Spire (Wiley, New York)

5.102 Lewin, R. (1982): *Science* **217**, 42–3

5.103 Chothia, C., Lesk, A. (1982): *J. Molec. Biol.* **160**, 309–23

5.104 Fox, S., ed. (1965): *The Origin of Prebiological Systems and of Their Molecular Matrices* (Academic Press, New York)

5.105 McAnliffe, C., ed. (1975): *Techniques and Topics in Bioinorganic Chemistry* (Macmillan, New York)

5.106 Eichhorn, G., ed. (1975): *Inorganic Biochemistry*, Vols. 1, 2 (Elsevier, Amsterdam)

5.107 Hughes, M. (1981): *The Inorganic Chemistry of Biological Processes* (John Wiley and Sons, Chichester, New York, Brisbane and Toronto)

5.108 Volkenstein, M. (1982): *Molek. Biol.* **16**, 901–29

Chapter 6

6.1   Kimura, M. (1968): *Nature* **217**, 624–6
6.2   Kimura, M., Crow, J. (1964): *Genetics* **49**, 725–39
6.3   Wright, S. (1931): *Genetics* **16**, 97–111
6.4   King, J., Jukes, T. (1969): *Science* **164**, 788–98
6.5   Kimura, M. (1983): *The Neutral Theory of Molecular Evolution* (Cambridge University Press, Cambridge)
6.6   Kimura, M. (1977): In *Molecular Evolution and Polymorphism*, ed. M. Kimura (National Institute of Genetics, Mishima)
6.7   Kimura, M. (1979): *Scientific American* **241**, 5, 94–104
6.8   Kimura, M. (1985): *New Scientist* **11**, July
6.9   Kimura, M. (1989): *Genome* **31**, 24–31
6.10   Nei, M. (1987): *Molecular Evolutionary Genetics* (New York: Columbia University Press)
6.11   Volkenstein, M. (1981): *Zhur. Obshchei Biol.* **42**, 680–6
6.12   Ptitsyn, O., Volkenstein, M. (1986): *J. Biomolec. Structure and Dynamics* **4**, 137–56
6.13   Richardson, J. (1981): *Adv. Protein Chem.* **34**. 167–203
6.14   Ptitsyn, O., Finkelstein, A. (1983): *Biopolymers* **22**, 15–23
6.15   Finkelstein, A. (1975): *Doklady Akad. Nauk SSSR* **223**, 744–7
6.16   Finkelstein, A. (1977): *Biopolymers* **16**, 525–37
6.17   Lewontin, R. (1974): *The Genetic Basis of Evolutionary Change* (Columbia University Press, New York)
6.18   Grant, V. (1977): *Organismic Evolution* (W. Freeman, San Francisco)
6.19   Dobzhansky, Th. (1955): *Cold Spring Harbor Symposia. Quant. Biol.* **20**, 1–15
6.20   Volkenstein, M., Esipova, N. (1988): *Molek. Biol.* **22**, 1673–7
6.21   Alexandrov, V. (1977): *Cells, Macromolecules and Temperature* (Springer-Verlag, Berlin)
6.22   Alexandrov, V. (1985): *Reactivity of Cells and Proteins* (Nauka, Leningrad) (in Russian)
6.23   Yutani, K., Ogasahara, K., Sugino, Y., Matsuhiro, A. (1977): *Nature* **267**, 274–5
6.24   Kimura, M. (1986): *Phil. Trans. Roy. Soc., London* **B312**, 343–54
6.25   Miyata, T., Yosunaga, T. (1981): *Proc. Nat. Acad. Sci. USA* **78**, 343–54
6.26   Ikemura, I. (1985): *Mol. Biol. Evol.* **2**, 13–34
6.27   Volkenstein, M. (1966): *Biochim. Biophys. Acta* **119**, 421–7
6.28   Volkenstein, M. (1994): *Biophysics* (Mir Publishers and The American Institute of Physics, Moscow and New York), 2nd ed
6.29   Schmalhausen, I. (1969): *Problems of Darwinism* (Nauka, Leningrad) (in Russian)
6.30   Crow, J. (1985): In *Population Genetics and Molecular Evolution*, eds. T. Ohta and K. Aoki (Japan Sci. Soc. Press, Tokyo; Springer-Verlag, Berlin) pp. 1–18
6.31   Kimura, M. (1985): *Ibid.*, 19–39
6.32   Kimura, M. (1985): *J. Genet.* (India) **64**, No. 7–19
6.33   Ohta, T. (1973): *Nature* **246**, 96–8
6.34   Mukai, T. (1985): In *Population Genetics and Molecular Evolution*, eds. T. Ohta and K. Aoki (Japan Sci. Soc. Press, Tokyo; Springer-Verlag, Berlin) pp. 125–45
6.35   Li, W.-H. (1985): *Ibid.*, 333–52
6.36   Mayr, E. (1982): *The Growth of Biological Thought. Diversity, Evolution and Inheritance* (Harvard University Press, Cambridge, Mass.)
6.37   Hartl, D., Dykhuizen, D. (1985): In *Population Genetics and Molecular Evolution* (Japan Sci. Soc. Press, Tokyo; Springer, Berlin) pp. 107–24
6.38   Zuckerkandl, E., Pauling, L. (1965): In *Evolving Genes and Proteins*, eds. V. Bryson and H. Vogel (Academic Press, New York)
6.39   Lee, Y., Fredman, D., Ayala, F. (1985): *Proc. Nat. Acad. Sci. USA* **82**, 824–8
6.40   Li, W.-H., Tanimura, M., Sharp, P. (1987): *J. Mol. Evol.* **25**, 330–42
6.41   Ohta, T. (1987): *J. Mol. Evol.* **26**, 1–6

6.42 Preparata, G., Saccone, G. (1987): *J. Mol. Evol.* **26**, 7–15
6.43 Syvanen, M. (1987): *J. Mol. Evol.* **26**, 16–23
6.44 Kimura, M. (1987): *J. Mol. Evol.* **26**, 24–33
6.45 Zuckerkandl, E. (1987): *J. Mol. Evol.* **26**, 34–46
6.46 Jukes, T. (1987): *J. Mol. Evol.* **26**, 87–98
6.47 Ochman, H., Wilson, A. (1987): *J. Mol. Evol.* **26**, 74–86
6.48 Eastel, S. (1988): *Proc. Nat. Acad. Sci. USA* **85**, 7622–6
6.49 Palmer, J., Herbon, L. (1989): *J. Mol. Evol.* **28**, 87–97
6.50 Sharp, P., Li, W.-H. (1989): *J. Mol. Evol.* **28**, 398–402
6.51 Dover, G. (1987): *J. Mol. Evol.* **26**, 47–58
6.52 Fitch, W. (1976): In *Molecular Evolution*, ed. F.J. Ayala (Sinauer Assoc., Inc., Sunderland, Mass.)
6.53 Ayala, F.J., Kiger, J. (1984): *Modern Genetics*, Chapter 26 (The Benjamin/Cummings Publishing Co., Inc., Menlo Park, Reading, London)
6.54 Gillespie, J.H. (1986): In *Evolutionary Processes and Theory* (Orlando)
6.55 Volkenstein, M. (1985): *Molek. Biol.* **19**, 55–66
6.56 Hochachka, P., Somero, G. (1984): *Biochemical Adaptation* (Princeton University Press, Princeton, New Jersey)
6.57 Volkenstein, M., Goldstein, B. (1986): *Molek. Biol.* **20**, 1645–54
6.58 Reich, J., Selkov, E. (1981): *Energy Metabolism of the Cell* (Academic Press, New York)
6.59 Goldstein, B. (1989): *Kinetical Graphs in Enzymology* (Nauka, Moscow) (in Russian)
6.60 Saifullin, S., Goldstein, B. (1985): *Molek. Biol.* **19**, 1092–9, 1273–7
6.61 Kaimachnikov, N., Selkov, E. (1976): *Biofizika* **21**, 220–4
6.62 Crick, F., Orgel, L. (1964): *J. Mol. Biol.* **8**, 161–70
6.63 Goldbeter, A., Koshland, Jr., D. (1981): *Proc. Nat. Acad. Sci. USA* **78**, 6840–4
6.64 Fersht, A. (1984): *Trends in Biochem. Sci.* **9**, 145–7
6.65 Lazdunski, M. (1974): *Progr. Bioorg. Chem.* **3**, 81–104
6.66 Spivak, V. (1986): *Molek. Biol.* **20**, 789–97
6.67 Wykoff, H. (1968): *Brookhaven Symp. Biol.* **21**, 252–7
6.68 Yanofsky, C., Horn, V., Thorpe, D. (1964): *Science* **146**, 1593–4
6.69 Volkenstein, M., Rass, T. (1987): *Doklady Akad. Nauk SSSR* **295**, 1513–6
6.70 Davitashvili, L. (1969): *The Causes of Extinction of Organisms* (Nauka, Moscow) (in Russian)
6.71 Rensch, B. (1954): *Neue Probleme der Abstammungslehre* (Stuttgart)
6.72 Ruzhentsev, V. (1962): *Fundamentals of Paleontology* (Nauka, Moscow) (in Russian)
6.73 Ruzhentsev, V., Shevyrev, A. (1965): *Transactions of the Paleontological Institute of the USSR Academy of Sciences* **108**, 47–57 (in Russian)
6.74 Gould, S., Eldredge, N. (1977): *Paleobiology* **3**, 115–51
6.75 Damuth, J. (1985): *Evolution* **39**, 1132–46
6.76 Chetverikov, S. (1983): In *Problems of General Biology and Genetics* (Nauka, Novosibirsk) (in Russian)
6.77 Schmalhausen, I. (1983): *Selected Works. Ways and Laws of the Evolutionary Process* (Nauka, Moscow) (in Russian)

Chapter 7

7.1 Lewin, B. (1985): *Genes* (John Wiley and Sons, New York)
7.2 Lewin, R. (1981): *Science* **212**, 28–32
7.3 Dawkins, R. (1983): *The Selfish Gene* (Granada, London, Toronto, Sydney, New York)
7.4 Doolittle, W., Sapienza, C. (1980): *Nature* **284**, 601–3
7.5 Orgel, L., Crick, F. (1980): *Nature* **284**, 604–7
7.6 Doolittle, W. (1982): In *Genome Evolution*, eds. G. Dover and R. Flavell (Academic Press, New York)

7.7   Dover, G. (1980): *Nature* **285**, 618–20
7.8   Normark, S. et al. (1983): *Overlapping Genes. Ann. Rev. Genetics* **17**, 499–525
7.9   Zuckerkandl, E. (1986): *J. Mol. Evol.* **24**, 12–27
7.10 Trifonov, E. (1989): *Bull. Math. Biol.* **51**, 417–31
7.11 Holmquist, G. (1989): *J. Mol. Evol.* **28**, 469–86
7.12 Westheimer, F. (1987): *Science* **235**, 1173–8
7.13 Sharp, P., Eisenberg, D. (1987): *Science* **238**, 729–30
7.14 Eigen, M. (1971): *Naturwissenschaften* **58**, 465–523
7.15 Eigen, M., Schuster, P. (1979): *The Hypercycle* (Springer-Verlag, Berlin, Heidelberg and New York)
7.16 Eigen, M. (1987): *Cold Spring Harbor Symposia Quant. Biol.* **52**, 307–20
7.17 Traut, T. (1988): *Proc. Nat. Acad. Sci. USA* **85**, 294–8
7.18 Schrödinger, E. (1945): *What is Life?* (Cambridge University Press, Cambridge)
7.19 Ayala, F., Kiger, J. Jr. (1984): *Modern Genetics* (The Benjamin/Cummings Publishing Co., Inc., Menlo Park, Reading)
7.20 McClintock, B. (1982): *Science* **226**, 792–801
7.21 Khesin, R. (1984): *The Non-Constancy of Genome* (Nauka, Moscow) (in Russian)
7.22 Georgiev, G. et al. (1977): *Science* **195**, 394–7
7.23 Rubin, G., Finnegan, D., Hogness D. (1976): *Progress in Nucleic Acids Research. Mol. Biol.* **19**, 221–6
7.24 Ohta, T. (1981): *Nature* **292**, 648–9
7.25 Georgiev, G. (1984): *Eur. J. Biochem.* **145**, 203–20
7.26 Georgiev, G. (1989): *Genes of Higher Organisms and Their Expression* (Nauka, Moscow) (in Russian)
7.27 Finnegan, D. et al. (1982): In *Genome Evolution*, eds. G.A. Dover and R.B. Flavell (Academic Press, London)
7.28 Kordium, V. (1982): *Evolution and Biosphere* (Naukova Dumka, Kiev) (in Russian)
7.29 Tatarinov, L. (1987): *Essays on the Theory of Evolution* (Nauka, Moscow) (in Russian)
7.30 Volkenstein, M. (1984): Molek. Biol. **18**, 858–9
7.31 Ohno, S. (1970): *Evolution by Gene Duplication* (Springer-Verlag, Berlin)
7.32 Raff, R., Kaufman, T. (1983): *Embryos, Genes and Evolution* (Macmillan, New York)
7.33 Dickerson, R. (1977): In *Molecular Evolution and Polymorphism*, ed. M. Kimura (National Institute of Genetics, Mishima)
7.34 Li, W.-H., Gojobori, T. (1983): *Mol. Bio. Evol.* **1**, 94–108
7.35 Li, W.-H. (1985): In *Population Genetics and Molecular Evolution*, eds. T. Ohta and K. Aoki (Japan Sci. Soc. Press, Tokyo; Springer-Verlag, Berlin)
7.36 Townes, T., Shapiro, S., Wernke, S., Lingrel, J. (1984): *J. Biol. Chem.* 259, 1896–900
7.37 Flavell, R. (1982): In *Genome Evolution*, eds. G.A. Dover and R.B. Flavell (Academic Press, London)
7.38 Ohta, T. (1980): *Evolution and Variation in Multigene Families* (Springer-Verlag, Berlin)
7.39 Ohta, T. (1985): In *Population Genetics and Molecular Evolution*, eds. T. Ohta and K. Aoki (Japan Sci. Soc. Press, Tokyo; Springer-Verlag, Berlin)
7.40 Ohta, T. (1983): *Theor. Pop. Biol.* **23**, 216–40
7.41 Ohta, T. (1985): *Genetics* **110**, 513–24
7.42 Dover, G. (1982): *Nature* **299**, 111–7
7.43 Dover, G., Brown, S., Coen, E., Dallas, J., Strachan, T., Flavell, R. (1982): In *Genome Evolution*, eds. G.A. Dover and R.B. Flavell (Academic Press, London)
7.44 Lewin, R. (1982): *Science* **218**, 552–53
7.45 Ohta, T. and G. Dover (1982): *Proc. Nat. Acad. Sci. USA* **80**, 4079–83
7.46 Dover, G. (1986): *Trends in Genetics* **2**(6), 159–65
7.47 Dover, G. (1986): In *Evolutionary Processes and Theory* (Academic Press, New York)
7.48 Dover, G. (1988): In *Prospects in Systematics*, ed. D. Hawksworth (Clarendon Press, Oxford)
7.49 Dover, G. (1989): *Trends in Genetics* **5**(4), 100–2
7.50 Goldsmith, M., Kafatos, F. (1984): *Ann. Rev. Genetics* **18**, 443–87

7.51   Jones, W., Kafatos, F. (1982): *J. Mol. Evol.* **19**, 87–103
7.52   Dover, G., Flavell, R. (1984): *Cell* **38**, 622–3
7.53   O'Hare, K., Rubin, G. (1983): *Cell* 34, 25–35
7.54   Karess, R., Rubin, G. (1984): *Cell* **38**, 135–46
7.55   Ohta, T., Dover, G. (1984): *Genetics* **108**, 501–23
7.56   King, M.-C., Wilson, A. (1975): *Science* **188**, 107–16
7.57   Volkenstein, M. (1987): *Biosystems* **20**, 289–304
7.58   Volkenstein, M. (1986): In *Self-organization by Nonlinear Irreversible Processes*, eds. W. Ebeling and H. Ulbricht (Springer, Berlin)

Chapter 8

8.1   Eigen, M. (1971): *Naturwissenschaften* **58**, 465–523
8.2   Eigen, M. (1973): *Self-organization of Matter and the Evolution of Biological Macromolecules* (Mir Publishers, Moscow) (in Russian)
8.3   Eigen, M. (1971): *Quarterly Reviews of Biophysics* 4, 149–65
8.4   Eigen, M., Schuster, P. (1979): *The Hyprecycle* (Springer, Berlin)
8.5   Eigen, M. (1986): *Chemica Scripta* **26B**, 13–26
8.6   Eigen, M. (1987): *Stufen zum Leben* (Piper, München, Zürich)
8.7   Eigen, M. (1987): *Cold Spring Harbor Symposia Quant. Biol.* **52**, 307–20
8.8   Eigen, M., Winkler-Oswatitsch, R., Dress, A. (1988): *Science* **85**, 5913–7
8.9   Eigen, M., McCaskill, J., Schuster, P. (1988): *J. Phys. Chem.* **92**, 6881–91; (1989) *Adv. Chem. Phys.* **75**, 149–263
8.10   Volkenstein, M. (1983): *General Biophysics*, Vol. 2 (Academic Press, New York)
8.11   Volkenstein, M. (1994): Biophysics, 2nd edn. (Mir Publishers and The American Institute of Physics, Moscow and York; translated by Artavaz Beknazarov)
8.12   Küppers, B.-O. (1985): *Molecular Theory of Evolution* (Springer, Berlin)
8.13   Volkenstein, M. (1990): *Entropie und Information* (Verlag Harri Deutsch, Frankfurt am Main)
8.14   Wigner, E. (1987): *Symmetries and Reflexions* (Indiana University Press, Bloomington, Indiana)
8.15   Volkenstein, M. (1973): *Uspekhi Fiz. Nauk* **109**, 499–515
8.16   Jones, B., Enns, R., Ragnekar, S. (1976): *Bull. Math. Biol.* **38**, 15–22
8.17   Cech, T. (1989): *Biochemistry International* **18**, 7–14
8.18   Ratner, V. (1985): In *Problems of Molecular Evolution*, ed. R. Salganik (Nauka, Novosibirsk) (in Russian)
8.19   Hamming, R. (1980): *Coding and Information Theory* (Prentice Hall, Englewood Cliffs)
8.20   Wright, S. (1982): *Evolution* **36**, 427–43
8.21   Swetina, J., Schuster, P. (1982): *Biophys. Chem.* **16**, 329–40
8.22   Leuthäusser, I. (1986): *J. Chem. Phys.* **84**, 1884–5; (1987) *J. Stat. Phys.* **48**, 343–50
8.23   Hill, T. (1956): *Statistical Mechanics* (McGraw-Hill, New York)
8.24   Rumer, Yu., Ryvkin, M. (1972): *Thermodynamics, Statistical Physics and Kinetics* (Nauka, Moscow) (in Russian)
8.25   Eigen, M. (1988): *In Phasensprunge und Stetigkeit in der natürlichen und kulturellen Welt*, eds. K. Hierhobzer and H. Wittmann (Wissenschaftliche Verlagsgesellschaft mbH, Stuttgart)
8.26   Spiegelman, S. (1970): In *The Neurosciences, Second Study Program*, ed. F. Smitt (Rockfeller University Press, New York)
8.27   Mills, D., Peterson, R., Spiegelman, S. (1967): *Proc. Nat. Acad. Sci. USA* **58**, 217–23
8.28   Levisohn, R., Spiegelman, S. (1969): *Proc. Nat. Acad. Sci. USA* **63**, 807–14
8.29   Safhill, R., Schneider-Bernlochz, H., Orgel, L., Spiegelman, S. (1970): *J. Mol. Biol.* **51**, 531–8
8.30   Mills, D., Kramer, F., Spiegelman, S. (1973): *Science* **180**, 216–21
8.31   Biebricher, C. (1987): *Cold Spring Harbor Symposia Quant. Biol.* **52**, 299–306
8.32   Biebricher, C. (1991): In *Biologically Inspired Physics*, ed. L. Peliti (Plenum Press, New York)

8.33   Maynard Smith, J. (1989): *Evolutionary Genetics* (Oxford University Press, Oxford)
8.34   Biebricher, C. (1986): *Nature* **321**, 89–92
8.35   Eigen, M. (1988): *Perspektiven der Wissenschaft* (Deutsche Verlags-Anstalt, Stuttgart)
8.36   Eigen, M. (1989): *Naturwissenschaften* **76**, 341–50
8.37   McCaskill, J. (1984): *Biol. Cybernetics* **50**, 63–70
8.38   Weinberger, E. (1987): *A Stochastic Generalization of Eigen's Model of Natural Selection.* Thesis (New York University Press, New York)
8.39   Weinberger, E. (1987): *J. Stat. Phys.* **49**. 1011–28
8.40   Weinberger, E. (1992): *Biol. Cybernetics* (in press)
8.41   Kauffman, S., Levin, S. (1987): *J. Theor. Biol.* **128**, 11–45
8.42   Binder, K., Young, A. (1986): *Rev. Mod. Phys.* **54**, 801–976
8.43   Anderson, P. (1983): *Proc. Nat. Acad. Sci. USA* **80**, 3386–90
8.44   Rokshar, D., Anderson, P., Stein, D. (1986): *J. Mol. Evol.* **23**, 119–26
8.45   Red'ko, B. (1986): *Biofizika* **31**, 511–4; (1990) **35**, 827–30
8.46   Amitrano, C., Peliti, L., Saber, M. (1989): *J. Mol. Evol.* **29**, 513–25
8.47   Derrida, B., Peliti, L. (1991): *Bull. Math. Biol.*
8.48   Eigen, M., Lindemann, B., Tietze, M., Winkler-Oswatitsch, R., Dress, A., von Haeseler, A. (1989): *Science* **244**, 673–9
8.49   Schuster, P. (1989): In *Optimal Structures in Heterogeneous Reaction Systems*, ed. by P. Plath. Springer Series in Synergetics. Vol. 44 (Springer, Berlin: pp. 101–122)
8.50   Fontana, W., Schnabl, W., Schuster, P. (1989): *Phys. Rev.* **A40**, 6, 3301–3321
8.51   Schuster, P. (1991): In *Optimization Dynamics on Value Landscapes. Modelling Molecular Evolution*, ed. by O. Solbrig and G. Nicolis, No. 6, pp. 115–161 (Paris)
8.52   Schuster, P. (1992): In *The Science and Theory of Information*, ed. by G. Wassermann, R. Kirby, and B. Rordoof. Labor and Fides, Geuf, pp. 45–57
8.53   Fontana, W., Griesmacher, T., Schnabl, W., Stadtler, P., Schuster, P. (1992): *Mh. Chemie* **122**
8.54   Kimura, M. (1983): *The Neutral Theory of Molecular Evolution* (Cambridge University Press, Cambridge)
8.55   Gojobori, T., Moriyama, E., Kimura, M. (1990): *Proc. Nat. Acad. Sci. USA* **87**, 10015–8
8.56   Wong, A. (1990): *J. Theor. Biol.* **146**, 523–43

Chapter 9

9.1   Losev, A. (1982): *Chaos. Myths of the Peoples of the World*, Vol. 2, pp. 579–81 (Soviet Encyclopedia, Moscow) (in Russian)
9.2   Toporov, V. (1982): *Chaos Primitive. Ibid.*, pp. 581–8
9.3   Asmus, V. (1973): *Immanuel Kant* (Nauka, Moscow) (in Russian)
9.4   Block, A. (1921): *About the Mission of a Poet*
9.5   Volkenstein, M. (1986): *Nauka i Zhizn'* **6**, 82–6
9.6   Volkenstein, M. (1969): *Dramatic Art* (Sovetsky Pisatel', Moscow) (in Russian)
9.7   Klimantovich, Yu. (1989): *Uspekhi Fiz. Nauk* **158**, 58–91, 350
9.8   Pool, R. (1988): *Science* **241**, 787–8; (1989) **243**, 25–28, 310–3; (1989) **245**, 26–8
9.9   Feigenbaum, M. (1983): *Uspekhi Fiz. Nauk* **141**, 343–74
9.10   Sinai, Ya, Krylov, N. (1979): *Works on the Foundations of Statistical Physics* (Princeton University Press, Princeton, New Jersey)
9.11   Kolmogorov, A. (1958): *Doklady Akad. Nauk SSSR* **119**. 861–5; (1959) **124**, 754–8
9.12   Ornstein, D. (1989): *Science* **243**, 182–7
9.13   Wolkenstein, M. (1990): *Entropie und Information* (Verlag Harri Deutsch, Frankfurt am Main)
9.14   Bishop, A. (1990): *Nature* **344**, 290–1
9.15   Brewer, R. et al. (1990): *Nature* **344**, 305–9

9.16   Hoffnagle, J. et al. (1988): *Phys. Rev. Lett.* **61**, 255–8
9.17   Klimantovich, Yu. (1982): *Statistical Physics* (Nauka, Moscow) (in Russian)
9.18   Gleick, J. (1988): *Chaos. Making a New Science* (Heinemann, London)
9.19   Lorenz, E. (1963): *J. Atoms. Sci.* **20**, 130–9
9.20   Polak, L., Mikhailov, A. (1983): *Self-Organization in the Non-equilibrium Physicochemical Systems* (Nauka, Moscow) (in Russian)
9.21   Nicolis, G., Prigogine, I. (1987): *Die Erforschung des Komplexen* (Piper, Munchen, Zurich)
9.22   Mandelbrot, B. (1986): In *The Beauty of Fractals*, eds. H.-O. Peitgen and P. Richter (Springer, Berlin)
9.23   Jürgens, H., Peitgen, H.-O., Saupe, D. (1990): *Scientific American* **263**, 2, 40–7
9.24   Mandelbrot, B. (1982): *The Fractal Geometry of Nature* (W. Freeman, San Francisco)
9.25   Barnsley, M. (1988): *Fractals Everywhere* (Academic Press, New York)
9.26   Schuster, H. (1988): *Deterministic Chaos* (VCH Verlagsgesellschaft mbH, Weinheim)
9.27   Ornstein, D. (1974): *Ergodic Theory, Randomness and Dynamical Systems* (Yale University Press, New Haven)
9.28   Hao Bai-Lin, ed. (1984): *Chaos* (World Scientific, Singapore)
9.29   Cvitanović, P. ed. (1984): *Universality in Chaos* (Hilger, Bristol)
9.30   Bohr, T., Cvitanović, P. (1987): *Nature* **329**, 391–2
9.31   Procaccia, I. (1988): *Nature* **333**, 618–23
9.32   Hess, B. (1990): *Fresenius J. Analyt. Chem.* **337**, 459–68
9.33   Bishop, A. (1990): *Nature* **344**, 290–1
9.34   Ebeling, W., Feistel, R. (1982): *Physik der Selbstorganization und Evolution* (Akademie Verlag, Berlin)
9.35   Conrad, M. (1986): In *Chaos*, ed. A. Holden (Manchester University Press, Manchester)
9.36   Arnold, V. (1983): *Uspekhi Fiz. Nauk* **141**, 571–90
9.37   Arnold, V. (1986): *Theory of Catastrophes* (Springer, Berlin)
9.38   Feigenbaum, M. (1980): "Universal behavior in nonlinear systems". Los Alamos Science **1**, 1, 4–27
9.39   Romanovsky, Yu., Stepanova, N., Chernavsky, D. (1984): *Mathematical Biopysics* (Nauka, Moscow) (in Russian)
9.40   Ivanitsky, G., Krinsky, V., Selkov, E. (1978): *Mathematical Biophysics of the Cell* (Nauka, Moscow) (in Russian)
9.41   Volkenstein, M. (1983): *General Biopysics*, Vol. 2 (Academic Press, New York)
9.42   Volkenstein, M. (1994): *Biopysics*, 2nd edn. (Mir Publishers and The American Institute of Physics, Moscow and New York; translated by Artavaz Beknazarov)
9.43   Volterra, V. (1931): *Leçons sur la Theorie Mathematique de la Lutte pour la Vie* (Gauthier-Villars, Paris)
9.44   Kolmogorov, A., Petrovsky, I., Piskunov, N. (1975): *Problems of Cybernetics* **12**, 3–12 (in Russian)
9.45   Lifshitz, E., Pitaevsky, L. (1979): *Theortical Physics*, Vol. 10, *Physical Kinetics* (Nauka, Moscow) (in Russian)
9.46   Polezhajev, A., Sidelnikov, D. (1992): *Biofizika* (in press)
9.47   Ayala, F., Kiger, J. Jr. (1984): *Modern Genetics* (The Benjamin/Cunnings Publishing Co., Menlo Park, Reading)
9.48   Segel, L. (1984): *Modeling Dynamic Phenomena in Molecular and Cellular Biology* (Cambridge University Press, Cambridge)
9.49   Moran, P. (1962): *The Statistical Processes of Evolutionary Theory* (Clarendon Press, Oxford)
9.50   Nei, M. (1987): *Molecular Evolutionary Genetics* (Columbia University Press, New York)
9.51   Belintsev, B., Volkenstein, M. (1977): *Doklady Akad. Nauk SSSR* **235**, 205–7
9.52   Volkenstein, M. (1987): *BioSystems* **20**, 289–304
9.53   Schlögl, F. (1972): *Zs. Phys.* **B253**, 147–560
9.54   Berg, L. (1969): *Nomogenesis or Evolution Determined by Law* (MIT Press, Cambridge, Mass.)
9.55   Volkenstein, M., Livshits, M. (1989): *BioSystems* **23**, 1–5

9.56 Livshits, M., Volkenstein, M. (1992): *BioSystems* (in press)
9.57 Poston, T., Stewart, I. (1978): *Catastrophe Theory and Its Applications* (Pitman, San Francisco and Melbourne)
9.58 Belintsev, B., Livshits, M., Volkenstein, M. (1981): *Zs. Phys. B: Condensed Matter* **44**, 345–51
9.59 Wright, S. (1968): *Experimental Results and Evolutionary Deductions. Evolution and the Genetics of Populations*, Vol. 1, pp. 211–72, 373–420; (1977) Vol. 3, pp. 443–73; (1978) Vol. 4, pp. 460–76 (University of Chicago Press, Chicago)
9.60 Wright. S. (1982): *Evolution* **36**, 427–43
9.61 Lande, R. (1985): *Proc. Nat. Acad. Sci. USA* **82**, 7641–5
9.62 Newman, C., Cohen, J., Kipnis, C. (1985): *Nature* **315**, 400–1
9.63 Lewin, R. (1986): *Science* **231**, 672–3
9.64 Livshits, M., Volkenstein, M. (1988): *Doklady Akad. Nauk SSSR* **301**, 731–4
9.65 Gardiner, C., Chaturvedi, S. (1977): *J. Stat. Phys.* **17**, 429–40
9.66 Feistel, R., Ebeling, W. (1982): *BioSystems* **15**, 291–6
9.67 Ebeling, W., Engel, A., Esser, B., Feistel, R. (1984): *J. Stat. Phys.* **37**, 369–84
9.68 Burger, R. (1986): *Evolution* **40**, 182–93
9.69 Glansdorff, P., Prigogine, I. (1971): *Thermodynamics of Structure, Stability and Fluctuations* (Wiley, New York)
9.70 Prigogine, I. (1980): *From Being to Becoming* (Freeman, San Francisco)
9.71 Nicolis, G., Prigogine, I. (1977): *Self-Organization in Nonequilibrium Systems* (Wiley, New York)
9.72 Nicolis, J. (1986): *Dynamics of Hierarchical Systems* (Springer, Berlin)
9.73 Prigogine, I., Stengers, I. (1979): *La Nouvelle Alliance* (Gallimard, Paris)
9.74 Enns, R. et al., eds. (1981) *Nonlinear Problems in Physics and Biology (Plenum Press, New York)*
9.75 Haken, H. (1977): *Synergetics* (Springer, Berlin)
9.76 Haken, H. (1989): *Advanced Synergetics* (Springer, Berlin)

Chapter 10

10.1   Schrödinger, E. (1945): *What is Life?* (Cambridge University Press, Cambridge)
10.2   Yaglom, A., Yaglom, I. (1962): *Probability and Information* (Dover Publications, New York)
10.3   Wolkenstein, M. (1990): *Entropie und Information* (Verlag Harri Deutsch. Thun, Frankfurt am Main)
10.4   Poplavsky, R. (1981): *Thermodynamics of Informational Processes* (Nauka, Moscow) (in Russian)
10.5   Romanovsky, Yu., Stepanova, N., Chernavsky, D. (1984): *Mathematical Biophysics* (Nauka, Moscow) (in Russian)
10.6   Ebeling, W., Volkenstein, M. (1990): *Physica* **A163**, 398–402
10.7   Quastler, H. (1964): *The Emergence of Biological Organization* (Yale University Press, New Haven and London)
10.8   Küppers, B.-O. (1989): *Information and the Origin of Life* (The MIT Press, Cambridge, Mass.)
10.9   Polanyi, M. (1968): *Science* **160**, 1308–11
10.10 Wiener, N. (1950): *Cybernetics* (Wiley, New York)
10.11 Blumenfeld, L. (1981): *Problems of Biological Physics* (Springer, Berlin)
10.12 Schmalhausen, I. (1968): *Cybernetic Problems in Biology* (Nauka, Novosibirsk) (in Russian)
10.13 Gatlin, L. (1972): *Information Theory and the Living Systems* (Columbia University Press, New York)
10.14 Wicken, J. (1987): *Evolution, Thermodynamics and Information* (Oxford University Press, Oxford)
10.15 Brooks, D., Wiley, E. (1988): *Evolution as Entropy* (University of Chicago Press, Chicago)

10.16 Weber, B., Depew, D., Smith, J. eds. (1988): *Entropy, Information and Evolution. New Perspectives in Physical and Biological Evolution* (Cambridge University Press, Cambridge, Mass.)

10.17 Ebeling, W., Feistel, R. (1982): *Physik der Selbstorganisation und Evolution* (Akademie Verlag, Berlin)

10.18 Volkenstein, M. (1994): *Biophysics*, 2nd edn. (Mir Publishers and The American Institute of Physics, Moscow and New York; translated by Artavaz Beknazarov)

10.19 Bongard, M. (1970): *Pattern Recognition* (Hayden Book Co., Spartan Books, Rochelle Park, New Jersey)

10.20 Kharkevitch, A. (1973): *Information Theory* (Nauka, Moscow) (in Russian)

10.21 Stratanovitch, R. (1975): *Information Theory* (Soviet Radio, Moscow) (in Russian)

10.22 Volkenstein, M., Chernavsky, D. (1978): *J. Soc. Biol. Structure* **1**, 95–106

10.23 Volkenstein, M., Chernavsky, D. (1984): In *Self-Organization Autowaves and Structures Far from Equilibrium*, ed. V. Krinsky (Springer, Berlin)

10.24 Saunders, P., Ho, M. (1976): *J. Theor. Biol.* **63**, 375–87; (1981) **90**, 515–26

10.25 Kolmogorov, A. (1965): *Problems of Information Transmission* **1**, 1, 3–8

10.26 Chaitin, C. (1975): *Scientific American* **232**(5), 47–58

10.27 von Neumann, J. (1966): *Theory of Self-Reproducing Automata* (University of Illinois Press, Urbana, London)

10.28 Dayhoff, M., ed. (1972): *Atlas of Protein Sequence and Structure* (Nat. Biomedical Research Foundation)

10.29 Bachinsky, A. (1976): *Zhur. Obshch. Biol.* **37**, 163–75

10.30 Bachinsky, A., Ratner, V. (1976): *Biomed. Zs.* **18**, 53–68

10.31 Volkenstein, M. (1979): *J. Theor. Biol.* **80**, 155–69

10.32 Woese, C. (1974): *J. Mol. Evol.* **3**, 109–20

10.33 Reichert, T., Yu, J., Christensen, R. (1976): *J. Mol. Evol.* **8**, 41–53

10.34 Yano, T., Hasegawa, M. (1974): *J. Mol. Evol.* **4**, 179–93

10.35 Conrad, M., Volkenstein, M. (1981): *J. Theor. Biol.* **92**, 293–9

10.36 Conrad, M. (1979): *Bull. Math. Biol.* **41**, 387–95

10.37 Eigen, M., Winkler, R. (1975): *Das Spiel* (Piper, München, Zürich)

10.38 Lasker, E. *Lehrbuch des Schachspiels*

10.39 Einstein, A. (1952): *Foreword to the book "Emmanuel Lasker"* by J. Hannack (Berlin)

10.40 Botvinnik, M. (1979): *From the Chessplayer to the Machine* (Fizkultura i Sport, Moscow) (in Russian)

10.41 Haken, H. (1988): *Information and Self-Organization* (Springer, Berlin)

10.42 Haken, H. (1987): *Biol. Cybernetics* **56**, 11–7

Author Index

Numbers in italics indicate that an author's work is referred to, although his name
is not cited in the text

Subject Index

Springer-Verlag
and the Environment

We at Springer-Verlag firmly believe that an international science publisher has a special obligation to the environment, and our corporate policies consistently reflect this conviction.

We also expect our business partners – paper mills, printers, packaging manufacturers, etc. – to commit themselves to using environmentally friendly materials and production processes.

The paper in this book is made from low- or no-chlorine pulp and is acid free, in conformance with international standards for paper permanency.